Animal Genetic Improvement

To Manolo Baselga,
master and friend

Animal Genetic Improvement

Agustín Blasco
Institute for Animal Science and Technology
Universitat Politècnica de València,
València, Spain

CABI

While the publisher has attempted to ensure that the full-text version of this eBook is accessible to all readers, rendering of native MathML is not fully functional in all support scenarios.

CABI is a trading name of CAB International

CABI
Nosworthy Way
Wallingford
Oxfordshire OX10 8DE
UK

Tel: +44 (0)1491 832111
E-mail: info@cabi.org
Website: www.cabi.org

CABI
200 Portland Street
Boston
MA 02114
USA

Tel: +1 (617)682-9015
E-mail: cabi-nao@cabi.org

A catalogue record for this book is available from the British Library, London, UK.

ISBN-13: 9781836990970 (hardback)
9781836990987 (ePDF)
9781836990994 (ePub)

DOI: 10.1079/9781836990994.0000

Commissioning Editor: Alexandra Lainsbury
Editorial Assistant: Theresa Regueira
Production Editor: Shankari Wilford
Translator: Agustín Blasco

Typeset by Straive, Pondicherry, India
Printed and bound in the UK by CPI Group (UK) Ltd, Croydon, CR0 4YY

Contents

Preface

Animal genetic improvement is a discipline that, despite having a biological foundation, fundamentally uses statistical methods for its application. Until the beginning of the 20th century, animal genetic improvement (animal breeding at that time) was an art in which reasonable proposals, such as evaluating males for their offspring, were mixed with superstitions such as maternal impregnation, telegony or the inheritance of acquired characters. The scientific foundations were originally provided by Mendel when he postulated the existence of what we now call *genes*, but his discovery remained unknown until it was rediscovered at the beginning of the 20th century. From then, research on the biological causes of inheritance began, leading in 1953 to the discovery of the mechanisms of inheritance by Watson and Crick, and culminating in research on the mechanisms of gene regulation and the techniques of gene editing. In parallel, the biometric school of Karl Pearson (who did not believe in the existence of genes) tried to explain the hereditary mechanisms by applying statistical analysis to evolution; despite the fact that, in their day, the members of this school ignored the existence of genes, their procedures have been very useful for animal breeders. The reason for this discrepancy between schools is that most of the productive traits are regulated by many genes with each one having a very small effect, so it appeared that the inheritance mechanisms proposed by Mendel did not apply to most of the observed traits. At the beginning of the 20th century, several models were proposed that tried to explain the continuity observed in the productive traits as a result of the joint action of many genes each with small effect; in fact, Mendel himself had already indicated that intermediate colours of the flowers could be due to the joint action of several of what we now call genes. It was Ronald Fisher, in 1918, who developed the model that has been used since then, approximating the joint effect of genes of small effect to a normal distribution, which allowed the application of linear models, common today in selection. From the 1960s on, an intense development of molecular genetics began, which not only led to a better understanding of the mechanisms of inheritance but also contributed a great deal of useful data to breeding programmes. The rapid development of molecular genetics led to the belief that it would replace the usual statistical methods of genetic improvement. However, it has not resulted in a substitution of one methodology for another, as they are not different paradigms (Fisher's model does not deny the existence of genes), but has resulted in an integration of molecular data in the statistical methods of quantitative genetics, so that nowadays it is common to use both methodologies in animal genetic improvement programmes.

The first associations for animal improvement were created in the 19th century, although the selection of animals was essentially based on morphological data. At the beginning of the 20th century, records began to be taken systematically, but there was still no established methodology for genetic improvement; early animal genetic improvement books (e.g. Shaw, 1901; Davenport, 1907; Marshall, 1912) did not contain what we would today consider useful information for selection, including Sewall Wright's book of 1920. It was in the mid-1930s when an Iowa professor, Jay Lush, used the knowledge developed by Fisher and other statistical geneticists such as Sewall Wright to apply it to genetic improvement. The development of artificial insemination in the 1930s and the freezing of bovine semen in the 1950s made it possible to have large amounts of data related to each other by kinship. The development of computing made it possible from the end of the 20th century to apply linear models that included all this information, and also to use new statistical methods, Bayesian methods, which were practically not applicable before. Modern full-sequencing techniques appeared at the beginning of the 21st century, and sequencing costs have been rapidly falling; single-nucleotide polymorphisms (SNPs), which are simple and inexpensive genetic markers, have been commercially developed in chips containing tens of thousands of them. This has led to the appearance of genomic selection, which has been integrated into traditional selection systems, increasing the response per unit of time in dairy cattle by more than 50%, and increasing the economic response from 20% to 30% in the genetic programmes of other species.

This book aims to fill a gap in the field of animal genetic improvement. At present, there are books containing different topics on quantitative genetics useful for animal genetic improvement, as well as books explaining specific aspects of rabbit, pig, poultry, sheep and cattle genetics. However, it is difficult to find a text that addresses the bases of quantitative genetics and the methodology for animal genetic improvement including Bayesian methods and the new genomic selection applications, and that also presents how genetic improvement companies are currently organized and what their selection programmes are. This book attempts to fill this gap. The book is aimed at students of animal science degrees; each chapter begins with a summary of the main topics to be covered, and ends with a list of important main concepts, problems to be solved (with their solutions) and a test to check whether the reader has understood the chapter. In the quantitative genetics chapters, I have tried to offer a rigorous approach, leaving the mathematical demonstrations in appendices, so as not to interrupt the exposition and facilitate its study in a first approximation. Nowadays, it is not uncommon for students to believe that by knowing how to apply a program in R or Python they already have all the knowledge they need to work; however, understanding the fundamental concepts of quantitative genetics is essential for their training, for their professional development, and for a good interpretation of the results of the selection programmes.

The first two chapters of the book are devoted to the quantitative genetics needed to apply genetic improvement methods, and they presuppose a certain knowledge of elementary statistics and classical genetics on the part of the reader. The first chapter deals with genetics considering a single gene with two alleles, and the second deals with Fisher's infinitesimal model. The third chapter deals with selection indices including family information, for one or more traits. Today, analyses using the linear model have displaced these indices, but they are still used in experiments and in some improvement programmes; furthermore, the study of indices is a good introduction to methods based on linear models. Chapters 4 and 5 deal with the use of the linear model in genetic improvement. Chapter 4 describes the selection method preferred today, best linear unbiased prediction (BLUP), and the estimation of variance components by residual maximum likelihood (REML), which is also the most widely used method for genetic parameters estimation. Chapter 5 deals with Bayesian methods, increasingly used in animal breeding, particularly since the implementation of genomic selection. This chapter introduces Bayesian theory and methods, and highlights the benefits of using Bayesian techniques in both simple and complex problems. Chapter 6 introduces genomic selection and gene detection, with a critical assessment of their use in different species. Chapter 7 deals with the concept of breed and conservation programmes. This chapter takes a critical perspective and discusses when to conserve a breed. The author of this book took a sabbatical at the headquarters of the Food and Agriculture Organization of the United Nations (FAO) in Rome, working on the FAO's breed conservation programme, so his position may be debatable but is well informed. Chapters 8 and 9 deal with genetic improvement schemes in prolific and low-prolific species (monogastric and ruminants, respectively). The main programmes of the genetic improvement companies in pigs, poultry, rabbits, cattle, sheep and goats are addressed. It is not possible to be encyclopaedic in a book with these characteristics, but it is intended to expose the most developed genetic improvement programmes; for these chapters, a certain prior knowledge of the reader about animal productions is assumed. The book ends with a chapter dedicated to the future of animal genetic improvement, dealing with topics such as gene editing and global warming. I have reduced the references to a minimum to facilitate reading.

In Islamic culture, it is believed that only Allah is perfect, and man should avoid creating perfect works. We know that, although the book has been fully reviewed by several people, errors are sure to appear. All errors will be noted as we become aware of them in the Errata.pdf file at a GitHub website that also will contain comments, extensions, and clarifications of the present text. The URL of the site is https://github.com/ablasco-upv/BlascoA

References

Davenport, E. (1907) *Principles of Breeding: A Treatise on Thremmatology*. Ginn & Co., Boston, Massachusetts.

Marshall, F.R. (1912) *Breeding Farm Animals*. The Breeder's Gazette, Chicago, Illinois.

Shaw, T. (1901) *Animal Breeding*. Orange Judd Co., New York.

Wright, S. (1920) *Principles of Livestock Breeding*. US Department of Agriculture, Washington, DC.

Acknowledgements

This book is dedicated to Manolo Baselga, my colleague and friend, to whom I owe my training in quantitative genetics, first as a professor and PhD supervisor and later through the scientific discussions that we have had for more than 30 years. I would like to thank Professors Baselga, Miguel Angel Toro and Juan Pablo Gutiérrez, as well as my former doctoral student Marina Martínez, for their detailed review of the book. My friendship with all of them has facilitated criticism and subsequent corrections. I would like to thank Armando Caballero for revising the first two chapters, Daniel Sorensen for revising Chapter 4, Andrés Legarra, Zulma Vitezica, Luis Varona and Cristina Casto for revising Chapter 6, Jean-Pierre Bidanel for revising Chapter 8 and Oscar González, Juan Antonio Jiménez and Dianelys González-Peña for revising Chapter 9. I am also grateful to other colleagues (Jesús Arango, Santiago Avendaño, Raúl Ponzoni, Eva Ugarte, Félix Goyache, Beatriz Villanueva, Ernesto Gómez, Daniel Gianola, Noelia Ibáñez, David Cavero and others too numerous to mention) for their advice and reviews of specific sections of this book. The book has benefited from the courses given in our Masters in Animal Genetic Improvement and Reproductive Biotechnology by geneticists from the main cattle, pig and poultry breeding companies and by our colleagues from various universities and research centres. Finally, I would like to thank Concha and our sons Alex and Ricardo for their patience when I was busy writing the book during the free time I should have spent with them.

1 The Genetics of Productive Traits

Abstract
This chapter covers what is known as population genetics. It examines the evolution of allele frequencies of one or two genes in an infinite population under random mating, and the frequencies of the genotypes to which they give rise. The concept of linkage disequilibrium is introduced. The coefficients of consanguinity and kinship are defined. The evolution of genotype frequencies in finite populations and in cross-breeding between populations is examined, defining the concept of heterosis. By assigning values to genotypes, the evolution of the mean and variance of a population is examined, with or without inbreeding. The concepts of genotypic value, additive value and dominant value of a genotype are introduced, and the breakdown of the variance of genotypic values into additive variance and dominant variance is examined.

1.1 Simple Genes, Genetic Markers and Linkage Disequilibrium

1.1.1 Major genes and minor genes

Until recent times, genetics developed not by looking at genes directly but by estimating their effects. In his famous 1866 article, Mendel spoke of directly observable 'characters', such as the smooth (A) or rough (a) shape, or the yellow (B) or green (b) colour of peas. The word 'gene' to name these inheritance factors was introduced by the Danish biologist Johannsen in 1909. The observation of the proportions of different colours in the seeds and flowers of the progeny, depending on the parental plants, allowed Mendel to infer the existence of the inheritance factors that caused them, even though their nature was a mystery. Mendel's famous laws explained how traits were transmitted to offspring and made it possible to predict their expected values for certain traits.

Mendel's works, although known in the field of plant hybrid producers, were ignored by biologists seeking the laws of inheritance until 1900, when they were independently rediscovered by Hugo de Vries and Carl Correns. The reason that Mendel's laws took so long to become established was that most of the observed traits did not appear to be governed by these laws but showed continuous variation, something that Mendel was perfectly aware of. In a letter to the eminent botanist Carl Nägeli, Mendel said, 'I realized that the results obtained by me did not easily reconcile with our contemporary scientific knowledge' (Stern and Sheerwood, 1966). Science was at that time dominated by Darwinism, and Darwin himself believed that individuals should be intermediate between the values of their parents. The problem with Darwin's theory was that, if true, the observed variability would quickly disappear and all individuals would be intermediate, which would go against Darwin's natural selection theory and, furthermore, contradict everyday experience.

However, Mendel himself pointed out that when intermediate flower colours were observed in a cross-breeding experiment, the cause was probably the combination of several of these inherited factors (Mendel, 1866). Although it was not perceived then, the

DOI: 10.1079/9781836990994.0001

fact that continuous variation could come from many independent factors allowed maintenance of the variation, as all the factors that intervened in a trait reappeared in the next generation. Subsequent work by three statisticians, Udny Yule in 1902, Karl Pearson in 1904 and especially Ronald Fisher in 1918, made possible to reconcile the existence of genes and the fact that continuous variation was normally observed. Thus, an explanatory model for the transmission of traits was finally accepted according to which larger genes with a large effect on a trait coexist with smaller genes having a negligible individual effect but which together produce substantial variability in this trait. An entertaining but rigorous history of these early steps in population and quantitative genetics is found in the classic book by Provine (2001).

> It is important to realize that the term *major gene* or *minor gene* does not refer to the size of the gene or its biological importance. When we speak of 'major' or 'minor', we mean the effect on a trait; the same gene can be 'major' for one trait and 'minor' for another. For example, the gene that regulates growth hormone may be biologically important but have a small effect on growth, as growth is influenced by many metabolic pathways regulated by many genes. There are also genes with intermediate effects, so the genetic determination of a trait can be complex.

The chromosomes of an animal cell, which are $2n$, are organized in n pairs of chromosomes, the chromosomes of each pair being called homologous chromosomes. Each member of the couple is contributed by a sperm and an egg. The genes, which are fragments of chromosomes made up of nucleotide sequences, are found in specific places on homologous chromosomes called *loci* (singular *locus*). At a locus, the nucleotide sequence of each homologous chromosome is called an *allele*. The same nucleotide sequence can be found at a locus on each homologous chromosome, in which case we say that the individuals are *homozygous*, or two different sequences can be found at a locus and we say that the individuals are *heterozygous*. Molecular genetics has progressed spectacularly in recent decades, allowing a better understanding of the functioning of genes. Today, we know that genes have fragments that can contribute to encode proteins (exons) and fragments that do not (introns), and that a gene can give rise to several proteins depending on which exons are activated in the transcription. All of this is regulated in a complex way by other nucleotide sequences of the chromosomes. We will, however, use a simple 'black box' approach and refer to genes by their effects; therefore, we will identify a locus with one gene and say that it has two alleles. For example, we know that the growth hormone gene encodes a polypeptide that is this hormone, but we will be interested in referring not to the hormone itself but to its effect on growth. This effect may be very small because there are many factors that influence growth – many metabolic pathways involving many genes. Today, it is considered that most productive traits are regulated by a large number of genes, more than 1000 in many cases.

When the number of genes determining a trait increases, the chances of new genotypes increase as well. Sometimes, if two genes are on the same chromosome, there are crossovers of the homologous chromosomes whereby part of the paternal chromosome exchanges with the maternal chromosome, giving rise to new gametes. Consider, for example, a trait controlled by four genes, each with two alleles, where we represent favourable alleles with uppercase and unfavourable alleles with lowercase. Figure 1.1

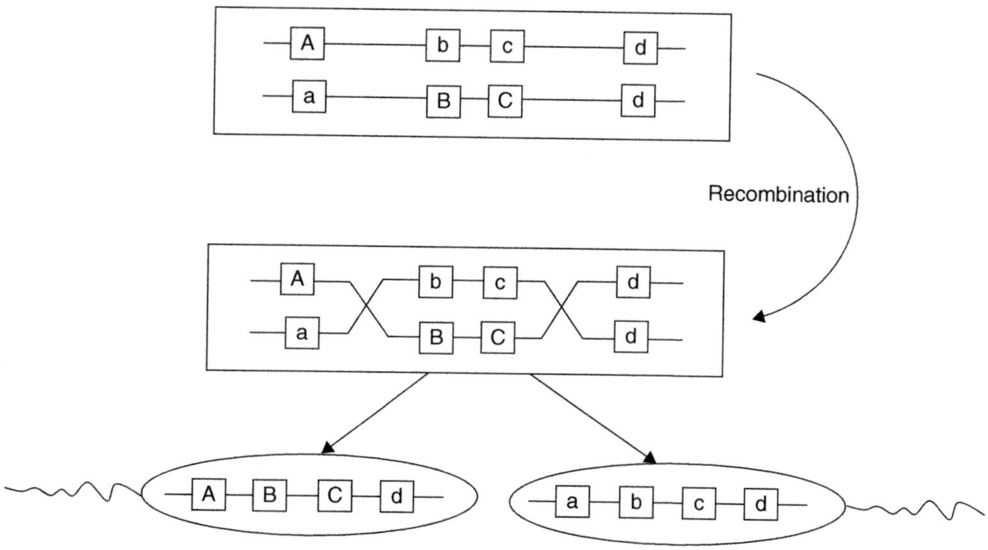

Fig. 1.1. Genetic determination of a trait, showing the effect of recombination of chromosomes.

shows how a parent with an intermediate genotype, through crossovers, can give rise to a sperm with many favourable genes and another with unfavourable genes. This is one of the causes of a well-known phenomenon in livestock – that good parents can occasionally have bad offspring.

A trait determined by n genes with two alleles each can give rise to 3^n different genotypes. Thus, most productive traits, which are determined by hundreds of genes, can give rise to billions of different genotypes. Furthermore, the number of possible alleles for the same gene can be greater than two, further increasing the number of possible genotypes.

1.1.2 Molecular markers

The recent development of molecular genetics has made it possible to sequence the genomes of animals at low cost. Since 2007 in dairy cattle, and later in other livestock species, chips containing a large number of genetic markers have been commercialized. A genetic marker is a point in the genome where we know that there are variants in the population. They can be composed of many bases or they can simply be what we call a single-nucleotide polymorphism (SNP), a simple nucleotide of which we know several versions, for example with two guanine bases (GG), two cytosine bases (CC) or one guanine and one cytosine (GC or CG), which allows it to be treated mathematically as if it were a gene with two alleles. Markers can be associated with major genes, such that one version of the marker is transmitted more frequently with the favourable version of the gene. In Fig. 1.2A, it can be seen how the marker (M, m) is associated with the major gene (Q, q), so that the M version of the marker is transmitted associated with the favourable Q version. For practical purposes, it is convenient to consider the M and m versions of the marker as if they were alleles of a gene. Here, the M version of the marker (which we treat as if it were an M 'allele') can be a simple base (e.g. guanine in a SNP) with no influence on the trait, while the Q allele of the larger gene is made up of many bases with great influence on the trait.

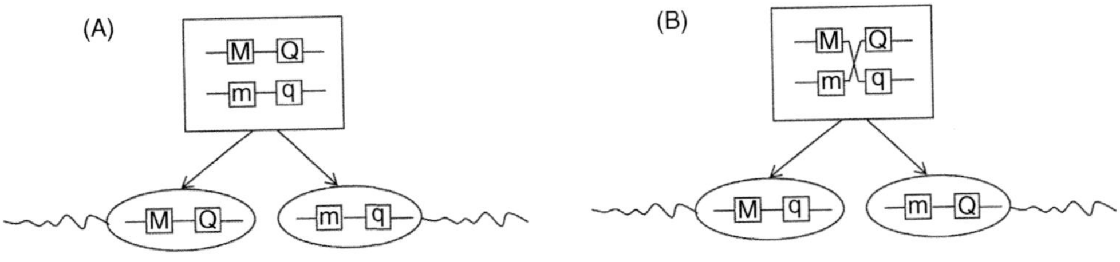

Fig. 1.2. Association of markers with genes. (A) Transmission without recombination. (B) Transmission after recombination.

We cannot observe Q, but we can observe M, so we observe that individuals with the M marker often have higher yields or favourable characteristics because M occurs more frequently associated with the Q allele. Ideally, if an individual has the M version of the marker, it should always have the Q version of the gene; this happens on some occasions when the marker is within the gene, for example it is a nucleotide or a set of nucleotides of the gene itself, but this is unusual. Generally, the marker is situated close to the gene and so the M version is usually transmitted with the Q allele. However, this is not always the case. Occasionally, a crossover can occur by recombination, as shown in Fig. 1.2B, and the M allele will instead be transmitted together with the q gene. As we only observe the M or m version of the marker, we do not know whether those with the M version have the Q or the q allele, but if the gene and marker are situated close to each other, then the Q allele will be more frequently associated with the M marker.

Thus, when two alleles are frequently transmitted together, they are usually physically close on the same chromosome and we say they are *linked*. Depending on the *recombination frequency* between these genes, these associations disappear with time. If the recombination frequency between them is very small, the association between M and Q will take time to disappear, but it will disappear in the long run, because once a recombination has occurred and M is associated with q, the recombination that would return again to M with Q will also be very infrequent. This is the reason why, in breeding programmes that use markers, the equations associating the productive traits with the markers have to be recalculated in each generation, or at least every three or four generations.

Markers can also be associated with genes even if they are not close to them or they are on different chromosomes. For example, if there is selection in a population, some genes on different chromosomes can influence the same trait, so that associations between alleles will appear even though there is no physical linkage between them. The mere fact that there are families in a population already implies that some individuals are more alike than others, so that associations between genes due to relationships may appear. Simply by chance, when a population has a small size, associations between genes can occur that would not occur in large populations, so we will find associations between genes and markers that are not due to physical linkage.

A logical way to assess the degree of association between genes, whatever its cause, is to calculate the correlation coefficient between alleles, as we will do in section 1.1.4. For this, we need to know their frequencies in a given population, as genes that are closely associated in one population may not be so closely associated in another; for example, the M-Q pair may be frequent in one population and the m-Q pair may be the most frequent in another population.

When markers are associated with certain genes in a population, this association may not be the same in another population, so the use of genetic markers to locate genes or to use them in genomic selection is better done in genetically close populations.

Below, we will consider the frequency with which alleles are found in a population, introducing the branch of genetics known as *population genetics*.

1.1.3 Equilibrium in the case of a gene

Let us consider a gene with two alleles, A and a, with frequencies p and q, respectively; for example, the population may have 60% A alleles and 40% a alleles. Thus, in this case, $p = 0.6$ and $q = 0.4$. Spermatozoids and ova will have these alleles in the same proportions of p and q (called *allelic* or *gametic frequencies*). When they meet at random, the frequencies of the genotypes will be as shown in Table 1.1, because the probability of two independent events is the product of their probabilities.

Table 1.1. Equilibrium frequencies of a gene with two alleles.

	A p	a q
○ A p	AA p^2	Aa $p \cdot q$
○ a q	aA $q \cdot p$	aa q^2

The genotypes Aa and aA are indistinguishable and we will simply call them Aa, with frequencies $pq + qp = 2pq$, so we will have three genotypes, AA, Aa and aa, with frequencies p^2, $2pq$ and q^2, respectively, which we will call homozygous AA, heterozygous Aa and homozygous aa. In Mendelian genetics, sometimes the A allele is called *dominant* and the a allele *recessive*, when the observed homozygote AA is the same or has the same value as the heterozygote Aa, but we will see that there are many intermediate situations. An example is shown in Table 1.2.

Table 1.2. Allelic and genotypic frequencies in equilibrium.

Gametes		Genotypes		
A	a	AA	Aa	aa
p	q	p^2	$2pq$	q^2
0.60	0.40	0.36	0.48	0.16

Note that even with a low frequency of aa genotypes in a population, the a allele will still be present in the heterozygotes. This is important because it is common for recessive a alleles to have undesirable effects (e.g. cause abnormalities) when they are homozygous (aa).

Although the frequency of these aa individuals in the population is small, many heterozygous Aa animals can be carriers, making it difficult to eliminate the a allele from the population because only the homozygous aa individuals show the trait. For example, in a population with only 1% recessive homozygous animals ($q^2 = 0.01$), the a allele has a frequency of $q = 0.1$, and the heterozygous Aa has a frequency $2pq = 2 \cdot 0.9 \cdot 0.1 = 0.18$; that is, 18% of the population carries the recessive allele without manifesting it.

The population described in Table 1.2 is said to be *in equilibrium* because generation after generation we will find the same genotypic frequencies if the mating occurs at random and if the frequencies of the alleles, p and q, are not modified. In order for the allele frequencies p and q not to change from generation to generation, several conditions must be met:

- *There should not be selection* and therefore all genotypes should have the same number of offspring. For example, if only the AA genotype has offspring, the entire population would be AA, so $p = 1$ and $q = 0$. The frequencies are also modified if, for example, the AA genotype produces more offspring than the other genotypes. When there are genotypes that produce more offspring than others, we say that there is *selection* in that population
- *There should not be mutations*. There are mutations when some A alleles become a, or vice versa. This would obviously modify the frequencies p and q. This mutation effect can occur simultaneously in both directions, with different frequencies.
- *The population should be in closed reproduction*. This means that no new members are introduced into the population that modify the allele frequencies. For example, an invasion of AA individuals (known as *migration*) would alter the allele frequencies.
- *The population should be infinite in size*. This is because if a non-infinite sample of individuals is taken as parents, it is likely that the A alleles will not be exactly 60% as in the parental population of our example, but there will be slightly more or slightly less. Therefore, for the frequencies of the alleles to maintain their p and q values exactly, the population has to be infinite in size. The sampling effect produced by the finite size of the population is called *genetic drift*, and is more pronounced the smaller the population (Fig. 1.3). *Note that alleles can be lost due to genetic drift* if no copy of that allele remains due to sampling. This is more likely to happen if they occur at low frequency; then one allele is lost and the other becomes *fixed*

To find the same frequencies of the genotypes Aa, Aa and aa generation after generation, one more condition is needed:

- The mating should occur at random. If alleles are not combined at random, the joining of alleles A and a is no longer an independent event and we cannot estimate the genotype frequencies as the product of the gametic frequencies, as we have done before.

When these conditions are met, generation after generation we find the same genotype frequencies, and we say that the population is in equilibrium. This law was formulated separately in 1908 by the British mathematician Hardy and the German physician Weinberg, which is why it is known as the *Hardy–Weinberg equilibrium* law. It states that

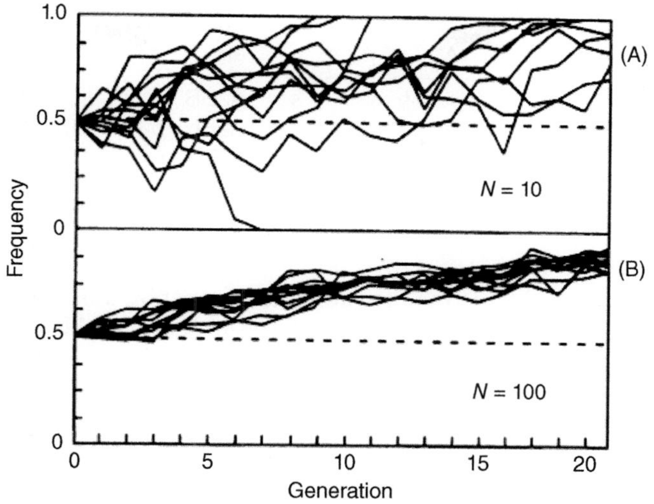

Fig. 1.3. Genetic drift. Selection over 20 generations was performed in 12 populations for a trait controlled by a gene with two alleles at initial frequencies of $p = q = 0.5$. (A) Small populations ($N = 10$). (B) Populations with $N = 100$. Differences between populations in allele frequencies in each generation are due to genetic drift. When $N = 10$, some populations fix or lose the allele by chance. From Falconer and Mackay (1996).

if there are no modifying forces of the frequencies (*selection*, *mutation*, *migration* or *genetic drift*), the frequencies are not modified, and, importantly, describes how to calculate these frequencies, based on the *statistical independence* of the union of alleles due to random mating. It was important at the time because it explained why genetic variability is maintained in populations.

Note that equilibrium is reached in the first generation of random mating, regardless of the initial genotype frequencies.

For example, in a population in which half of the genotypes are AA and the other half are aa (no heterozygotes), then $p = q = 0.5$, and the next generation of random mating will produce the allele frequencies shown in Table 1.1, and equilibrium will be reached with the frequencies of AA, Aa and aa being p^2, $2pq$ and q^2 as in Table 1.2, but as $p = q = 0.5$, the genotype frequencies will be 0.25, 0.50 and 0.25, and thus, in all subsequent generations, the genotype frequencies will remain constant at the values 0.25, 0.50 and 0.25.

1.1.4 Linkage disequilibrium

Let us take a trait regulated by two genes with two alleles each (A, a, and B, b) with frequencies of p and q, and r and s, respectively. The nine genotypes in our population will be: AABB, AABb, AAbb, AaBB, AaBb, Aabb, aaBB, aaBb and aabb. If the alleles were statistically independently, the frequencies of the genotypes would be the product of their probabilities. Table 1.3 shows the equilibrium frequencies for the genotypes.

Table 1.3. Gametic and genotypic frequencies of two genes in equilibrium.

♂ / ♀	AB $p \cdot r$	Ab $p \cdot s$	aB $q \cdot r$	ab $q \cdot s$
AB $p \cdot r$	AABB $p^2 \cdot r^2$	AABb $p^2 \cdot r \cdot s$	AaBB $p \cdot q \cdot r^2$	AaBb $p \cdot q \cdot r \cdot s$
Ab $p \cdot s$	AABb $p^2 \cdot r \cdot s$	AAbb $p^2 \cdot s^2$	AaBb $p \cdot q \cdot r \cdot s$	Aabb $p \cdot q \cdot s^2$
aB $q \cdot r$	AaBB $p \cdot q \cdot r^2$	AaBb $p \cdot q \cdot r \cdot s$	aaBB $q^2 \cdot r^2$	aaBb $q^2 \cdot r \cdot s$
ab $q \cdot s$	AaBb $p \cdot q \cdot r \cdot s$	Aabb $p \cdot q \cdot s^2$	aaBb $q^2 \cdot r \cdot s$	aabb $q^2 \cdot s^2$

If the equilibrium conditions that we described in section 1.1.3 hold, the genotypic frequencies will always be the same, generation after generation. In this case, the population is said to be in *linkage equilibrium*, and otherwise in *linkage disequilibrium*. This expression is particularly unfortunate, as there may be linkage disequilibrium but no linkage between genes (e.g. because they are on different chromosomes), or genes very close to each other on the chromosome may be in equilibrium in some populations and not in others, as we will now see in the examples below.

> Although it is common for genes that are physically close on the chromosome to be correlated, the correct way to express the relationship between genes should be to indicate whether or not there is *statistical independence* between them, without using the word *linkage*, because there may be correlations between alleles even if they are not on the same chromosome. However, it is common to use the expression *linkage disequilibrium* as a synonym for statistical dependence.

EXAMPLE 1.1

Let us take a population of 200 animals with two genes with two alleles each (A and a, and B and b) *on different chromosomes* (thus not physically linked), with the genotypes distributed as indicated in Table 1.4.

Table 1.4. Genetic structure of a population, shown as the number of animals for each genotype.

	BB	Bb	bb	
AA	50	42	6	98
Aa	20	44	20	84
aa	2	10	6	18
	72	96	32	200

The allele frequencies are:

$$freq(\mathrm{A}) = p = \frac{98 + \frac{84}{2}}{200} = 0.70 \quad freq(\mathrm{B}) = r = \frac{72 + \frac{96}{2}}{200} = 0.60$$

$$freq(\mathrm{a}) = q = 1 - p = 0.30 \quad freq(\mathrm{b}) = s = 1 - r = 0.40$$

Note that if we derive genotype frequencies from allelic frequencies, the results are different from those predicted in Table 1.4. For example, if we calculate the frequency of the AaBB genotype from the allelic frequencies (Table 1.3), we get:

$$freq_{EQ}(\text{AaBB}) = freq_{EQ}\left[\text{AB}_{sp}\ \text{aB}_{ov}\right] + freq_{EQ}\left[\text{aB}_{sp}\ \text{AB}_{ov}\right] =$$
$$= p \cdot q \cdot r^2 + q \cdot p \cdot r^2 = 2 \cdot p \cdot q \cdot r^2 = 2 \cdot 0.7 \cdot 0.3 \cdot 0.6^2 = 0.15$$

where *sp* is the spermatozoid, *ov* is the ovum and $freq_{EQ}$ is the equilibrium frequency, while in Table 1.4, the observed frequency of the AaBB genotype is:

$$freq(\text{AaBB}) = \frac{20}{200} = 0.10$$

In this population, there is no *statistical independence* between alleles, so the population is not in equilibrium. Note that in this example, the two genes that are in '*linkage disequilibrium*' (A, a, and B, b) are on different chromosomes; they do not need to be physically linked on the same chromosome to be in 'linkage disequilibrium'.

> Note that for an equilibrium between two genes, one more condition than in the case of a single gene is required. Here, it is not enough that the gametes mate randomly to achieve statistical independence between alleles; it is also necessary that *within each gamete there should be independence between alleles.*

For example, if gamete AB does not occur with a frequency $p \cdot r$, the genotypes in which it occurs (first row and first column of Table 1.3) will no longer have the equilibrium frequencies. When there is independence between alleles within each gamete, it is said that there is *gametic equilibrium*. The equilibrium frequencies of the gametes are shown in Table 1.5.

Table 1.5. Gametic equilibrium.

♂	A p	a q
B r	AB $p \cdot r$	aB $q \cdot r$
b s	Ab $p \cdot s$	ab $q \cdot s$

♀	A p	a q
B r	AB $p \cdot r$	aB $q \cdot r$
b s	Ab $p \cdot s$	ab $q \cdot s$

EXAMPLE 1.2

We can find out if the population in the previous example is in gametic equilibrium or not. For example, the equilibrium frequency of gamete AB (be it sperm or egg), should be:

$freq_{EQ}(\text{AB}) = p \cdot r = 0.7 \cdot 0.6 = 0.42$

To calculate the frequency of AB gametes in the population, consider that AB gametes are produced by AABB, AABb, AaBB and AaBb individuals, although not all of them produce the AB gametes equally: AABB individuals only produce AB gametes but for AABb and AaBB individuals, only half of their gametes are AB, and for AaBb only a quarter of their gametes are AB. The frequency of the gametes will be, according to the data in Table 1.4:

$$freq(\text{AB}) = \frac{50 + \frac{42}{2} + \frac{20}{2} + \frac{44}{4}}{200} = 0.46$$

which is different from the equilibrium frequency 0.42, and therefore the population is not in equilibrium because we do not have *gametic equilibrium*, although the population is mating at random and the allele frequencies (p, q, and r, s), are not modified. Therefore, the frequency of the genotypes will not be the product of the allele frequencies. This association between genes A, a and B, b is statistical, not physical, although it is true that physically close alleles on the same chromosome are often associated.

> Note that although there is no equilibrium between two genes, there can be equilibrium within each gene.

In fact, in our population, both genes are in equilibrium separately, because:

$$freq(\text{AA}) = \frac{98}{200} = 0.49 = 0.7^2 = p^2$$

$$freq(Aa) = \frac{84}{200} = 0.42 = 2 \cdot 0.7 \cdot 0.3 = 2pq$$

$$freq(aa) = \frac{18}{200} = 0.09 = 0.3^2 = q^2$$

and the same happens with the other gene (B, b).

The 'linkage disequilibrium' is measured as the difference between the observed frequency of a gamete and what it should be if it were in equilibrium (i.e. under conditions of statistical independence). In our example, the linkage disequilibrium, D, of gamete AB is:

$$D(\text{AB}) = freq(\text{AB}) - freq_{\text{EQ}}(\text{AB}) = 0.46 - 0.42 = 0.04$$

The frequency of the association between A and B in our population is 0.46, higher than if the alleles were completely independent (0.42). If we perform the same calculations for the remaining gametes, we will obtain:

$freq(\text{Ab}) = 0.24 \qquad freq(\text{aB}) = 0.14 \qquad freq(\text{ab}) = 0.16$

$$D(\text{Ab}) = freq(\text{Ab}) - freq_{\text{EQ}}(\text{Ab}) = 0.24 - (0.7 \cdot 0.4) = -0.04$$

$$D(\text{aB}) = freq(\text{aB}) - freq_{\text{EQ}}(\text{aB}) = 0.14 - (0.3 \cdot 0.6) = -0.04$$

$$D(\text{ab}) = freq(\text{ab}) - freq_{\text{EQ}}(\text{ab}) = 0.16 - (0.3 \cdot 0.4) = 0.04$$

As can be seen, the value is the same and only the sign varies, and furthermore, gamete AB has the same linkage disequilibrium as gamete ab, and Ab has the same as aB. This is a general property of linkage disequilibrium (see Appendix 1.1). We can represent the gametic frequencies under gametic disequilibrium. Using D as the linkage disequilibrium with an absolute value and f_{AB}, f_{Ab}, f_{aB} and f_{ab} as the gametic frequencies (spermatozoids or ova), we have the results shown in Table 1.6.

Table 1.6. Gametic disequilibrium.

	A	A	
B	$f_{AB} = p \cdot r + D$	$f_{aB} = q \cdot r - D$	r
b	$f_{Ab} = p \cdot s - D$	$f_{ab} = q \cdot s + D$	s
	p	q	

Linkage disequilibrium is a measure of statistical association; therefore, it must be related to the correlation between the effects of the gamete alleles. We will assume that alleles A and B have a value of k for a trait (e.g. k grams if the trait is weight), with the values of alleles a and b being null. This is simply a change in the origin of coordinates and does not imply a loss of generality. We will also assume that the gametes are not in equilibrium, and therefore their frequencies will not be the product of the frequencies of the alleles within the gametes but rather will be different, f_{AB}, f_{Ab}, f_{aB} and f_{ab}. To calculate this correlation between the effects of the alleles, we will generate two variables, the variable x will have two values, k or zero depending on whether the gamete has allele A or a, and the variable y will also have two values, k or zero depending on whether the gamete has allele B or b (Table 1.7).

Table 1.7. Computing the correlation between the effects of the alleles of a gamete.

x \ y		$y_1 = k$ $freq(y_1) = r$	B	$y_2 = 0$ $freq(y_2) = s$	b
$x_1 = k$ $freq(x_1) = p$	A	$x_1 \cdot y_1 = k^2$ $freq(x_1,y_1) = f_{AB}$	AB	$x_1 \cdot y_2 = 0$ $freq(x_1,y_2) = f_{Ab}$	Ab
$x_2 = 0$ $freq(x_2) = q$	a	$x_2 \cdot y_1 = 0$ $freq(x_2,y_1) = f_{aB}$	aB	$x_2 \cdot y_2 = 0$ $freq(x_2,y_2) = f_{ab}$	ab

It is important to realize that Table 1.7 does not represent the value of the gametes AB, Ab, aB and ab but rather the product of the variables x and y for those gametes. For example, the value of gamete AB is $k + k = 2k$, but the product of variables $x \cdot y = k \cdot k = k^2$ appears in the box for gamete AB, because this product is the one used to calculate the covariance. The same happens with the other gametes; Table 1.7 has a purely operational mission to facilitate the calculation of the covariance:

$$\text{Mean}(x) = \bar{x} = k \cdot p + 0 \cdot q = k \cdot p$$

$$\text{Mean}(y) = \bar{y} = k \cdot r + 0 \cdot s = k \cdot r$$

$$\text{cov}(x,y) = \Sigma x_i y_j f_{ij} - \bar{x} \cdot \bar{y} = k \cdot k \cdot f_{AB} + 0 + 0 + 0 - kp \cdot kr = k^2 (f_{AB} - pr)$$

$$\text{var}(x) = \Sigma x_i^2 f_i - \bar{x}^2 = k^2 p + 0 - k^2 p^2 = k^2 p(1-p) = k^2 pq$$

$$\text{var}(y) = \Sigma y_i^2 f_i - \bar{y}^2 = k^2 r + 0 - k^2 r^2 = k^2 r(1-r) = k^2 rs$$

$$r_D = \frac{\text{cov}(x,y)}{\sqrt{\text{var}(x) \cdot \text{var}(y)}} = \frac{k^2 (f_{AB} - pr)}{\sqrt{k^2 pq \cdot k^2 rs}} = \frac{D_{AB}}{\sqrt{pqrs}}$$

where D_{AB} is the linkage disequilibrium between alleles A and B. The same expression can be obtained for the linkage disequilibria of the other alleles.

As the 'linkage disequilibrium' D is always the same with one or the other sign, it is normally represented, to avoid the sign problem, as:

$$r_D^2 = \frac{D^2}{pqrs}$$

Note that when there is linkage equilibrium, $D = 0$ and the correlation between gamete alleles is then $r_D = 0$, as expected.

Linkage disequilibrium is a common situation in populations due to the large number of factors that can cause it. The same causes that lead to Hardy–Weinberg disequilibrium in a gene also cause linkage disequilibrium: *selection*, *mutation*, *migration* and *genetic drift* cause the gametes not to have their alleles in equilibrium, and furthermore, *if the mating is not random*, the genotypes will not be in equilibrium. Any factor altering the independence of alleles within a gamete generates linkage disequilibrium.

As we have said before, these associations between alleles disappear over time. For example, if there is a physical association between alleles A and B (i.e. they are on the same chromosome), there are crossovers that associate allele A with allele b, which leads to a decrease in the frequency of AB gametes, approaching the equilibrium frequency. Simultaneously, and also by crossover, new AB gametes are generated, until the situation is balanced. The speed with which this process occurs depends on the recombination fraction, c; if there is frequent recombination between alleles A and B the process is faster, but if not (i.e. if alleles A and B are 'closely linked'), the process is slower. Figure 1.4 shows how equilibrium is reached with different recombination fractions.

In our example, where the genes are on different chromosomes, $c = 0.5$ and equilibrium is reached quickly. We can calculate the frequencies of the gametes in the next generation.

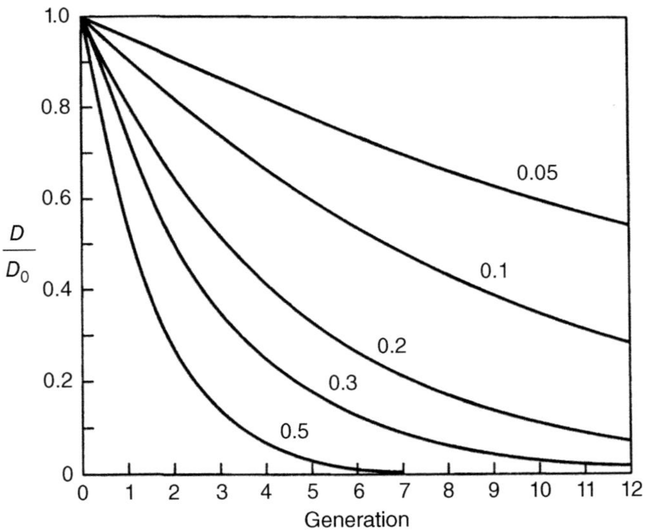

Fig. 1.4. Evolution of the *gametic disequilibrium*, *D*, with reference to D_0 (gametic disequilibrium in generation zero), for two genes with different recombination fractions. From Falconer and Mackay (1996).

First, it is necessary to calculate the frequencies of the genotypes produced by the gametes; for example, the frequency of the AABB genotype, if mating is random, will be:

$$freq(\text{AABB}) = freq(\text{AB}) \cdot freq(\text{AB}) = 0.46 \cdot 0.46 = 0.21$$

which is not the frequency of equilibrium, as the gametes are not in *gametic equilibrium*, despite having random mating; the equilibrium frequency should be:

$$freq_{EQ}(\text{AABB}) = p^2 \cdot r^2 = 0.7^2 \cdot 0.6^2 = 0.49 \cdot 0.36 = 0.18$$

If we calculate the frequencies of the remaining genotypes and, as before, from them we calculate the frequencies of the gametes, we will find that:

$$freq(\text{AB}) = 0.44$$
$$D(\text{AB}) = freq(\text{AB}) - freq_{EQ}(\text{AB}) = 0.44 - 0.42 = 0.02$$

The linkage disequilibrium for genes that are not physically linked has been reduced by half. The same will happen in the next generation, so the association between alleles A and B will quickly disappear. The consequences for the use of markers in the selection are important.

- The associations between genes and markers are lost over time, so it is necessary to re-estimate the relationships between markers from time to time, at least every three or four generations.
- The relationship between genes and markers depends on each population. Genes and closely associated markers in one population may be loosely associated in another population; *the linkage disequilibrium is not the same in all populations.* This implies that caution must be taken when using markers in a population distant from the population used to detect associations, such as detecting markers in a breed and trying to used them in other breeds, or in populations of the same breed but genetically distant (e.g. when using markers of North America Angus beef cattle in other Angus populations that are genetically distant).

1.2 Inbreeding and its Consequences

1.2.1 Coefficient of kinship and coefficient of inbreeding

One of the equilibrium conditions that we saw earlier is random mating. We will now see what happens to a population when there is no random mating. It is common to avoid mating relatives to avoid the detrimental effects derived from this sort of mating.

To study mating involving related individuals, it is convenient to have some measure of kinship. When individuals are closely related, their alleles are likely to be the same more often than if they are unrelated. One way to measure the kinship of two individuals is to randomly extract an allele from each one and see if they are identical because they were transmitted by a common ancestor; that is, to determine whether they are *identical by descent*. We describe the *kinship coefficient* of two individuals as the probability that two alleles taken at random, one from each individual, are identical by descent. An offspring of these two individuals will randomly receive one allele from each parent, so the probability that these two alleles are identical by descent, which we call the *inbreeding coefficient*, is the kinship coefficient of their parents.

- The *kinship coefficient*, r_{XY}, between two individuals, *X* and *Y*, is the probability that when taking an allele at random from each individual, these two alleles are identical by descent.
- The *inbreeding coefficient*, *F*, of an individual is the probability that the two alleles of that individual are identical by descent.
- The inbreeding coefficient of an individual coincides with the kinship coefficient of the parents.

In infinite populations, mating between relatives leads to the alleles of the offspring being identical more often than they would be if the mating were random; that is, the mating between relatives modifies the frequencies of the genotypes. Allele frequencies remain unchanged as long as there are no other reasons to modify them (e.g. selection or mutation).

To calculate the new genotype frequencies, we will divide the population into two groups; in one are all the individuals whose alleles are identical by descent, and in the other are the rest (Table 1.8). In the first group, there is a frequency p of AA individuals and q of aa individuals; the allelic and genotypic frequencies coincide because there are no heterozygotes and the number of AA genotypes with respect to the total number of individuals is equal to the number of A alleles with respect to the total number alleles. In the second group are the individuals that come from a random mating, and the frequencies of the genotypes will be those of equilibrium (p^2, $2pq$ and q^2).

As F is the probability of having the two identical alleles by descent, F will be the fraction of individuals in the population that belongs to the first subpopulation and $1 - F$ the fraction that belongs to the second subpopulation.

Table 1.8. Genotypic frequencies with inbreeding.

	Group 1: Frequency of alleles identical by descent (F)	Group 2: Frequency of alleles at random ($1 - F$)	Both groups
freq (AA)	p	p^2	$pF + p^2(1 - F) = p^2 + pqF$
freq (Aa)	0	$2pq$	$2pq(1 - F)$
freq (aa)	q	q^2	$qF + q^2(1 - F) = q^2 + pqF$

Note that as inbreeding increases, the percentage of homozygotes increases and that of heterozygotes decreases. Inbreeding is therefore a measure of homozygosity; that is, a measure of the proportion of homozygotes in a population.

A natural way to check if the two alleles of one individual are identical to those of another individual more frequently than if they were identical by chance is to calculate their correlation coefficient. In a population in equilibrium, an individual can have allele A with probability p and allele a with probability q. Let us consider two alleles of an individual and a variable x for the first allele, which can take the value k if it is the allele A and zero if it is the allele a, and consider a variable y for the other allele with the same characteristics (Table 1.9). The correlation is calculated as in section 1.1.4 (see Table 1.7):

Table 1.9. Computing the correlation between the effects of the alleles of a gamete under inbreeding.

x \ y		$y_1 = k$ $freq(y_1) = p$	A	$y_2 = 0$ $freq(y_2) = q$	a
$x_1 = k$ $freq(x_1) = p$	A	$x_1 \cdot y_1 = k^2$ $freq(x_1,y_1) = p^2 + pqF$	AA	$x_1 \cdot y_2 = 0$ $freq(x_1,y_2) = pq(1-F)$	Aa
$x_2 = 0$ $freq(x_2) = q$	a	$x_2 \cdot y_1 = 0$ $freq(x_2,y_1) = pq(1-F)$	Aa	$x_2 \cdot y_2 = 0$ $freq(x_2,y_2) = q^2 + pqF$	aa

$$\mathrm{Mean}(x) = \bar{x} = k \cdot p + 0 \cdot q = k \cdot p$$
$$\mathrm{Mean}(y) = \bar{y} = k \cdot p + 0 \cdot q = k \cdot p$$
$$\mathrm{cov}(x,y) = \Sigma x_i y_j f_{ij} - \bar{x} \cdot \bar{y} = k \cdot k \cdot (p^2 + pqF) + 0 + 0 + 0 - kp \cdot kp = k^2 pqF$$
$$\mathrm{var}(x) = \Sigma x_i^2 f_i - \bar{x}^2 = k^2 p + 0 - k^2 p^2 = k^2 p(1-p) = k^2 pq$$
$$\mathrm{var}(y) = \Sigma y_i^2 f_i - \bar{y}^2 = k^2 p + 0 - k^2 p^2 = k^2 p(1-p) = k^2 pq$$
$$r(x,y) = \frac{\mathrm{cov}(x,y)}{\sqrt{\mathrm{var}(x) \cdot \mathrm{var}(y)}} = \frac{k^2 pqF}{k^2 pq} = F$$

> Note that the inbreeding coefficient, F, is the correlation between the alleles of the same individual. It measures the association between the two alleles that is not due to chance but to having identical alleles by descent produced by mating relatives, instead of alleles that are identical but come from random mating.

From the definition of kinship coefficients, the kinship coefficients of close relatives are easily deduced, for example:

$$r(\text{Father, Son}) = \frac{1}{4}; \qquad r(\text{Full sibs}) = \frac{1}{4}; \qquad r(\text{Half sibs}) = \frac{1}{8}$$

and the kinship coefficient of an individual X with themself would be:

$$r_{XX} = \frac{1}{2} + \frac{1}{2} \cdot F_X = \frac{1 + F_X}{2}$$

The coefficients of consanguinity and kinship can be calculated recurrently. If we know that individual X is the offspring of A and B, and that individual Y is the offspring of C and D, it can easily be shown that:

$$r_{XY} = r_{A \times B, C \times D} = \frac{1}{4}(r_{AC} + r_{AD} + r_{BC} + r_{BD})$$

which allows us to calculate the coefficients of an entire genealogy by substituting A, B, C and D for their parents, and so on.

When pedigree information is not available, such as in natural populations or in populations in which kinship is not controlled, inbreeding can be estimated using molecular markers. As inbreeding is related to the proportion of heterozygotes, by using markers one can get an idea of the mean heterozygosity for each gene in the population, although here no distinction is made between identity by descent or by chance, so the coefficients tend to be higher. Moreover, it is assumed that the foundational population is unrelated, which is usually not the case, and mutations are ignored although they change the ratio of homozygous to heterozygous individuals over time. To examine the relationships between molecular kinship and the one we have defined in this section, the reader can consult Caballero (2020).

1.2.2 Control of inbreeding

It has been known for a long time that mating of related individuals produces a decrease in fertility and the frequent appearance of offspring with malformations. In addition, in certain characteristics, normally related to reproduction, there is also a decrease in performance, such as smaller litter sizes. Table 1.10 shows some experimental examples of yield loss due to inbreeding in pigs, poultry and dairy cattle, although it can be generalized to all domestic and non-domestic species (de Rose and Roff, 1999) and to plants (Austerlitz *et al.*, 2012).

Table 1.10. Inbreeding depression in some livestock species (from several sources).

Poultry	Egg number to 42 weeks	Percentage hatched of fertile eggs
Inbreeding depression of the mean for 10% F		
Line EN	−1.75	0.60
Line EW	−3.84	−2.95
Line EM	−0.31	−2.10

Adapted from Sewalem *et al.* (1999).

Dairy cattle	Milk (kg)	Fat (kg)	Protein (kg)
Inbreeding depression of the mean for 1% F	−95	−3.25	−2.93

Adapted from Smith *et al.* (1998).

Pig	Total born	Born alive	Weaned
Inbreeding depression of the mean for 10% F			
Landrace	−0.21	−0.19	−0.16
Large white	−0.16	−0.12	−0.21

Adapted from Köck *et al.* (2009).

In animal breeding programmes, inbreeding increases for various reasons. In dairy cattle, it is due to the reduced number of males used (see section 9.1.1.B.3), while in pigs or poultry, it is due to the relatively small population selected in the farm nucleus (see section 8.1.1). Furthermore, selection tends to increase inbreeding, as individuals having favourable genes also have relatives with the same alleles, and it is likely that both will be selected. This happens especially when family information is used for selection, which is common in intensive livestock breeding programmes.

Although it is difficult to generalize, it is generally recommended that inbreeding should not exceed 0.5% per generation. An approximate way of estimating the increase in inbreeding per generation is to calculate it assuming that we start a population from one *generation zero*, in which the founders are not related and also have zero inbreeding. In the next generation, there will be related individuals. To calculate the kinship coefficient of any two individuals, X and Y, let us take a random allele of X; the probability that it comes from its father is ½. The same happens if we take an allele of Y; the probability that it comes from its father is again ½. If there are N_m males, the probability that the father of X and Y is the same is $1/N_m$. But the father has two alleles; the probability that the allele of X and that of Y is the same is ½. The probability of all this happening will be:

$$\frac{1}{2}\cdot\frac{1}{2}\cdot\frac{1}{N_m}\cdot\frac{1}{2}=\frac{1}{8N_m}$$

The same happens with the allele from the mother, so if there are N_f females, the probability that two individuals taken at random have an identical allele by descent from the mother is $1/8N_f$, and the probability that they have an identical allele by descent, coming from the father or the mother, is:

$$r_{XY}=\frac{1}{8N_m}+\frac{1}{8N_f}$$

As the inbreeding coefficient of a descent is the kinship coefficient of its parents, and we have started from a population with zero inbreeding, the increase in inbreeding in the following generation due to the limited size of the population is precisely the kinship coefficient of two individuals taken at random, $\Delta F = r_{XY}$. In farm populations, the number of females is much higher than the number of males, so the second term is usually negligible:

$$\Delta F \approx \frac{1}{8N_m}$$

In the next generation, things get complicated because some of the males and females are already related, several parents share identical alleles by descent, and the probability that two individuals, X and Y, have an allele identical by descent is somewhat greater, but as the relationship between individuals of the population is still very small, this is a good approximation of the increase in inbreeding by generation. Thus, for example, a population with 15 males and 300 females suffers an increase in inbreeding per generation of approximately 1/120 (i.e. <1%). If it is intended that ΔF is less than 0.5%, the number of males must be at least 25. Note, however, that these numbers are only indicative.

An intuitive way to indicate the size of a population needed for maintaining a certain level of inbreeding is to calculate the *effective population number*, which is the size of an ideal population mating at random, including self-fertilization, producing an increase in inbreeding of ΔF. To calculate the increase in inbreeding in a population of these characteristics with N individuals, consider that the probability of getting an allele at random in an individual is ½, and the probability that their other allele is identical by descent to the previous one is $1/N$, so the increase in inbreeding from generation zero to the next is:

$$\Delta F=\frac{1}{2}\cdot\frac{1}{N}=\frac{1}{2N}$$

Therefore, we define the effective population number as:

$$N_e=\frac{1}{2\cdot\Delta F}$$

For example, if a population should not increase its inbreeding beyond $\Delta F = 0.005$ per generation (0.5%), the *effective number* or *effective population size* should not be less than $N_e = 100$. Note that a population of 25 males already has approximately that effective number, so in livestock, *the effective number is approximately equal to four times the number of males*. Calculation of the effective number can be complicated by calculating it for several generations, for selected populations in which certain individuals leave more offspring than others, for variable population sizes in each generation, etc.; the interested reader can consult Caballero (2020) for these details. The effective number is an intuitively interesting concept to compare the increase in inbreeding in our population with respect to an ideal population, but it is not defined for some specific cases; for example, when mating the least related individuals, inbreeding decreases. In this case, the effective number would be negative, which does not make sense.

Controlling inbreeding is necessary in all breeding programmes. However, in prolific species (e.g. pigs, poultry, rabbits), inbreeding does not usually affect the farmer, who frequently uses cross-bred animals with zero inbreeding, because when crossing, the male and female are from different lines and are not related (see section 8.1.1). In prolific species, inbreeding has to be controlled in the farm nuclei of selection because these nuclei have lines in closed reproduction. Inbreeding also has to be controlled in the case of dairy cattle, in which artificial insemination utilizes a very small number of males to serve a world population, and in the case of small populations in closed reproduction, typical of local breeds of ruminants (e.g. sheep, goats, beef cattle).

There are several methods to control increased inbreeding, for example:

- *Mating individuals avoiding kinship*. This can be achieved using simple methods such as avoiding common grandparents, or using more sophisticated methods such as linear programming and other methodologies that optimize inbreeding control. A technique sometimes used is to subdivide the population into small subpopulations that interbreed on a rotating basis, although this technique harms genetic progress by reducing the chances of selecting the best animals, as now the selection must be made within each small subpopulation. The problem with all these techniques is that, in the long run, inbreeding increases. By mating non-relatives, more related individuals are eventually mated in the long run than they would be if the population were randomly mated. This, however, is not a problem given that, for the sizes of populations usually used in animal breeding, this increase in inbreeding takes a long time to happen, and we have seen that after many generations, unfavourable alleles are lost due to genetic drift, and new heterozygotes appear due to mutations.
- *Introducing new males into the population*. This is effective in reducing the average inbreeding in a population, as the new males are not related to the existing ones, but they should not be genetically inferior to those of the recipient population to avoid losing genetic progress.
- *Increasing the number of males compared with those used in a production farm*. As we have seen, the increase in inbreeding in a population depends more on the number of males than that of females, as fewer males are always used on farms. The cost of keeping more males is relevant, because, due to the widespread use of artificial insemination, the number of males required on a farm is small.

An important question is how far one can go back in generations to calculate an individual's inbreeding coefficient, as the more generations that are considered, the more the coefficient of inbreeding will increase. In general, except in very small populations where close kinship is common, calculating the inbreeding coefficient for three or four generations is enough for practical purposes. In the long run, the inbreeding coefficient loses its original meaning because two phenomena occur:

- After many generations (around 20 in animals), the appearance of mutations means that inbreeding is no longer a direct measure of homozygosity, so by increasing the number of heterozygotes due to mutations, the negative effects of inbreeding are mitigated (see section 10.2.1).
- Sometimes, none of the gametes carries copies of a particular allele and the allele is lost. This implies that certain recessive alleles that are lethal in homozygosity or have negative effects, are lost; a *purge of alleles* detrimental to reproduction is said to have occurred. This phenomenon is more pronounced in small populations, in which genetic drift has a higher effect. As the harmful recessives disappear, the effects of inbreeding are mitigated. This makes some authors distinguish between *new inbreeding* and *old inbreeding* (Doekes *et al.*, 2019).

The inbreeding coefficient loses its meaning when calculated beyond approximately 20 generations, and studies in mice indicate that it probably loses its meaning much earlier. As the important inbreeding in livestock is '*new inbreeding*', it seems reasonable to pay more attention to its recent evolution, for example in the last five generations (see section 9.1.2.B3.

1.2.3 Changes in mean and variance due to inbreeding

As we have seen earlier, inbreeding does not change the allele frequencies (p and q), but it does change the genotypic frequencies, decreasing the percentage of heterozygotes. We have seen previously that inbreeding produces a decrease in yields for many traits, particularly those related to reproduction.

Suppose that the homozygous AA genotype increases the value of a characteristic by an amount k, the homozygous aa genotype reduces it by an amount $-k$ and the heterozygote has a quantity d over the average of the homozygous mean (which in our example is zero; Fig. 1.5). In this situation, we will say that there is *dominance* of the heterozygote, also called Mendelian dominance, or d (Table 1.11). Generally, $d > 0$ and $d < k$, but in some rare cases $d > k$ and we say there is *overdominance*.

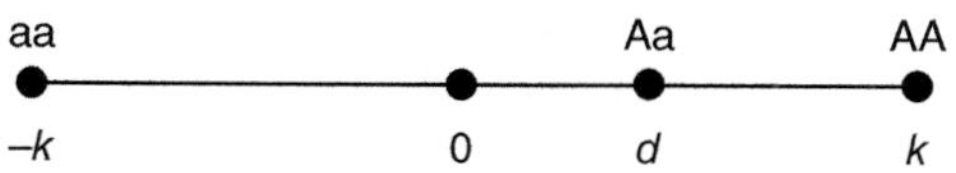

Fig. 1.5. Genotypic values.

Table 1.11. Genotypic values, and genotypic frequencies with or without inbreeding.

	AA	Aa	aa
Frequencies without inbreeding	p^2	$2pq$	q^2
Frequencies with inbreeding	$p^2 + pqF$	$2pq(1 - F)$	$q^2 + pqF$
Genotypic value	k	d	$-k$

In a population with no inbreeding, the mean will be:

$$M_0 = k \cdot p^2 + 2pqd - k \cdot q^2 = k(p^2 - q^2) + 2pqd = k(p+q)(p-q) + 2pqd = k(p-q) + 2pqd$$

With inbreeding, the mean is:

$$M_F = k(p^2 + pqF) + d \cdot 2pq(1-F) - k(q^2 + pqF) = k(p^2 - q^2) + 2pqd - 2pqdF = M_0 - 2pqdF$$

When there is inbreeding and positive dominance, the mean decreases because the number of heterozygotes decreases. Note that when $d = 0$, *when there is no dominance, the mean does not change when inbreeding increases*. This explains why some traits are sensitive to inbreeding, but others are hardly altered even if inbreeding increases. In general, the traits related to growth and carcass quality (e.g. percentage of fat, or percentage of expensive retail cuts of the carcass) do not change or change very little with inbreeding, while those related to reproduction (e.g. litter size, fertility, milk production, egg production) decrease with increasing inbreeding.

To see the effect on the variance, we will examine a simple example, without dominance ($d = 0$). If there were no inbreeding, the variance would be:

$$V_0 = k^2p^2 + (-k)^2q^2 - (M_0)^2 = k^2(p^2 + q^2) - k^2(p-q)^2 = 2pq \cdot k^2$$

If there is inbreeding but no dominance ($d = 0$), we saw that $M_F = M_0$ and the variance becomes:

$$V_F = k^2\left(p^2 + pqF\right) + (-k)^2\left(q^2 + pqF\right) - (M_0)^2 = k^2\left(p^2 + q^2\right) + 2pqk^2F - k^2(p-q)^2 =$$
$$= 2pqk^2 + 2pqk^2F = V_0 + V_0F = V_0(1+F)$$

The effect of inbreeding is to decrease the proportion of heterozygotes. By decreasing the proportion of heterozygotes, which have intermediate values, the population tends to be concentrated at the extremes, increasing the frequency of AA and aa homozygotes, and *therefore the variance increases with inbreeding*. When there is Mendelian dominance ($d > 0$), the effect is the same, but the formula is more complex and has a quadratic term of inbreeding (F^2).

1.2.4 Heterosis

When two populations that have remained isolated in reproduction (e.g. two breeds) are crossed, *the crossed individuals present better yields than the average of the parents' populations* for the traits that are affected by inbreeding. This phenomenon is called *heterosis*. As the effect of inbreeding is to decrease the proportion of heterozygotes, the effect of crossing should be to increase the proportion of heterozygotes, as we will see below.

Suppose that the frequencies of the two alleles of a locus, A and a, in one population are p and q, and in the other population are r and s. Frequencies will normally have been modified by mutation and drift, and sometimes by selection, and they are not expected to be the same in both populations. The frequency of the AA genotype in the two parental populations together, before crossing them, will be

$$freq(\mathrm{AA})_{\mathrm{PARENTAL}} = \frac{p^2 + r^2}{2}$$

Assuming that males from the first population mate at random with females from the second population and vice versa, the probability that a gamete with allele A from the first population meets a gamete with allele A from the second population is $p \cdot r$, and thus the frequency of the genotype AA in the cross is:

$$freq(\text{AA})_{\text{CROSS}} = p \cdot r$$

To compare both frequencies, we will add and subtract the quantity $p \cdot r$ from the frequency in the parental population, which will not change:

$$freq(\text{AA})_{\text{PARENTAL}} = \frac{p^2 + r^2}{2} + pr - pr = pr + \frac{p^2 + r^2 - 2pr}{2} = pr + \frac{(p-r)^2}{2}$$

The difference between both frequencies is a positive quantity:

$$freq(\text{AA})_{\text{PARENTAL}} - freq(\text{AA})_{\text{CROSS}} = pr + \frac{(p-r)^2}{2} - pr = \frac{(p-r)^2}{2} > 0$$

Therefore, the frequency of the homozygous AA genotype is higher in the population of parents. The argument is the same for the homozygous aa genotype, and thus we can conclude that there are more homozygotes in the parental populations and fewer homozygotes in the crossed population, or that *there are more heterozygotes in the crossed population*. This is greater if the distance between the populations is greater; that is, if the difference between the allele frequencies of the population ($p - r$, or $q - s$) is large. If there is dominance ($d > 0$), the cross population has higher yields than the mean of the parental populations. This explains why it is preferred to cross populations that have been separated for a long time, for example two breeds, because they are expected to have very different allele frequencies. Sometimes, this is not possible because one breed has a much higher performance than any other (e.g. Leghorn chickens for white eggs). In this case, different lines of the same breed that were separated in reproduction for a long time are used.

> Note that heterosis does not guarantee that the offspring are, on average, superior to both parental populations, but rather *to the mean of the parental populations*. If the litter size of two pig breeds is 10 and 14 respectively, and heterosis is 10%, the expected mean of the parents is 12, and the expected mean of the cross is 10% higher than 12; that is, 12 + 1.2 = 13.2, lower in this case than one of the two parental populations.

If the crossed population is maintained in closed reproduction, that is, *if the subsequent generations are the product of crossed males with crossed females*, then *heterosis is gradually lost*. In the crossed population, the frequencies of the gametes are:

$$freq(\text{A})_{\text{CROSS}} = p_{CROSS} = \frac{p+r}{2} \quad ; \quad freq(\text{a})_{\text{CROSS}} = q_{CROSS} = \frac{q+s}{2}$$

In next generation (CROSS1), the genotypic frequency of the homozygous AA genotype will be, adding and subtracting $p \cdot r$ as before:

$$freq(\text{AA})_{\text{CROSS1}} = p^2_{\text{CROSS}} = \left(\frac{p+r}{2}\right)^2 = \frac{p^2 + r^2 + 2pr}{4} + pr - pr = pr + \frac{p^2 + r^2 - 2pr}{4} = pr + \frac{(p-r)^2}{4}$$

The difference between the frequency of AA in the cross and in the original parental population is:

$$freq(\text{AA})_{\text{PARENTAL}} - freq(\text{AA})_{\text{CROSS1}} = pr + \frac{(p-r)^2}{2} - pr - \frac{(p-r)^2}{4} = \frac{(p-r)^2}{4}$$

Note that the distance to the parental generation has been halved.

Maintaining a crossed population in close reproduction is quite common for creating what is called synthetic lines, common in monogastric animals and sometimes in beef cattle. The heterosis of the first cross disappears with time.

1.3 Breeding Value

1.3.1 Partition of the genotypic value

In a trait regulated by one gene with two alleles, A and a, each genotype (AA, Aa or aa) has a value determined by the individual effects of each allele, and their interaction.

EXAMPLE 1.3

Let us consider a population in equilibrium with allele frequencies $p = 0.4$ and $q = 0.6$, with the genotypic values shown in Table 1.12.

Table 1.12. Genotypic values and frequencies of a trait regulated by a gene with two alleles.

	AA	Aa	aa
Frequency	0.16	0.48	0.36
Genotypic value	0.75	2	–3

These genotype values are constant: all individuals with genotype AA will have the value of 0.75, and the same applies for the other genotypes. However, the genotypes do not pass directly to the offspring, as gametes only have one allele, A or a. It is important to calculate the average effect of an allele, as it is this effect that will pass to the offspring. Table 1.13 shows the genotypic values, with the frequencies in parentheses. To calculate the *average effect* of each allele in this population, we will calculate the mean of the rows (or columns) of Table 1.13.

Table 1.13. Average effects of the alleles (frequencies in parentheses).

	A	a	Average effect
A	0.75 (0.16)	2 (0.24)	1.5 (0.40)
a	2 (0.24)	–3 (0.36)	–1 (0.60)
Average effect	1.5 (0.40)	–1 (0.60)	0

$$\text{Average effect}(\text{A}) = \frac{0.75 \cdot 0.16 + 2 \cdot 0.24}{0.16 + 0.24} = 1.5$$

$$\text{Average effect}(\text{a}) = \frac{2 \cdot 0.24 - 3 \cdot 0.36}{0.24 + 0.36} = -1$$

$$\text{Mean} = 1.5 \cdot 0.4 - 1 \cdot 0.6 = 0$$

These average effects of each allele are associated with the gametes, and on average they are transmitted to the offspring. The *breeding value* of an animal is the sum of the effects of its alleles, and it is also called the *additive value*. Note that the sum of the allele effects does not coincide with the genotypic value (Table 1.12). We call the difference between the genotypic value and the sum of the individual effects of the alleles the *dominant value* or *dominant deviation* (Table 1.14).

> Note that the *dominant value* appears in the three genotypes, not only in the Aa genotype, and therefore it should not be confused with the dominance *d* that we have defined before (the so-called *Mendelian dominance*).

Table 1.14. Components of the genotypic value.

	AA	Aa	aa
Frequency	0.16	0.48	0.36
Genotypic value	0.75	2	−3
Breeding value	1.5 + 1.5 = 3	1.5 − 1 = 0.5	−1 − 1 = −2
Dominant value	0.75 − 3 = −2.25	2 − 0.5 = 1.5	−3 − (−2) = −1

For the general case, we will take the genotypic values of Fig. 1.5. We assume that the population is in Hardy–Weinberg equilibrium (Table 1.15).

Table 1.15. Genotypic values and frequencies of a trait regulated by a gene with two alleles.

	AA	Aa	aa
Frequency	p^2	$2pq$	q^2
Genotypic value	k	d	$-k$

Table 1.16 shows the genotypic values with the frequencies in parentheses. We calculate the *average effect* of each allele as in the example.

Table 1.16. Average effects of the alleles.

	A	a	Average effect
A	$k(p^2)$	$d(pq)$	$kp + dq$
a	$kp + dq$	$-k(q^2)$	$-kq + dp$
Average effect	$kp + dq$	$-kq + dp$	m

$$\text{Average effect}(\text{A}) = \frac{kp^2 + dpq}{p^2 + pq} = \frac{p(kp + dq)}{p(p + q)} = kp + dq$$

$$\text{Average effect (a)} = \frac{dpq - kq^2}{pq + q^2} = \frac{q(dp - kq)}{q(p + q)} = -kq + dp$$

$$m = (kp + dq)p + (-kq + dp)q = k(p - q) + 2dpq$$

We can express the average effect as an absolute value, but it is usually referred to as the *mean of the population*. In our example, the mean was zero, but this is not normally the case; thus, the centred average effects are:

$$\alpha_A = \text{Average effect}(A) - m = kp + dq - k(p-q) - 2dpq =$$
$$= q(d + k - dp - dp) = q(k + d(1-p) - dp) = q(k + d(q-p))$$

$$\alpha_a = \text{Average effect}(a) - m = -kq + dp - k(p - q) - 2dpq = -p(k + d(q - p))$$

If $\alpha = k + d(q - p)$, we have:

$$\alpha_A = q \cdot \alpha$$
$$\alpha_a = -p \cdot \alpha$$
$$\alpha_A - \alpha_a = q \cdot \alpha - (-p \cdot \alpha) = (q + p)\alpha = \alpha$$

The *breeding value* or *additive value*, A, of a genotype is the sum of the effects of its alleles:

$$A_{AA} = \alpha_A + \alpha_A = q \cdot \alpha + q \cdot \alpha = 2q \cdot \alpha$$
$$A_{Aa} = \alpha_A + \alpha_a = q \cdot \alpha - p \cdot \alpha = (q - p) \cdot \alpha$$
$$A_{aa} = \alpha_a + \alpha_a = -p \cdot \alpha - p \cdot \alpha = -2p \cdot \alpha$$

The importance of the breeding value or additive value is that it is the one that is transmitted to the offspring, which is not the case with the dominant or genotypic values, as the offspring receive the two alleles separately, each in a gamete. The dominant effects are interactions between the alleles of a gene that disappear when the gametes are formed, and therefore they are not transmitted to the offspring. The breeding value (or additive value) of an animal is the sum of the effects of its alleles.

We can also reference the genotypic value, G, to the mean of the population, and it can be verified after performing some operations:

$$G_{AA} = k - m = 2q(\alpha - dq)$$
$$G_{Aa} = d - m = (q - p)\alpha + 2pqd$$
$$G_{aa} = -k - m = -2p(\alpha + dp)$$

The dominant value, D, will be the difference between the genotypic value and the additive value. After doing some operations, we can verify that:

$$D_{AA} = G_{AA} - A_{AA} = -2q^2 d$$
$$D_{Aa} = G_{Aa} - A_{Aa} = 2pqd$$
$$D_{aa} = G_{aa} - A_{aa} = -2p^2 d$$

Therefore, and in a general way, we can write for each of the genotypes, using the centred genotypic values:

$$G = A + D$$

Note that when the value of the heterozygote is located exactly between the homozygotes ($d = 0$), there are no dominant interactions in any of the three genotypic values, the genotypic values are equal to the additive ones, and the effect of substitution of one allele for another α is just the quantity k.

Note that *the additive and dominant values depend on the frequencies of the alleles in the population*. Two populations with the same genotype values can have different additive and dominant values. If an individual migrates from one population to another population that has different allele frequencies, its breeding value changes, because the allele frequencies of the new population are different. The partition of genotype values into additive and dominant values is statistical, and these values only make sense in a specific population with certain allele frequencies; they are not like genotype values, which are fixed for each genotype and do not depend on frequencies.

Note that if we reference the genotype values to the population mean, then they do depend on the allele frequencies, because the mean depends on these frequencies.

Finally, note that the mean of the additive values and the mean of the dominant values (the sum of the effects multiplied by their frequencies) is in both cases zero.

The fact that the same expression, 'dominant', is used to refer to the *dominant effect of the heterozygote, d*, and to the *dominant values of the genotypes*, D_{AA}, D_{Aa} and D_{aa} is a source of confusion. We can have a population with a strictly dominant genotypic determination, and yet have additive and dominant values, as we will see in the following example.

EXAMPLE 1.4

We will take the genotype values and genotype frequencies of Table 1.17, where heterozygotes have a value of 1 and homozygotes have a value of zero. The population is in equilibrium with allele frequencies of $p = 0.1$ and $q = 0.9$.

Table 1.17. Genotypic values and frequencies in a population completely dominant in a Mendelian sense.

	AA	Aa	aa
Frequency	0.01	0.18	0.81
Genotypic value	0	1	0

The mean of this population is:

$$m_0 = 0.01 \cdot 0 + 0.18 \cdot 1 + 0.81 \cdot 0 = 0.18$$

If we practice selection of the best individuals, all of them will be heterozygous, and thus in the next generation we would have genotype frequencies AA, Aa and aa of 0.25, 0.50 and 0.25. The mean, calculated as before, will be m_1=0.50, and thus we would have obtained a response to selection despite the fact that the genotypic determination of the trait is strictly dominant. As the only values transmitted to the offspring are the additive values, this means that the genotype Aa should have an additive value; indeed, all genotypes have additive values and dominance values. In this example:

$$\alpha = k + d(q - p) = q - p = 0.8$$

Therefore, the additive and dominant values are:

$$
\begin{aligned}
A_{AA} &= 2q \cdot \alpha = 1.44 & D_{AA} &= -2q^2 d = -1.62 \\
A_{Aa} &= (q-p) \cdot \alpha = 0.64 & D_{Aa} &= 2pqd = 0.18 \\
A_{aa} &= -2p \cdot \alpha = -0.16 & D_{aa} &= -2p^2 d = -0.02
\end{aligned}
$$

and, as we can see:

$$
\begin{aligned}
G_{AA} &= m_0 + A_{AA} + D_{AA} = 0.18 + 1.44 - 1.62 = 0 \\
G_{Aa} &= m_0 + A_{Aa} + D_{Aa} = 0.18 + 0.64 + 0.18 = 1 \\
G_{aa} &= m_0 + A_{aa} + D_{aa} = 0.18 - 0.16 - 0.02 = 0
\end{aligned}
$$

The model can be generalized to several genes with several alleles. A genotypic value (e.g. that of AaBB) is determined by the values of the individual effects of alleles A, a, B and B, and by their interactions. If we could measure all the possible genotypic values, we could estimate the value of these effects and their interactions as in the conventional analysis of variance with a factorial design with four effects. We would then have a model with four effects with double, triple and quadruple interactions. We have called the interactions between alleles of the same locus the *dominant value*, and we have said that it should not be confused with Mendelian dominance, *d*. The remaining interactions are called *epistasis*. With two loci, the interactions between the effects of the alleles of different loci are called *epistatic additive per additive value* (*AA*). The interactions between the effects of one allele and the dominant effect of the other locus are called *epistatic additive per dominant value* (*AD*). Finally, the interactions between dominant effects of the two loci are called *epistatic dominant per dominant value* (*DD*). The partition of the genotypic value becomes:

$$G = A + D + AA + AD + DD$$

This notation can be generalized to more than two genes, but we will not do so as it is not feasible to estimate epistatic values of such a high order. In fact, epistasis can rarely be estimated, other than in laboratory experiments or using pure lines and their crosses in plants. The calculation of average values is done exactly the same way as in Example 1.4.

1.3.2 Partition of genotypic variance

The variance of the additive (breeding) values, which we call *additive variance*, will be, taking into account that the mean of the additive values is zero:

$$\sigma_A^2 = (A_{AA})^2 p^2 + (A_{Aa})^2 2pq + (A_{aa})^2 q^2 = (2q\alpha)^2 p^2 + (q-p)^2 \alpha^2 2pq + (-2p\alpha)^2 q^2 = 2pq\alpha^2$$

The variance of the dominant values is calculated in the same way:

$$\sigma_D^2 = (D_{AA})^2 p^2 + (D_{Aa})^2 2pq + (D_{aa})^2 q^2 = (-2q^2 d)^2 p^2 + (2pqd)^2 2pq + (-2p^2 d)^2 q^2 = (2pqd)^2$$

When there is Hardy–Weinberg equilibrium, the covariance between the additive and dominant values is zero and they are independent:

$$
\begin{aligned}
\operatorname{cov}(A,D) &= (A_{AA} D_{AA}) p^2 + (A_{Aa} D_{Aa}) 2pq + (A_{aa} D_{aa}) q^2 = \\
&= (2q\alpha)(2q^2 d) \cdot p^2 + (q-p)\alpha(2pqd) \cdot 2pq + (-2p\alpha)(-2p^2 d)^2 \cdot q^2 = 0
\end{aligned}
$$

As this covariance is zero, the variance of the genotype values can be broken down into the sum of both variances:

$$\sigma_G^2 = \sigma_A^2 + \sigma_D^2 = 2pq\alpha^2 + (2pqd)^2$$

The importance of the additive variance lies in it being the variance of the breeding values, used for predicting genetic progress when selection is made (see section 2.3.3.A). If there is no additive variance, all genotypes have the same additive value and there is no genetic progress.

Note that in a Mendelian additivity situation (i.e. when $d = 0$), the dominant variance is zero ($\sigma_D^2 = 0$) and the genetic variance is the same as the additive variance ($\sigma_G^2 = \sigma_A^2$). In these cases, the variance is maximum for intermediate allele frequencies, $p = q = 0.5$.

Note that *if there were a complete advantage of the heterozygote* ($k = 0$, $d > 0$), *there would still be additive variance*, except for intermediate frequencies, $p = q = 0.5$, in which $\alpha = d\ (q - p)$ would be null. This is a frequent source of confusion, because at first glance it appears that a purely Mendelian dominant model should not have additive variance. As in the case of dominant values, the confusion comes from the fact that Mendelian dominance is not the same as the statistical dominance that we are dealing with in this chapter. *Additive variance is not necessarily produced by genes with additive effect* (i.e. when $d = 0$), *but can be produced by genes with any degree of dominance.* The same occurs with epistasis; as in the case of dominance, it may happen that certain combinations of genes can have high value and be selected. In this case, the frequencies of the alleles that make up that combination increase and there is genetic progress, so there is additive variance. This is the meaning of the somewhat mysterious statement that 'epistasis generates additive variance'.

Note that we have defined the additive value as the sum of the effects of the alleles. In the case where they are in Hardy–Weinberg equilibrium, this sum coincides with the sum of the average effects of each allele. In the case where the population is not in equilibrium, the average effects of the alleles no longer give rise to additive values independent of the dominant values. The additive effects should then be estimated by the least-squares method, which ensures that the additive effects will be independent of the dominant effects (see Appendix 1.2).

A final comment regarding the nomenclature: in statistics, we usually speak of *estimation* of a parameter, and of *prediction* of an unobserved value or unobserved data. For example, if you want to *predict* carcass weight from live weight from a regression equation, you can *estimate* the regression coefficient and *predict* a carcass weight *not observed* in an individual, from its live weight. In our case, we will speak from now on about estimation of the variances but about the prediction of the breeding values, because this nomenclature is frequently used among animal breeders, although the expression 'estimation of breeding values' is sometimes used (e.g. Lynch and Walsh, 1998, pp. 745–778).

Appendix 1.1

There is a useful property of determinants where, by substituting a column for a linear combination of other columns including the one being substituted, the result does not change; for example, substituting a column for the sum of the columns, the determinant does not change:

$$\begin{vmatrix} a & x \\ b & y \end{vmatrix} = ay - bx = ay - bx + ab - ab = a(b+y) - b(a+x) = \begin{vmatrix} a & a+x \\ b & b+y \end{vmatrix}$$

and the same happens if a row is replaced by the sum of the rows – the determinant does not change.

In the case of gametic frequencies:

$$\begin{vmatrix} f_{AB} & f_{Ab} \\ f_{aB} & f_{ab} \end{vmatrix} = \begin{vmatrix} f_{AB} & f_{AB}+f_{Ab} \\ f_{aB} & f_{aB}+f_{ab} \end{vmatrix} = \begin{vmatrix} f_{AB} & p \\ f_{aB} & q \end{vmatrix} = \begin{vmatrix} f_{AB} & p \\ f_{AB}+f_{aB} & p+q \end{vmatrix} = \begin{vmatrix} f_{AB} & p \\ r & 1 \end{vmatrix} = f_{AB} - pr$$

$$\begin{vmatrix} f_{AB} & f_{Ab} \\ f_{aB} & f_{ab} \end{vmatrix} = \begin{vmatrix} f_{AB}+f_{Ab} & f_{Ab} \\ f_{aB}+f_{ab} & f_{ab} \end{vmatrix} = \begin{vmatrix} p & f_{Ab} \\ q & f_{ab} \end{vmatrix} = \begin{vmatrix} p+q & f_{Ab}+f_{ab} \\ q & f_{ab} \end{vmatrix} = \begin{vmatrix} 1 & s \\ q & f_{ab} \end{vmatrix} = f_{ab} - qs$$

and therefore $f_{AB} - pr = f_{ab} - qs$, and in the same way it can be shown that $f_{Ab} - ps = f_{aB} - qr$ with the opposite sign.

Appendix 1.2

When the population is not in equilibrium, the additive values estimated as the mean effect are correlated with the dominant values. To avoid this, we can estimate the additive effects using a multiple regression model of the type $y = m + b_1x_1 + b_2x_2 + e$ in which y represents the genetic values, the variable x_1 is the number of A alleles in the genotype and x_2 is the number of a alleles. The residual e is the dominant value, and thus the equation, using $y = G$, $x_1 = N_A$, $x_2 = N_a$, $b_1 = \alpha_A$, $b_2 = \alpha_a$ and $e = D$, is:

$$G = m + \alpha_A N_A + \alpha_a N_a + D$$

where we know that the number of alleles cannot be higher than two: $N_A + N_a = 2$ (Table 1.18).

Table 1.18. Genotypic values and genotypic frequencies with no Hardy–Weinberg equilibrium.

	AA	Aa	aa
Frequencies	f_{AA}	f_{Aa}	f_{aa}
Genotypic values	G_{AA}	G_{Aa}	G_{aa}
N_A	2	1	0
N_a	0	1	2
Additive values	$2\alpha_A$	$\alpha_A+\alpha_a$	$2\alpha_a$
Dominant values	$D_{AA} = G_{AA} - 2\alpha_A$	$D_{AA} = G_{Aa} - \alpha_A - \alpha_a$	$D_{aa} = G_{aa} - 2\alpha_a$

It is convenient to change the notation to simplify the calculations. *We will put the origin of coordinates in the mean*. This means that the mean is null and the equation is:

$$G = \alpha_A N_A + \alpha_a N_a + D$$

We will estimate the regression coefficients by least squares. In a classical regression equation, this means minimizing $E(e^2) = \Sigma f_i e_i^2$ with respect to the unknowns (the regression coefficients α_1 and α_2), which in our case means minimizing:

$$\sum_1^3 f_i \cdot D_i^2 = \sum_1^3 (G - \alpha_A N_A - \alpha_2 N_a)_i^2 = f_{AA}(G_{AA} - 2\alpha_A)^2 + f_{Aa}(G_{Aa} - \alpha_A - \alpha_a)^2 + f_{aa}(G_{aa} - 2\alpha_a)^2$$

$$\frac{\partial}{\alpha_A} \sum_1^3 f_i \cdot d_i^2 = 2(G_{AA} - 2\alpha_A)(-2)f_{AA} + 2(G_{Aa} - \alpha_A - \alpha_a)(-1)f_{Aa} = 0$$

$$\frac{\partial}{\alpha_a} \sum_1^3 f_i \cdot d_i^2 = 2(G_{aa} - 2\alpha_a)(-2)f_{aa} + 2(G_{Aa} - \alpha_A - \alpha_a)(-1)f_{Aa} = 0$$

Solving the equations, we get:

$$\alpha_A = \frac{q\left(G_{AA} f_{AA} + \frac{1}{2} G_{Aa} f_{Aa}\right)}{2pq - \frac{1}{2} f_{Aa}} \qquad \alpha_a = \frac{p\left(G_{aa} f_{aa} + \frac{1}{2} G_{Aa} f_{Aa}\right)}{2pq - \frac{1}{2} f_{Aa}}$$

Adding these two equations, and taking into account that, using our scale, $m = 0$, we get:

$$G_{AA} f_{AA} + G_{aa} f_{aa} + G_{Aa} f_{Aa} - 2\alpha_A f_{AA} - 2\alpha_a f_{aa} - (\alpha_A + \alpha_a) f_{Aa} = 0$$

$$m = G_{AA} f_{AA} + G_{aa} f_{aa} + G_{Aa} f_{Aa} = \alpha_A f_{AA} + \alpha_a f_{aa} + (\alpha_A + \alpha_a) f_{Aa} = 0$$

$$\alpha_A\left(f_{AA} + \frac{1}{2} f_{Aa}\right) + \alpha_a\left(f_{aa} + \frac{1}{2} f_{Aa}\right) = \alpha_A p + \alpha_a q = 0$$

$$\alpha = \alpha_A - \alpha_a = \alpha_A + \frac{p}{q}\alpha_A = \left(1 + \frac{p}{q}\right)\alpha_A = \frac{\left(G_{AA} f_{AA} + \frac{1}{2} G_{Aa} f_{Aa}\right)}{2pq - \frac{1}{2} f_{Aa}}$$

$$\alpha_A = q \cdot \alpha$$

$$\alpha_a = -p \cdot \alpha$$

In Appendix 3.1, it is shown that, in a multiple regression equation, residuals e are not correlated with data y. In our case, the residuals are dominant values D, and the data are the genotypic values G; therefore:

$$\text{cov}(A, D) = 0$$

$$\sigma_G^2 = \sigma_A^2 + \sigma_D^2 \quad \rightarrow \quad \sigma_D^2 = \sigma_G^2 - \sigma_A^2$$

$$\sigma_G^2 = G_{AA}^2 f_{AA} + G_{Aa}^2 f_{Aa} + G_{aa}^2 f_{aa}$$

$$\sigma_A^2 = 4\alpha_A^2 f_{AA} + (\alpha_A + \alpha_a)^2 f_{Aa} + 4\alpha_a^2 f_{aa} = \cdots = 2\alpha^2\left(2pq + \frac{1}{2} f_{Aa}\right)$$

Main Concepts

- *Major gene:* A gene whose effect on a trait is relevant. If a gene has little effect on a trait, we call it a *minor gene*.
- *Genetic marker:* A point of the genome where its position is known. Genetic markers can be made up of many bases or they can simply be what we call a single-nucleotide polymorphism (SNP), a single nucleotide of which there are several versions.
- *Hardy–Weinberg equilibrium:* A situation in which the frequencies of the genotypes of a gene (AA, Aa and aa) remain constant from generation to generation.
- *Modifying forces of allelic frequencies:* These are selection, mutation, migration and genetic drift, which modify the frequencies of the alleles of a gene (A, and a).
- *Genetic drift:* The sampling effect that occurs in finite populations, giving rise to gametes with alleles that do not occur in the same frequency as in the parental population.
- *Linkage disequilibrium:* The difference between the observed frequency of a gamete (e.g. gamete AB) and what it should be if it were in equilibrium (the product of the frequencies of alleles A and B).
- *Relationship coefficient between two individuals:* The probability that when extracting an allele from a gene at random in each individual, these alleles are identical by descent.
- *Inbreeding coefficient of an individual:* The probability that the two alleles of a gene are identical by descent.
- *Effective number of a population:* The size of an ideal population mating randomly, including selfing, and that produces the observed increase in inbreeding in the population.
- *Heterosis:* In a cross between two populations, heterosis is the difference between the value of a trait in the offspring and the mean of the two parental populations.
- *Genotypic value:* In a gene with two alleles, genotypic value is the value of each of the genotypes. It can be generalized to the value of genotypes with several genes.
- *Mendelian additivity:* In a gene with two alleles, when heterozygotes have intermediate value between the two homozygotes, we say that there is Mendelian additivity.
- *Additive value:* In a gene with two alleles of a population, additive value is the sum of the effects of each allele. It is coincident with the mean allele effect in the population, but only if the population is in Hardy–Weinberg equilibrium; otherwise, it should be estimated by least squares. *It depends on the allele frequencies in the population*, so the same genotypes can have different additive values in different populations, depending on the allele frequencies. It can be generalized to the sum of the values of several genes or to all the genes of an individual.
- *Mendelian dominance:* In a gene with two alleles, when a heterozygote has a value higher or lower than the mean of the homozygotes, we say that there is Mendelian dominance.
- *Dominant value:* In a gene with two alleles, this is the difference between the genotypic value and the additive value. Each of the genotypes AA, Aa and aa has a dominant value, which can be null if the genotypic value of the heterozygotes is intermediate between the homozygotes (i.e. if there is no Mendelian dominance). It depends on the allele frequencies in the population, so the same genotypes can have different dominant values in different populations.

Problems

1.1. The Hardy–Weinberg law predicts the stability of populations, but populations evolve and some genes become fixed and others are lost. What are the causes of this evolution?

1.2. An AA male interbreeds with 999 aa females, and their offspring mate randomly. In which generation is Hardy–Weinberg equilibrium reached?

1.3. Will a finite population mating randomly with no selection, mutation or migration end up being homozygous for all its genes?

1.4. Does inbreeding affect all productive traits?

1.5. A male with an inbreeding coefficient of 0.4 is mated to an unrelated female who also has an inbreeding coefficient of 0.4. What is the inbreeding coefficient of the offspring?

1.6. In an infinite population, mating between siblings takes place for three generations. In the fourth generation, they mate randomly. What will be the genotype frequencies in the fifth generation?

1.7. Calculate how many males a pig farm should have so that inbreeding does not increase by more than 1% per generation.

1.8. A gene (AA, Aa, aa) has genotypic values of 0, d and 0, with equilibrium frequencies of 0.09, 0.42 and 0.49, respectively. Does this population have additive values? Can we practice selection with success?

1.9. Some pig breeds in China show high prolificacy despite being raised under high inbreeding conditions. How can this phenomenon be explained?

1.10. A marker (MM, Mm, mm) is on a different chromosome from a gene with important effects on a trait (QQ, Qq, qq). Is it possible that selecting MM increases the frequency of QQ?

Self-assessment Questions (True or False)

1. A major gene for one trait may be a minor gene for another trait.

2. A marker close to a major-effect allele is always linked to it.

3. Finite populations without selection, mutation or migration retain their allele frequencies from generation to generation.

4. A gene with two alleles in a population that has no modifying forces on allele frequencies and in which mating is random gradually reaches equilibrium.

5. Two genes in linkage disequilibrium in a population that have no modifying forces on allele frequencies and in which mating is random progressively reach equilibrium.

6. Two genes that are not in equilibrium together may separately be in Hardy–Weinberg equilibrium.

7. Two genes that are not separately in Hardy–Weinberg equilibrium may be in equilibrium together.

8. Inbreeding affects all productive traits.

9. A trait with complete Mendelian dominance (i.e. in which the homozygotes have a value of zero and only the heterozygote has a value d) has no additive variance.

10. Heterosis implies that the crossed individuals are superior to their parental populations.

References

Austerlitz, F., Gleiser, G., Teixeira, S. and Bernasconi, G. (2012) The effects of inbreeding, genetic dissimilarity and phenotype on male reproductive success in a dioecious plant. *Proceedings of the Royal Society B* 279, 91–100. DOI: 10.1098/rspb.2011.0652

Caballero, A. (2020) *Quantitative Genetics*. Cambridge University Press, Cambridge, UK.

de Rose, M.A. and Roff, D.A. (1999) A comparison of inbreeding depression in life-history and morphological traits in animals. *Evolution* 53, 1288–1292. DOI: 10.1111/j.1558-5646.1999.tb04541.x

Doekes, H.P., Veerkamp, R.F., Bijma, P., de Jong, G., Hiemstra, S.J. and Windig, J.J. (2019) Inbreeding depression due to recent and ancient inbreeding in Dutch Holstein–Friesian dairy cattle. *Genetics Selection Evolution* 51: 54. DOI:10.1186/s12711-019-0497-z

Falconer, D.S. and Mackay, T.F.C. (1996) *Introduction to Quantitative Genetics*. Longman, Harlow, UK.

Fisher, R.A. (1918). The correlation between relatives on the supposition of Mendelian inheritance. *Transactions of the Royal Society of Edinburgh* 52, 399–433. DOI:10.1017/S0080456800012163

Johannsen, W. (1909) *Elemente der exakten Erblichkeitslehre*. Gustav Fischer, Jena, Germany.

Köck A., Fürst-Waltl B. and Baumung, R. (2009) Effects of inbreeding on number of piglets born total, born alive and weaned in Austrian Large White and Landrace pigs. *Archiv für Tierzucht* 52, 51–64. DOI: 10.5194/aab-52-51-2009

Lynch, M. and Walsh, B. (1998) *Genetics and Analysis of Quantitative Traits*. Sinauer, Sunderland, Massachusetts.

Mendel, G. (1866) Versuche über Plflanzenhybriden. *Verhandlungen des naturforschenden Vereines in Brünn, Bd. IV für das Jahr 1865, Abhandlungen, 3–47*. English translation: Druery, C.T. and Bateson, W. (1901) Experiments in plant hybridization. *Journal of the Royal Horticultural Society* 26, 1–32.

Pearson, K. (1904) On a generalized theory of alternative inheritance, with special reference to Mendel's Laws. *Philosophical Transactions of the Royal Society A* 203, 53–86. DOI: 10.1098/rsta.1904.0015

Provine, W.B. (2001) *The origins of theoretical population genetics*. University of Chicago Pres, Chicago, Illinois.

Sewalem, A., Johansson, K., Wilhelmson, M. and Lillpers, K. (1999) Inbreeding and inbreeding depression on reproduction and production traits of White Leghorn lines selected for egg production traits. *British Poultry Science* 40, 203–208. DOI: 10.1080/00071669987601

Smith, L.A., Cassell, B.G. and Pearson, R.E. (1998) The effects of inbreeding on the lifetime performance of dairy cattle. *Journal of Dairy Science* 81, 2729–2737. DOI: 10.3168/jds.S0022-0302(98)75830-8

Stern, C. and Sherwood, E.R. (1966) *The Origins of Genetics*. Freeman, San Francisco, California.

Yule, G.U. (1902) Mendel's Laws and their probable relations to intra-racial heredity. *New Phytologist* 1, 193–207, 222–238. DOI: 10.1111/j.1469-8137.1902.tb07336.x

Solutions

Solutions to the problems

1.1. The allele frequencies p and q are modified by selection, migration, mutation or genetic drift, which leads the population to change its genotype frequencies.

1.2. In the first generation, all will be heterozygotes, and from the first generation of random mating, the genotypic frequencies will remain constant (0.25, 0.50 and 0.25), assuming that the Hardy–Weinberg equilibrium conditions are fulfilled.

1.3. Being finite, the population will have consanguinity that will increase, so the heterozygotes will progressively decrease even with mating being random, so the population will end up being homozygous for AA and aa.
1.4. Only those that are not strictly additive and where the heterozygote is not intermediate between the homozygotes ($d \neq 0$).
1.5. Zero. The parents are not related.
1.6. The equilibrium frequencies are p^2, $2pq$ and q^2. The population is infinite, so if mating is at random, as inbreeding has not modified the allele frequencies, by mating randomly we return to the equilibrium situation.
1.7. Approximately 13.
1.8. Yes, because the additive value or breeding value depends on the allele frequencies, and unless $p = q$, there will be an additive value for the three genotypes. In this case, $\alpha = k + (q - p)d = 0 + (0.7 - 0.3) \cdot 1 = 0.4$, so the additive and dominant values for each genotype can also be calculated.
1.9. It may happen that due to genetic drift all the alleles that harm litter size in homozygosity have been purged (lost), so that litter size is not affected by inbreeding.
1.10. If there is linkage (i.e. if they are correlated), it is possible. They may be correlated (in linkage disequilibrium), even though they are on different chromosomes.

Solutions to self-assessment questions

1. T, 2. F, 3. F, 4. F, 5. T, 6. T, 7. F, 8. F, 9. F, 10. F.

2 Methods of Genetic Improvement

Abstract

We begin this chapter by examining the action of genes and the environment on phenotypes. The infinitesimal model (model of infinite genes, each with an infinitesimal effect) is then presented and this model is used to examine changes in the mean and variance of a selected population. The covariance between relatives, both genetic and environmental, is deduced, and the main genetic parameters are presented: repeatability, heritability and genetic correlations. The 'breeder's equation' is explained and the parameters of selection are defined: selection pressure and selection intensity, selection differential and expected response. The selection procedure is explained in general terms and then applied to various selection methods based on different criteria: selection based on individual data or on data from relatives, or indirect selection based on data from a correlated characteristic.

2.1 Phenotype and the Effect of Genes and Environment

2.1.1 The environmental effect

The values observed in an individual for productive traits (e.g. growth rate, milk production, litter size) are due to the effect of both genes and the environment. The observed value, called phenotypic value *P*, has two components, a genetic component *G*, and an environmental component *E*, responsible for the differences among individuals. For individuals with higher observed values of *P* than the population mean, this can be either because they have better genes or because the environment has favoured them, for example because they are on a farm with better management or because they have been measured in a favourable season (e.g. winter favours growth), or simply by unidentified random environmental factors.

There may be interactions between the genotype and environmental effects; the situation can arise where the sum of the genotypic and environmental values does not give rise to the phenotypic value. It is known that the best genotypes in certain environments are not necessarily the best in other environments. For example, Friesian cows are much more productive in temperate climates than local breeds in South-East Asia, but they produce less milk in hot, humid climates. The feed conversion ratio (FCR) for milk measured on farms in one part of the world can be used for bull testing, and the semen of these bulls can be sent to other parts of the world where the environment can be very different, so the best bulls in one place may not be the best in another place. Imported breeds of sheep from countries with abundant pasture may perform worse in drier climates than local breeds. Likewise, in pigs, poultry and rabbits, animals raised in selection farms, with a good environment, may not grow as well on commercial farms. Taking this into account, the phenotypic value is represented as:

$$P = m + G + E + G * E$$

where *P* is the phenotypic value, *m* is the population mean, *G* is the genotypic value, *E* is the environmental value, and the term G*E indicates the interaction between the genotype

DOI: 10.1079/9781836990994.0002

and the environment, and is sometimes represented as G×E, or as GE to avoid confusion with the × or ∗ signs, which here do not indicate multiplication. Note that here the genotypic value is centred, with reference to the mean, so that the animals can have positive or negative genotypic values, and the same happens with the environmental value and with the interaction.

Genotype–environment interaction can be measured in plants, where it is easy to have clones with identical genotypes and grow them in different environments. In animals, where this is not feasible in practice (see section 10.1.1), several methods are used to approximate these interactions, such as using semen from the same sires in two environments and estimating the correlations between performances obtained in each environment as if they were different traits. In intensive animal production, the facilities and the care of the animals are very similar everywhere, so the genotype–environment interaction is not a relevant problem. This is particularly noticeable in prolific species (poultry, pigs and rabbits), and also in dairy cattle raised under intensive conditions. In species more closely linked to environmental conditions such as sheep, beef cattle and goats, these interactions may be more important. In difficult environments, caution should be taken before introducing a new breed, due to these interactions.

If we do not consider genotype–environment interactions, the observed variability can be expressed as:

$$P = m + G + E$$

$$\sigma_P^2 = \sigma_G^2 + \sigma_E^2 + 2\,\mathrm{cov}(G, E)$$

where the term cov(G, E) is the covariance between the genotypic and environmental values. The term cov(G, E) is non-zero when there is an association between genotypic and environmental values. For example, dairy cattle farmers who buy expensive semen from bulls of high genetic value frequently raise the daughters with particular care. Thus, the best environmental values are provided to the daughters with the best genotypic value, creating a positive covariance. As this is not taken into account when evaluating the animals, the breeding value of the sire is overestimated (this phenomenon, known as non-declared preferential treatment, is one of the problems of genetic prediction in dairy cattle). Estimates of heritability can also be affected by the presence of this covariance. *Genotype–environment covariance* should not be confused with *genotype–environment interaction*, because they are two completely different concepts. The covariance indicates that the best genotypes are in the best environments, whereas the interaction indicates that the best genotypes are not the same in one environment as in another.

In the same way that the genotypic value can be partitioned into additive and dominant fractions and into various classes of epistasis (see section 1.3.1), the environmental value can also be broken down into fractions attributable to known causes. For example, if we have the weaning weights of piglets from the same litter, these weights will resemble each other not only because the siblings share genes but also because they have been gestated together and have been fed by lactation from the same female; this is known as the *common litter environmental effect*. Another example comes from siblings from other litters of the same female, which will also have weaning weights more similar than if we compare them with the weights of animals taken at random, because the amount of milk produced and the care given by this female will be similar for all litters that she produces. This is known as a *permanent environmental effect* and it is another example of a common environment; we will see other examples later in this chapter.

2.1.2 The genetic effect

A. The infinitesimal model

Some productive traits depend on a single gene, but most depend on many genes, hundreds or thousands of them in most cases, each one with a very small effect.

This does not mean that genes are irrelevant. Some major genes can control the production of important hormones or proteins, but these hormones or proteins may each have only a small effect on the trait. For example, litter size involves many metabolic pathways including ovulation, implantation, fetal development, etc. These pathways are governed by many metabolites, each having a small effect on the number of piglets in a litter. We do not work directly with genes but with the effects of genes. For this reason, Fisher (1918) proposed a model in which traits are determined by *infinite independent genes with each having an infinitesimal effect*, so that the genetic value follows a normal distribution, according to the Central Limit Theorem. This theorem states that the sum (or the mean) of many independent variables of small effect each follows a normal distribution. To ensure independence, the population must be in linkage equilibrium (i.e. there must be statistical independence between genes, whether or not they are on the same chromosome), and there must be no major genes or important epistatic effects in order to ensure small values for the variables. Populations are not usually in linkage equilibrium, but except for extreme cases, this does not affect the normality of the distribution much (Bulmer, 1985, p. 160). The importance of finding a system in which the variables are normally distributed is that in this case the relationships between the variables are linear, which greatly simplifies all the calculations.

In this model, the additive and dominant values are the sum of the corresponding values for each of the genes:

$$G = A + D + I = A_1 + A_2 + A_3 + A_4 + \cdots + D_1 + D_2 + D_3 + D_4 + \cdots + I$$

where I contains the epistatic effects (AA, AD, DD, etc.; see section 1.3.1), which should be very small in order to be able to apply the Central Limit Theorem (in practice, they are usually ignored).

In the infinitesimal model, the offspring has different additive value from the parents, as each full sibling would have inherited a different combination of alleles from their parents. Each additive value, A_I, of the offspring can be represented as the mean of the parents' values (A_S and A_D), plus a quantity A_e, which is the lack or excess to adjust to the additive value that each individual actually has:

$$A_I = \frac{1}{2}A_S + \frac{1}{2}A_D + A_e$$

In the infinitesimal model, the additive values are continuous variables and their distribution is normal with mean zero and variance σ_A^2. The additive value A_e is a residual, and is independent of the additive values of the parents; A_e is known as the *Mendelian sampling effect* or intrafamily deviation, because all full siblings have the same parents' additive values *of* A_S and A_D.

In this breakdown of the genotypic value, the additive value (i.e. the sum of the additive effects of the genes) has a special importance, because as we saw in section 1.3.1, the additive value is the part of the genotypic value that passes to the offspring. Parents with high additive values can give rise to some offspring with bad additive values because some A_e can by chance be strongly negative (see Fig. 1.1), but if we take the mean of all the possible offspring of two parents, who carry all the possible combinations of their alleles, this mean will also be high, because the mean of all A_e will be zero.

B. Selection under the infinitesimal model

B1. CHANGES IN THE MEAN As the gene effects are very small (infinitesimal), we can feel a certain perplexity because it is not intuitive how much the sum of infinite infinitesimals gives, how the gene frequencies will change or how will it be possible to increase the mean of the population by selection. We will see below that the changes in gene frequencies are also very small (infinitesimal) and the effects of the genes are also infinitesimal, but despite all of this, it is possible to improve the mean of the population by selection. Let us consider a simple example in which all genes have the same effect and all have same gene frequencies (p and q). Suppose they are strictly additive genes ($d = 0$) with an infinitesimal additive effect k. If n is the number of genes involved in the trait, the mean will be (section 1.3.1):

$$m = \sum_{1}^{n} m_i = n \cdot k(p-q) = nk(p-1+p) = nk(2p-1)$$

By selecting, we go from an allele frequency p to a frequency p'. The difference between means due to the change in frequencies by selection is:

$$m' - m = nk(2p' - 1) - nk(2p - 1) = 2nk(p' - p)$$

As $m' - m$, which is the sum of n quantities ($n \to \infty$), cannot be infinite, the quantities k and $p' - p$ have to be infinitesimal in the order of $1/\sqrt{n}$, as the product of both infinitesimals is in the order of $1/n$ and compensates for the sum of n quantities. Thus, the change of the population mean by selection could be finite, despite being the sum of infinite quantities, and the effects of each gene could be infinitesimal as well as the change in gene frequencies. This model, like all models, is false, but it is an approximation to the real situation, in which the traits usually depend on many genes each of small effect.

> Note that in the infinitesimal model, there is no fixation or loss of genes; they simply change their frequencies infinitesimally, although they can asymptotically approach 0 or 1.

B2. CHANGES IN THE VARIANCE The variance will be the sum of the variances of each of the genes. There are no covariances between genes because the population is in equilibrium (the effects of the genes are statistically independent). Ignoring epistasis, we have:

$$\sigma_G^2 = \sigma_A^2 + \sigma_D^2 = \sigma_{A_1}^2 + \sigma_{A_2}^2 + \sigma_{A_3}^2 + \cdots \sigma_{D_1}^2 + \sigma_{D_2}^2 + \sigma_{D_3}^2 + \cdots$$

Selection changes gene frequencies, so *linkage equilibrium* (the independence between genes; see section 1.1.4) is lost and one of the conditions holding the Central Limit Theorem for normality of the distribution of additive values is lost as well. As the changes in gene frequencies due to selection are infinitesimal, this loss of normality is not important, but the loss of linkage equilibrium affects the additive variance, as we will now see. Taking the previous example of n genes with gene frequencies p and q for all of them, and an additive model ($d = 0$) with an infinitesimal additive effect k in each gene, the sum of the additive variances of all genes is (see section 1.3.2):

$$\sum_1^n \sigma_A^2 = n \cdot 2pqk^2 = 2np(1-p)k^2 = 2n(p-p^2)k^2$$

If we select and move from a gene frequency p in generation zero (G0) to a frequency p' in generation 1 (G1), the difference in the sums of variances between both generations will be:

$$\left[\sum_1^n \sigma_A^2\right]_{G1} - \left[\sum_1^n \sigma_A^2\right]_{G0} = 2n(p'-p'^2)k^2 - 2n(p-p^2)k^2 = 2nk^2\left[(p'-p)+p^2-p'^2\right]$$

where $p' - p$ and k are infinitesimal in the order of $1/\sqrt{n}$, as we saw in the previous section, so $p^2 - p'^2$ is infinitesimal in the order of $1/n$. Although $n \to \infty$, this is compensated by multiplying by k^2, which is infinitesimal in the order of $1/n$, but the product $2nk^2$ $((p' - p) + (p^2 - p'^2))$ is infinitesimal because $(p' - p) + (p^2 - p'^2)$ is also infinitesimal; that is, change of the sum of variances from G0 to G1 is an infinitesimal value, which means that the change in the sum of the variances of the genes is so small that there is no change in practice (a more formal derivation can be found in Crow and Kimura, 1970, pp. 238–239).

The additive values of the parents are independent when the mating is carried out at random, so the additive variance of the offspring is:

$$\sigma_{A_I}^2 = \frac{1}{4}\sigma_{A_S}^2 + \frac{1}{4}\sigma_{A_D}^2 + \sigma_{Ae}^2$$

Without selection, the offspring, sires and dams are random samples of the population and their additive variance is the same as it is the additive variance of the population, so:

$$\sigma_A^2 = \frac{1}{4}\sigma_A^2 + \frac{1}{4}\sigma_A^2 + \sigma_{Ae}^2 \quad \to \quad \sigma_{Ae}^2 = \frac{1}{2}\sigma_A^2$$

However, if there is selection, the selected parents will resemble each other and will have more similar additive values A_S^* and A_D^*. Then, the variances $\sigma_{A_S^*}^2$ and $\sigma_{A_D^*}^2$ will be less than those of the unselected parents, so the additive variance of the offspring $\sigma_{A_I^*}^2$ will be less than the additive variance before selecting. Therefore, the additive variance is reduced due to selection (see Appendix 2.2).

We have seen, however, that the sum of the additive variances of each of the genes is not appreciably reduced due to selection. The only explanation is that there are negative covariances between the different genes:

$$\sigma_A^2 = \sigma_{A_1}^2 + \sigma_{A_2}^2 + \sigma_{A_3}^2 + \cdots + 2\operatorname{cov}(A_1,A_2) + 2\operatorname{cov}(A_1,A_3) + \cdots$$

In linkage equilibrium, the genes are independent and the covariances between their effects are zero. When there is selection, certain alleles increase in frequency and others decrease in frequency, breaking the statistical independence between alleles and therefore producing covariances between them. As the variance is reduced, and the sum of the variances does not change appreciably, *these covariances must be negative*. This is strongly counterintuitive, because it seems that when selecting a trait, we would tend to select for the dominant alleles A and B, not their corresponding recessives, so the covariance would be expected to be positive. We will give an example to show that this is not the case.

In section 1.1.4, Table 1.7, we calculated the correlation between two genes (AA, Aa, aa, and BB, Bb, bb) in which having allele A or B contributes a quantity k, and having allele a or b contributes zero. The frequencies of the alleles A or a are p and q, and of alleles B or b are r and s, respectively. If the population is not in equilibrium, the gamete frequencies cannot be calculated from the allele frequencies, so we will call these frequencies f_{AB}, f_{Ab}, f_{aB} and f_{ab}. Giving allele A a value of x and allele B a value of y, the covariance between the values of the gametes is (section 1.1.4, Table 1.7):

$$\mathrm{cov}(x,y) = k^2(f_{AB} - pr)$$

If there is gametic equilibrium, $f_{AB} = p \cdot r$, so the covariance is zero. Now we will assume that there is selection and that the selection favours alleles A and B, so that we have reached the following frequencies:

$$f_{AB} = 0.6 \quad f_{Ab} = 0.2 \qquad f_{aB} = 0.2 \qquad f_{ab} = 0$$

The allele frequencies are:

$$p = f_{AB} + f_{Ab} = 0.6 + 0.2 = 0.8 \qquad q = 1 - p = 0.2$$
$$r = f_{AB} + f_{aB} = 0.6 + 0.2 = 0.8 \qquad s = 1 - r = 0.2$$

and the covariance is:

$$\mathrm{cov}(x,y) = k^2(0.6 - 0.8 \cdot 0.8) = -0.04 \cdot k^2$$

The covariance is negative, as we expected. This will always happen, because when there is selection, the AB gamete is favoured, but also the Ab and aB gametes, and the product $p \cdot r$ will be higher than the frequency of the AB gamete. When selecting, the frequency of recessives will decrease until it approaches zero, and the sum of the frequencies of all the other gametes ($f_{AB} + f_{Ab} + f_{aB}$)will approach 1, thus:

$$p.r = (f_{AB} + f_{Ab})(f_{AB} + f_{aB}) = f^2_{AB} + f_{AB}\, f_{Ab} + f_{AB}\, f_{aB} + f_{Ab}\, f_{aB} = f_{AB}\,(f_{AB} + f_{Ab} + f_{aB}) + f_{Ab} f_{aB} \simeq f_{AB} + f_{Ab} f_{aB} > f_{AB}$$

This can also be seen through linkage disequilibrium. If selection decreases the frequency of the recessive f_{ab}, soon $q \cdot s > f_{ab}$, which implies that $p \cdot r > f_{AB}$, as can be seen in section 1.1.4 and in Table 1.6.

Bulmer (1971) drew attention to this fact, which is why it has been known since then as the 'Bulmer effect'. When selection is relaxed, the linkage equilibrium gradually recovers, as we saw in Fig. 1.3, and consequently the covariance becomes zero and the additive variance increases. In practice, about two or three generations of random mating are enough for the additive variance to recover (Bulmer, 1985, p. 155).

> Selection generates linkage disequilibrium and negative covariances between alleles of different genes. The additive variance decreases with selection, and is recovered when selection is relaxed and mating is at random. This happens when the infinitesimal model holds.

The infinitesimal model holds for most productive traits, at least in the short and medium term. In the long term, consanguinity increases and genes are fixed and lost due to genetic drift, but new genes are generated due to mutation, so the additive variance is maintained. This explains why no limits to selection are generally found in selection programmes, as we will see in section 10.2.

2.1.3 Correlation between relatives

A. Environmental covariance between relatives

Relatives look alike for genetic reasons but also for environmental reasons (e.g. from sharing the same litter). Let us take two related individuals, *X* and *Y*, from a population in equilibrium. We will assume that their phenotypic values are the sum of their genetic and environmental values, and there is no genotype–environment interaction. We will also consider that the genotype–environment covariances are null, or we will try to make them so. In this case, the phenotypic covariance is simply the sum of the genetic and environmental covariance. Taking into account that m is constant and all covariances with m are zero:

$$\mathrm{cov}(P_X,P_Y) = \mathrm{cov}\ (m + G_X + E_X, m + G_Y + E_Y) = \mathrm{cov}\ (G_X,G_Y) + \mathrm{cov}\ (E_X,E_Y)$$

An example of environmental covariance is the case of full siblings, particularly those that come from the same litter. Full siblings have shared the female's uterine environment, normally they share maternal care, and they are fed on milk produced by their mother. These common environmental effects are known as *maternal effects*. Some traits such as weaning weight will be very sensitive to maternal effects and others will not (e.g. muscular pH). When the trait is affected by maternal effects, the phenotypic value can be broken down into the following, if we assume that there are no genotype–environment interactions:

$$P = m + G + E_M + E_0$$

where E_M is the maternal effect and E_0 the environmental effect on the trait, which we call the *environmental direct effect*, which is independent from the maternal effect ($\mathrm{cov}(E_0, E_M) = 0$). For example, if two sows have the same litter size and one produces more milk than the other, the weaning weight of her piglets will be higher, so the sow will have a positive maternal effect (E_M) on her piglets; the direct effect (E_0) brings together all the other environmental causes affecting the trait. Note that, although *the maternal effect is environmental*, a female produces more milk as a result of both genetic and environmental causes, so it could be said that females with genes to produce more milk exert a positive environmental maternal effect on the weaning weight of her offspring. This allows us to break down the maternal environmental effect into an environmental part due to the mother's genes,E_{GM}, and another environmental part, E_{EM}, which results from the environment acting on the mother. Note that there are genes responsible for the mother producing more milk, butE_{GM}*is an environmental effect on the weaning weight of the offspring*, and is measured in grams of weight. The phenotypic value will thus be:

$$P = m + G + E_{GM} + E_{EM} + E_0$$

We could say, for example, that a piglet weighing 10 kg at weaning is 1 kg more than the farm average because its mother has favourable genes to produce milk, produce a good uterine environment, etc.; the part of the maternal effect attributable to the mother's genes is 1 kg. In the same way, we could say that this mother had a suitable environment, which led her to produce more milk than usual, and therefore this 10 kg piglet has 400 g more weight at weaning than the average of the farm; this is the E_{EM} component. Thus, the piglet is 1.4 kg larger than the average of the farm. The rest of the effects are the genetic effect G and the direct environmental effect E_0 that we have described earlier. Genetic effects (G) and maternal genetic effects (E_M) can be correlated. For example, some of the genes that make the mother produce more milk can also affect weaning weight in the offspring. Consider, for example, the genes regulating growth hormone; this hormone has an effect on the weaning weight of the piglet and also on the mother's milk production. In this case, the G and E_M effects are correlated and their covariance is not zero. In principle, this information could be used to select animals increasing the frequencies of genes favourable for weaning weight (G) and for the dams to produce a favourable environment for weaning weight (E_M). In practice, a large amount of data is needed to predict these effects, and breeding programmes only consider this option in schemes such as dairy cattle, where there is a massive amount of data.

There are other examples of a common environment, such as when family groups are reared together, such as families of fish each reared in a different pond. In general, care must be taken when analysing the data to try to take into account all possible environmental effects that can be corrected. We will discuss this in detail in Chapter 4.

B. Genetic covariance between relatives

As we have seen before, there is covariance between relatives that is due to genetic causes. If we consider that epistasis is not very important, in a population in equilibrium the genetic covariance between two individuals, X and Y, will only have the additive and dominant components, which are unrelated (see section 1.3.2, so this covariance will be:

$$\mathrm{cov}(G_X, G_Y) = \mathrm{cov}(A_X + D_X,\ A_Y + D_Y) = \mathrm{cov}(A_X,\ A_Y) + \mathrm{cov}(D_X,\ D_Y)$$

In section 2.1.2.A, we saw that the additive value of each individual can be expressed as the mean of the additive value of the parents, A_S and A_D, plus a random term A_e. We also assume that the parents are not related, as usually happens in order to avoid increased inbreeding, and that there is random mating so that the additive values of sires and dams are independent. We will see several examples of covariance between relatives. Let us consider two relatives, X and Y.

B1. ADDITIVE COVARIANCE BETWEEN SIRE AND OFFSPRING

$$A_X = A_S \quad ; \quad A_Y = \frac{1}{2}A_S + \frac{1}{2}A_D + A_e$$

The offspring have half the additive value of the sire, and this is the only part in common that they have, so the general formula for the additive covariance is:

$$\mathrm{cov}(A_X, A_Y) = \mathrm{cov}\left(A_S, \frac{1}{2}A_S\right) = \frac{1}{2}\cdot\sigma^2_{As} = \frac{1}{2}\sigma^2_A$$

as the parents are random samples of the population and their variance is the additive variance if there is no selection.

B2. ADDITIVE COVARIANCE BETWEEN HALF-SIBLINGS

$$A_X = \frac{1}{2}A_S + \frac{1}{2}A_{DX} + A_{eX} \quad ; \quad A_y = \frac{1}{2}A_S + \frac{1}{2}A_{DY} + A_{eY}$$

Half-sibs share a common sire, but their mothers are different, so:

$$\text{cov}(A_X, A_Y) = \text{cov}\left(\frac{1}{2}A_S, \frac{1}{2}A_S\right) = \frac{1}{2}\cdot\frac{1}{2}\cdot\sigma_{As}^2 = \frac{1}{4}\sigma_A^2$$

B3. ADDITIVE COVARIANCE BETWEEN FULL SIBLINGS

$$A_X = \frac{1}{2}A_S + \frac{1}{2}A_D + A_{eX} \quad ; \quad A_Y = \frac{1}{2}A_S + \frac{1}{2}A_D + A_{eY}$$

Full siblings have the same father and mother. We assume that the parents are not related and that there is random mating; therefore covariance between A_S and A_D is null, so:

$$\text{cov}(A_X, A_Y) = \text{cov}\left(\frac{1}{2}A_S + \frac{1}{2}A_D, \frac{1}{2}A_S + \frac{1}{2}A_D\right) = \frac{1}{2}\cdot\frac{1}{2}\cdot\sigma_{A_S}^2 + \frac{1}{2}\cdot\frac{1}{2}\cdot\sigma_{A_D}^2 = \frac{1}{4}\sigma_A^2 + \frac{1}{4}\sigma_A^2 = \frac{1}{2}\sigma_A^2$$

B4. COVARIANCE BETWEEN ANY TWO RELATIVES The general formula for the additive covariance between relatives is:

$$\text{cov}(A_X, A_Y) = 2r_{XY}\sigma_A^2$$

where r_{XY} is the *kinship coefficient* that we saw in section 1.2.1.

Covariance between dominant effects is deduced similarly. For the dominance effect to be the same in two relatives, the alleles of a gene have to be identical by descent in both relatives; that is, they must have arrived not by chance but precisely due to having common ancestors. The dominant covariance will be:

$$\text{cov}(D_X, D_Y) = u_{XY}\sigma_D^2$$

where u_{XY} is the probability that when extracting the two alleles of a locus in individual X and comparing them with those of individual Y, both are respectively identical by descent; is known as the *identity coefficient*. In our previous examples, only full siblings can have both alleles identical by descent. In practice, the dominant variance is much smaller than the additive one, and it only affects certain relationships (e.g. full siblings), so it tends to be ignored.

2.2 Genetic Parameters

2.2.1 Repeatability

There are traits that can be recorded more than once, such as litter size or traits for milk production in dairy cattle and sheep. In these cases, the resemblance between several measurements of the same individual is due, on the one hand, to the fact that the same genes control these measurements, and on the other hand, because there is an environment that affects those measurements in the same direction. For example, if a sow has cysts in the uterus, she will have a low litter size throughout all her births, and this is an environmental cause not shared by other females; the same can happen in milk production, where a disease could affect milk production throughout all lactations of the cow. We can decompose the phenotypic value of each measure as:

$$P = m + G + E_p + E_e$$

where E_p is the part of the phenotype attributable to the effect of the *permanent environment* on the trait, and E_e is the general effect of the remaining environment.

It is interesting to know the correlation between several measurements of a trait. For example, if the correlations between milk litter size of several births in pigs is small, one data measurement from the female is not enough to evaluate the productivity of the animal. Let us take an example in dairy cattle. Consider the records of milk production of a cow in the first lactation, P_1, and in the second lactation, P_2. These measurements have in common the genotypic value of the female and the permanent effect that affects all her lactations:

$$P_1 = m + G + E_p + E_{e1} \quad ; \quad P_2 = m + G + E_p + E_{e2}$$

Assuming that genetic and environmental effects are independent, and permanent and general environmental effects are also independent, the correlation between both measurements is:

$$\mathrm{corr}(P_1,P_2) = \frac{\mathrm{cov}(P_1,P_2)}{\sigma_{P_1}\sigma_{P_2}} = \frac{\sigma_G^2 + \sigma_{E_p}^2}{\sigma_P^2}$$

where we consider that $\sigma_{P1} = \sigma_{P2} = \sigma_P$ because the standard deviation of milk from both lactations is approximately the same and approximately equal to the standard deviation of the trait calculated with all lactations.

This correlation coefficient is known as the *repeatability* of the trait and depends not only on the trait but also on the population, because there are populations with different genetic variability, and both the general environment and the permanent environment may also be more variable in some populations. For example, on a well-kept farm, there is less environmental variability than on a farm with poorer management.

Sometimes, *repeatability* is defined as the ratio between the variance due to permanent effects and the phenotypic variance ($\sigma_{E_p}^2/\sigma_P^2$), which can be a source of confusion.

2.2.2 Heritability

We have already seen that additive values are those that pass from one generation to the next. In a population, there are individuals with higher and lower values, so it is interesting to know what proportion of the observed variation is due to additive effects, which are the effects that are transmitted to the descents. We define heritability as the ratio between the additive and the phenotypic variance:

$$h^2 = \frac{\sigma_A^2}{\sigma_P^2}$$

In a simple model (but quite close to reality), the phenotype is simply broken down into the sum of the additive value and the environment, ignoring dominance and epistasis, and genotype–environment interactions:

$$P = m + A + E$$

The phenotypic variance is then broken down into the additive and the environmental variances:

$$\sigma_P^2 = \sigma_A^2 + \sigma_E^2 \qquad h^2 = \frac{\sigma_A^2}{\sigma_A^2 + \sigma_E^2}$$

If a trait has high heritability, the variability of the trait observed in the population has genetic causes and the environment has little influence on the trait, so that an individual with high observed values will generally also be an individual with high additive genetic values. However, if heritability is low, an individual with high additive genetic values will be difficult to detect, as the observed variability among individuals is essentially due to environmental causes (Fig. 2.1). As environmental effects vary among populations, and the genetic variability of a trait should also be different in each population, heritability values should differ among populations; however, in general, reproductive traits tend to have low heritability values (< 0.1), growth, milk production and composition traits moderate values (~0.3–0.4), and carcass fat content high values (~0.5–0.6). In practice, the heritability values of a trait do not differ much from population to population, because the facilities of the breeding companies are good and relatively similar, and also because all companies try to have as much additive genetic variability as possible in order to have success in selection.

Note that heritability depends on the environment; therefore, reducing the environmental variance increases heritability. For example, keeping the composition of the food constant, avoiding large variations in temperature, and maintaining the most uniform management possible reduces the environmental variability and increases the heritability.

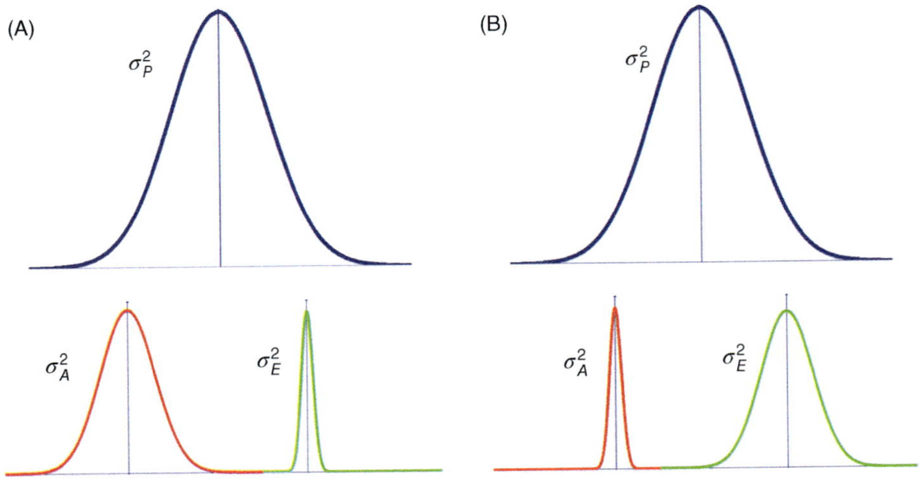

Fig. 2.1. Phenotypic variability (blue) **and underlying additive** (red) **and environmental** (green) **variabilities.** (A) When the heritability is 0.9. (B) When the heritability is 0.1.

This is why the farms in which the selection is made for prolific species (see section 8.1.1) are very well-conditioned farms, with the most uniform management possible.

> Most of the productive traits have relatively low heritability values, between 0.1 and 0.3, which implies that the phenotypic value is not generally a good indicator of the additive genetic value of the animal. These low heritability values also explain why cloning a highly productive animal does not imply that the clones will perform as well as the cloned animal.

For example, if we made a clone of a sow because it had a farrowing of 20 piglets, the sows of the clone would have on average a much lower litter size, because the heritability of litter size is very low (see section 10.1.1) and the observed value of 20 piglets was likely to be due to environmental reasons. The same will happen to that sow, which will probably have much smaller litter sizes in her successive parities, as repeatability of litter size is also very low.

2.2.3 Genetic correlations

Sometimes, the same group of genes influences several traits simultaneously. For example, genes controlling growth hormone influence the height and weight of individuals simultaneously. This effect is known as *pleiotropy*, and it is the main cause of *genetic correlation* between traits. Environmental factors can also produce correlations between traits; for example, environmental effects of diet or management will produce individuals that are more or less heavy and that will also be more or less tall. This correlation between environmental effects is known as *environmental correlation*. We are interested in knowing the correlation between additive genetic effects, because if we select for a trait, we are interested in knowing how the other genetically correlated traits will be modified, and because we could select for several traits at the same time by taking into account their genetic correlations, as we shall see in section 3.2.2.

We define the phenotypic, genetic and environmental correlation between two traits, X and Y, of the same individual, as:

$$r_P = \frac{\text{cov}(P_X, P_Y)}{\sigma_{P_X} \cdot \sigma_{P_Y}} \qquad r_A = \frac{\text{cov}(A_X, A_Y)}{\sigma_{A_X} \cdot \sigma_{A_Y}} \qquad r_E = \frac{\text{cov}(E_X, E_Y)}{\sigma_{E_X} \cdot \sigma_{E_Y}}$$

> We cannot modify the genetic correlation between traits by modifying the environment; it is a parameter strictly dependent on the genetic determination of the population.

The sign of these three correlations is often the same, and the value of the genetic correlation is generally similar to the value of the phenotypic correlation (Cheverud, 1988; Koots and Gibson, 1996). However, this is not always the case, so caution must be exercised in the case of traits for which no prior information is available.

Note that *the phenotypic correlation is not the sum of the genetic and environmental ones*, but:

$$r_P = \frac{\text{cov}\,(P_X, P_Y)}{\sigma_{P_x} \cdot \sigma_{P_Y}} = \frac{\text{cov}\,(A_X, A_Y)}{\sigma_{P_x} \cdot \sigma_{P_Y}} + \frac{\text{cov}\,(E_X, E_Y)}{\sigma_{P_x} \cdot \sigma_{P_Y}} = \frac{\text{cov}\,(A_X, A_Y)}{\sigma_{A_x} \cdot \sigma_{A_Y}} \cdot \frac{\sigma_{A_x} \cdot \sigma_{A_Y}}{\sigma_{P_x} \cdot \sigma_{P_Y}} + \frac{\text{cov}\,(E_X, E_Y)}{\sigma_{E_x} \cdot \sigma_{E_Y}} \cdot \frac{\sigma_{E_x} \cdot \sigma_{E_Y}}{\sigma_{P_x} \cdot \sigma_{P_Y}} =$$
$$= r_A \sqrt{h_X^2 \cdot h_Y^2} + r_E \sqrt{\left(1 - h_X^2\right)\left(1 - h_Y^2\right)}$$

The estimation of genetic parameters requires certain hypotheses that are not always fulfilled. Often populations are not in equilibrium, and if they have been selected, additive variances have been reduced due to the Bulmer effect (see section 2.1.2.B2). If the population is not closed, the introduction of new genes produces not only linkage disequilibrium but also changes in gene frequencies. Associations between genes also appear, but they will disappear as equilibrium is reached (see section 1.1.4 and Fig. 1.4); thus, *when crossing two populations, genetic correlations between traits appear due to linkage disequilibrium, which will disappear over time.* In Chapters 4 and 5, we will discuss the estimation of genetic parameters and how to avoid the effect of selection on the estimation.

2.3 Selection

2.3.1 The breeder's equation

When representing the values of parents and offspring on coordinate axes, we observe that on average the group of best parents gives rise to the group of best offspring. Figure 2.2 represents the relationship between parents and offspring for a quantitative trait, such as the weight at slaughter age in pigs. Each point represents a pair of data (x-axis is phenotypic mean in the parents, y-axis is phenotypic mean in the offspring), and we assume that the population is not selected and therefore the global mean m of parents and offspring is the same.

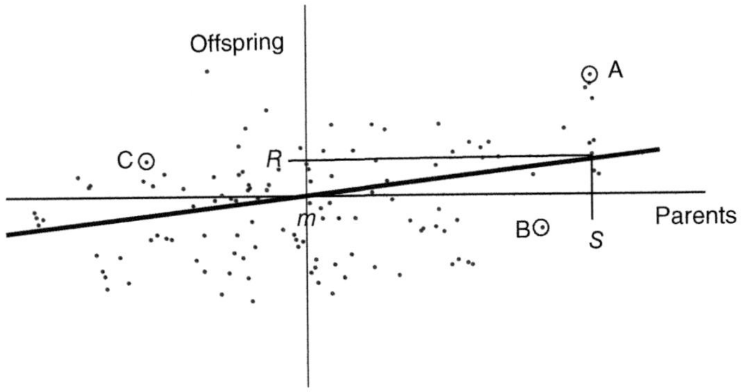

Fig. 2.2. Relationship between the means of the offspring and their parents for a trait with heritability $h^2 = 0.20$. In general, the offspring are more similar to the population mean (m) than to the value of the parents. In specific cases, good parents can have offspring who are superior (A) or inferior (B) to the mean. Bad parents can occasionally have offspring who are above the mean (C). S: selection differential (mean of the selected parents); R: response to the selection (mean of the offspring of the selected parents).

The regression line relating the mean of the parents, $\bar{P}$, and the mean of the offspring $\bar{H}$, has the heritability of the trait as the regression coefficient (see Appendix 2.1). The regression equation is:

$$\left(\bar{H}-m\right)=h^2\left(\bar{P}-m\right)$$

This is what is known as the 'breeder's equation', and it is the fundamental equation in animal breeding. In the case where the value of only one of the parents is represented on the x-axis, the equation is:

$$\left(\bar{H}-m\right)=\frac{1}{2}h^2\left(P-m\right)$$

Good parents can have good or bad offspring (see section 1.1.1 and Fig. 1.1). In Fig. 2.2, parents above the mean generally have offspring more similar to the population mean than to the mean of their parents, and cases such as 'A' are exceptional. It also happens that some parents well above the population mean may have offspring below the mean (Fig. 2.2, case 'B'), the opposite of what happens to the parents in case 'C', but these are exceptional cases. The reason is that what we observe is not the means of the additive values but the phenotypic means, and these are influenced by the environment, as we have seen earlier. Only if $h^2 = 1$ would the offspring be on average the same as the parents' mean; however, as heritability values are usually lower than 0.4, most of the variability that we observe is due to the environment.

> The exceptional values observed in some parents essentially have environmental and non-genetic causes; in general, we cannot expect the offspring to have on average phenotypic values as high as those we observe in their parents.

From the breeder's equation for one parent, we have:

$$\bar{H}=m+\frac{1}{2}h^2\left(P-m\right)=\frac{1}{2}h^2\cdot P+\left(1-\frac{1}{2}h^2\right)\cdot m$$

Let us give an example for FCR in pigs. Suppose that the population mean has a conversion rate of $m = 3$, but an exceptional male with a FCR of $P = 2$ has arrived at an artificial insemination station. Such a male would be immensely popular with farmers, who expect that this male's offspring will also have exceptional FCRs; however, as the heritability of the FCR is around 0.2, the expected mean of its offspring would be:

$$\bar{H}=\frac{1}{2}0.2\cdot 2+\left(1-\frac{1}{2}0.2\right)\cdot 3=2.9$$

which is practically the same as the mean of the population. Even mating this male with an exceptional female also having a FCR of 2, the expected mean of the offspring would be, according to the breeder's equation:

$$\bar{H}=h^2\cdot\bar{P}+\left(1-h^2\right)\cdot m=0.2\cdot 2+\left(1-0.2\right)\cdot 3=2.8$$

which is very far from the expectations of the farmers. Thus, we can make an important observation:

> Unless the heritability of a trait is very high (>0.5), the expected value of the offspring is closer to the population mean than to the values observed in the parents.

2.3.2 Selection parameters

A. Response to selection, selection differential, selection intensity and selection pressure

If we select the best parents, the mean of all their offspring will be higher than the population mean. For example, in Fig. 2.2, we have selected a set of parents whose phenotypic values have a mean value of S. Their offspring has much lower values, with a mean value of R. This is because the heritability in this example is $h^2 = 0.20$. If the heritability were $h^2 = 1$, the mean of the offspring would be equal to that of the parents, but this is never the case, as the environment has a strong influence on the productive traits. The mean of selected parents is called *selection differential* (S), and the mean of the offspring of selected parents is called *response to selection* (R), and in both cases we assume that R and S are centred. This means that S measures the superiority of the parents over the population mean and that R indicates the progress of the population mean after selection. Applying the breeder's equation to these values, we obtain:

$$R = h^2 \cdot S$$

This equation allows prediction of the response to selection if the superiority of the selected parents over the population mean is known, but also allows estimation of heritability if data from the next generation are available and the response R achieved is known. Then:

$$h^2 = \frac{R}{S}$$

This estimate of heritability is called the *realized heritability* and is important because it is not based on correlations between relatives but rather is the heritability that has effectively led to a response.

The response to the selection R depends not only on the heritability of the trait but also on the possibilities of selecting parents with high records, allowing S to be high. In a population in which all individuals are alike, it will not be possible to find large values of S. This becomes evident if we standardize the selection differential by dividing it by the standard deviation of the trait. We call the standardized selection differential the *selection intensity* (i):

$$i = \frac{S}{\sigma_P}$$

so that the expected response R to the selection becomes:

$$R = h^2 \cdot i \cdot \sigma_P$$

> *It can be seen that it is not enough that the heritability of the trait in the population is large to obtain a good response to selection.* Response also depends on the variability of the trait and on the selection intensity applied.

There are traits, such as litter size, that have low heritability but their variability is great, so that an appreciable response can be obtained despite their low heritability. To compare the variability between traits, the *coefficient of variation* (CV) is used, which has no dimensions. This is the standard deviation divided by the mean of the trait:

$$CV = \frac{\sigma_P}{m}$$

It is not a percentage but is sometimes represented multiplied by 100 and with the percentage sign. It is not defined for traits that can take negative values (in these cases, dividing by the mean can give the wrong idea about the variability of the trait). It is convenient because it has no dimensions and allows the variability of traits of different types to be compared; for example, traits related to growth might have a CV of around 0.10 and those related to litter size might have a CV of around 0.30. The formula for the expected response would then be:

$$R = h^2 \cdot i \cdot CV \cdot m$$

If the selection intensity i is different in males than in females, the selection differential is the mean of the selection differentials between males and females, and the selection intensity is also the arithmetic mean of the intensities.

As selection intensity has no dimension and is standardized, it is possible to compare the intensity of selection of several experiments or selection programmes. Selection intensity is usually determined by the facilities available. For example, it is not possible to keep more animals than available cages. Sometimes, the problem is biological: for example, if cows have three parities on average during their productive life, we only can have 1.5 heifers per cow for replacement, which makes selection via mothers very ineffective. In other cases in which it is possible to apply strong intensity of selection, this may be contraindicated because it can lead to a strong increase in consanguinity; for example, in fish it would be possible to apply high selection intensities given the large number of eggs they produce, but this would lead to the next generation coming from only a few parents.

Before selection, we know the percentage of animals we can keep of each sex. This would be useful to calculate the *selection intensity* and to be able to predict the response. Suppose that the trait is normally distributed and that individuals are selected when they exceed a certain *selection threshold*, which we will call c (Fig. 2.3). If p is the proportion of selected individuals or *selection pressure*, it can be shown (Appendix 2.2) that the mean of these selected individuals is, if we standardize as before:

$$i = \frac{f(c)}{p}$$

where $f(c)$ is the value of the standard normal distribution for c. The response to selection is:

$$R = h^2 \cdot \frac{f(c)}{p} \cdot \sigma_P$$

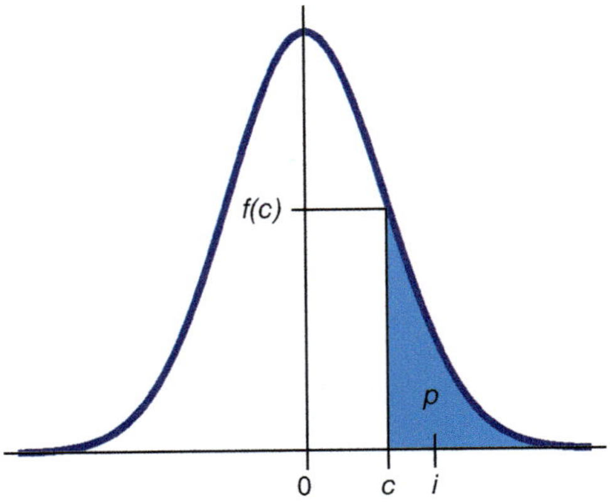

Fig. 2.3. Selection pressure *p* and selection intensity *i* for a standardized trait. *c*: Selection threshold; *f*(*c*): standard normal distribution for *c*.

where we can see that by decreasing the selection pressure p, the response increases. The term *selection pressure* is somewhat misleading, because in colloquial language it is sometimes said that pressure increases when p decreases, which leads to confusion.

EXAMPLE 2.1

In a population of rabbits of 100 females (dams, D) and 20 males (sires, S), it is decided to select by growth rate between 4 weeks (weaning) and 9 weeks, when they reach commercial weight. The mean is 35 g/day, and the coefficient of variation of the trait is 0.15. The heritability of the trait is, in this population, $h^2 = 0.20$. We have to decide the selection threshold and predict the expected response to selection. In order to reduce the generational interval, dams will only have one parity, having ten kits on average, of which we consider that only four kits are candidates, as the others may die before puberty, or may be culled due to diseases or other causes.

We have in the first parity 400 candidate rabbits for selection, of which on average 200 are male and 200 are female. After selection, the original population of 100 females and 20 males has to be reconstituted, and thus the selection pressure should be $p_D = 100/200 = 50\%$ in females and $p_S = 20/200 = 10\%$ in males. The thresholds c and values $f(c)$ of females and males for the respective pressures of 50% and 10% are:

$$p = 50\% \qquad p = 10\%$$
$$c_D = 0 \qquad c_S = 1.28$$
$$f(c_D) = 0.40 \quad f(c_S) = 0.18$$

The threshold values u for non-standardized data are, for females (u_D) and males (u_S):

$$\frac{u-m}{\sigma_P} = c \rightarrow u = m + c \cdot \sigma_P = m + c \cdot \mathrm{CV} \cdot m \rightarrow \begin{cases} u_D = 35 + 0 \cdot 0.15 \cdot 35 = 35 \\ u_S = 35 + 1.28 \cdot 0.15 \cdot 35 = 42 \end{cases}$$

Therefore, males having grown more than 42 g/day and females having grown more than 35 g/day will be selected. The selection intensities will be:

$$i = \frac{f(c)}{P} \quad \rightarrow \quad i_D = \frac{0.4}{0.5} = 0.8; \quad i_S = \frac{0.18}{0.10} = 1.8$$

The expected response R and the mean in next generation m' will be:

$$R = h^2 \cdot \frac{i_D + i_S}{2} \cdot CV \cdot m = 0.20 \cdot \frac{0.8 + 1.8}{2} \cdot 0.15 \cdot 35 = 1.4\text{g / day}$$

$$m' = m + R = 35 + 1.4 = 36.4\text{g / day}$$

B. Generation interval

The effectiveness of a selection process is determined by the response obtained per unit of time. If the sows of a line have a mean litter size of ten piglets, and the candidates for selection come from four parities, we will have 40 candidates per sow (20 males and 20 females) to carry out the replacement, but if they come from the first two parities, we will only have 20 candidates. The response will be greater in the first case, but the time needed for obtaining this response will also be greater, so the *generation interval* L should be calculated in both cases to predict the *response per unit of time*. The response per unit of time is:

$$R = h^2 \cdot \frac{i}{L} \cdot \sigma_P$$

EXAMPLE 2.2

We will calculate the response per unit of time of the previous example, in the case where candidates for selection come from two parities. The generational interval with one parity is $L_1 = 6$ months, and with two parities is $L_2 = 7.5$ months. With two parities, we duplicate the number of candidates, so selection pressures are reduced by half: $p_D = 25\%$ and $p_S = 5\%$. The values of the normal density for these selection pressures are $f(c_D) = 0.32$ and $f(c_S) = 0.10$. The selection intensities are:

$$i = \frac{f(c)}{p} \quad \rightarrow \quad i_D = \frac{0.32}{0.25} = 1.28; \quad i_S = \frac{0.10}{0.05} = 2$$

The expected response with two parities and the responses per unit time with one parity (R_1) and with two parities (R_2) are:

$$R = 0.20 \cdot \frac{1.28 + 2}{2} \cdot 0.15 \cdot 35 = 1.7 \qquad \begin{cases} R_1 = \dfrac{1.4}{6} \cdot 12 = 2.8\text{g /year} \\ R_2 = \dfrac{1.7}{7.5} \cdot 12 = 2.7\text{g /year} \end{cases}$$

Although the response using two parities is higher, enlarging the generation interval does not compensate for this, as the response per year is lower. In practice, other criteria are considered; for example, high selection pressure on males is avoided because it increases inbreeding.

Sometimes, the arguments are not genetic; for example, the high cost of replacing dams with only one parity can create financial problems, and thus it is preferred to leave them to have more parities.

When the intensity of selection is different between sexes and the generational interval is different as well, or when the generation interval is different for males and females because males are substituted by males just when they have provided semen to be stored (e.g. in dairy cattle), it can be shown that the response per unit of time is:

$$R = h^2 \cdot \frac{\Sigma i}{\Sigma L} \cdot \sigma_P$$

2.3.3 Selection methods

A. Selection criteria

In the previous example, selection was based on the mean of the phenotypic value of the parents, the *selection criterion*, but this is not always the case; sometimes the information of relatives is used. In these cases, Fig. 2.2 is the same, but the *x*-axis represents the selection criterion C used instead of the phenotypic mean of the parents. The breeder's equation is then:

$$\left(\bar{H} - m\right) = b\left(C - m_C\right)$$

where b is the regression coefficient of the offspring mean of the selection criterion, m is the mean of the selected trait and m_C is the mean of the selection criterion, which is not necessarily the same as the mean of the selected trait. In the case of individual selection (see section 2.3.3.B), we will see that b coincides with the heritability of the trait.

The expected response is:

$$R = b \cdot S = \frac{\operatorname{cov}(A,C)}{\sigma_C^2} \cdot i \cdot \sigma_C = \frac{\operatorname{cov}(A,C)}{\sigma_C \cdot \sigma_A} \cdot i \cdot \sigma_A = \rho_{A,C} \cdot i \cdot \sigma_A$$

where $\rho_{A,\,C}$, called the *precision*, is the correlation between the additive value and the selection criterion, so the formula for the expected response is:

$$R = \rho_{A,\,C} \cdot i \cdot \sigma_A$$

The expected response to selection depends on the effect of the additive genetic variability, measured by σ_A, on the intensity of selection i and on the precision $\rho_{A,\,C}$. Both the intensity of selection and the additive genetic variability are difficult to modify, the first, as we have seen before, due to limitations in the facilities, and the second because it depends on the trait and its genetically determined in our population. However, we can choose a criterion that increases the precision, thereby increasing the response. The square of the correlation coefficient $\rho_{A,C}^2$ is called the *reliability* and is a parameter commonly used in dairy cattle breeding, as we will see in Chapter 9.

To predict the additive values, we will start by using a simple model but one that is a good approximation to reality:

$$P = m + A + E$$

Prediction of the additive value can be done by regression on the selection criteria:

$$A = b(C - m_C) + e$$

The precision of this prediction is determined by the standard deviation of the error, later used to calculate the confidence interval. From the previous equation and taking into account that the criterion and the error are independent, we have:

$$\sigma_A^2 = b^2\sigma_C^2 + \sigma_e^2$$

$$\sigma_e^2 = \sigma_A^2 - b^2\sigma_C^2 = \sigma_A^2 - \left(\frac{\text{cov}(A,C)}{\sigma_C^2}\right)^2 \sigma_C^2 \cdot \frac{\sigma_A^2}{\sigma_A^2} = \sigma_A^2\left(1-\rho_{A,C}^2\right) = h^2\sigma_P^2\left(1-\rho_{A,C}^2\right)$$

where σ_e^2 is the *prediction error variance*. The standard error is the square root of the variance of the prediction error. As the errors are normally distributed, the length of the 95% confidence interval of our prediction will be approximately twice the standard error. The length of the *confidence interval* (*CI*) will be:

$$CI_{95\%} = \pm 2\sigma_e = \pm 2 \cdot h \cdot \sigma_P \sqrt{\left(1-\rho_{A,C}^2\right)}$$

Sometimes, the prediction of the additive value ± the standard error is given, but this is confusing because the standard error is not the 95% confidence interval. It is better to avoid the ± sign for standard errors and instead give them in parentheses next to the prediction. Confidence intervals are preferred because they better express the uncertainty around the true value.

B. Individual selection

Here, the selection criterion C is the phenotypic value of the individual, so $C = P$. The parameters are:

$$b = \frac{\text{cov}(A,P)}{\sigma_P^2} = \frac{\sigma_A^2}{\sigma_P^2} = h^2$$

$$\rho_{A,P} = \frac{\text{cov}(A,P)}{\sigma_A\sigma_P} = \frac{\sigma_A^2}{\sigma_A\sigma_P} = h$$

$$\sigma_e^2 = \sigma_P^2 \cdot h^2\left(1-h^2\right)$$

$$CI_{95\%} = \pm 2\sigma_e = \pm 2 \cdot h \cdot \sigma_P \sqrt{\left(1-h^2\right)}$$

The predicted additive value and predicted response to selection are:

$$\hat{A} = h^2(P-m) \pm 2h\sigma_P\sqrt{1-h^2}$$

$$R = h \cdot i \cdot \sigma_A$$

Now we will assume that we know h and σ_A, and in Chapters 4 and 5, we will discuss how to estimate them.

EXAMPLE 2.3

We will make the prediction of the additive value of an extraordinary individual, whose phenotypic value is P = 50 g/day, with data from the previous example. For this, we need the phenotypic and additive standard deviations:

$$\sigma_P = CV \cdot m = 0.15 \cdot 35 = 5.3\,; \qquad \sigma_A = h \cdot \sigma_P = \sqrt{0.20} \cdot 5.3 = 2.4$$

$$\hat{A} = 0.20(50-35) = 3.0$$

$$\sigma_e^2 = 5.3^2 \cdot 0.20 \cdot (1-0.20) = 4.5 \rightarrow \sigma_e = \sqrt{4.5} = 2.1 \rightarrow IC_{95\%} = \pm 4.2$$

The predicted additive value is:

$$\hat{A} = 3.0 \pm 4.2$$

The additive value is in the interval [−1.8, 7.2] Adding the population mean, we get $m + A = 38 \pm 4.2$, and the interval is [33.8, 42.2].

The precision and reliability of the prediction are, respectively:

$$\rho_{A,P} = h = \sqrt{0.20} = 0.45 \qquad \rho_{A,P}^2 = h^2 = 0.20$$

Note that an exceptional individual, with a phenotypic value of P = 50 g/day, has a predicted additive value, adding the mean, of 38 g/day, which is close to 35 g/day of the population mean. Note also the lack of precision of the prediction – we cannot state that this individual is higher than the population mean, because according to the confidence interval, they could have an additive value, adding the mean, of 34 g/day. There is a tendency to think that the true value must be in the middle of the confidence interval, but this is not the case; we cannot be sure if the true value is in the centre, or at one end or the other of the confidence interval. We will not discuss this point because it is a strictly statistical subject, but the reader can consult, for example, Blasco (2017, Chapter 1), where these statistical concepts are extensively discussed.

Sometimes an individual has several records of the same trait, such asseveral litter sizes in a sow or several lactations in a cow or sheep. Here, the selection criterion C is the mean of n phenotypic values of the individual, $C = \bar{P}$, that have a phenotypic correlation r_P. The parameters are (see Appendix 2.3):

$$b = \frac{n}{1+(n-1)r_P} h^2$$

$$\rho_{A,\bar{P}} = h\sqrt{\frac{n}{1+(n-1)r_P}}$$

$$\sigma_e^2 = \sigma_A^2\left(1-\rho_{A,\bar{P}}^2\right)$$

$$CI_{95\%} = \pm 2\sigma_e = \pm 2 \cdot h \cdot \sigma_P \sqrt{\left(1-\rho_{A,\bar{P}}^2\right)}$$

The mean of several records can be used to increase both the precision and the response to selection, although collecting several records on the same individual also increases the generational interval, and may not compensate if the response per unit time is lower (see section 2.3.2.B).

C. Indirect selection

Indirect selection is a particular case of individual selection. Here, the additive value of a trait is predicted by measuring another trait. This is common when the trait of interest is difficult or expensive to measure, such as in the case of FCR, which is economically important but expensive to measure because feed consumption has to be recorded individually. FCR has a medium to high correlation with growth rate, which is easy to measure, so FCR is often selected indirectly through growth rate. We will call I the trait selected indirectly by directly measuring the trait D. Here, the additive value of the indirectly selected trait $\hat{A}_I$ is predicted by the phenotype of the measured trait P_D; the criterion C is the phenotypic value of the measured trait, $C = P_D$. Taking into account that:

$$\mathrm{cov}(A_I,P_D) = \mathrm{cov}(A_I,A_D + E_D) = \mathrm{cov}(A_I,A_D)$$

the parameters are:

$$b = \frac{\mathrm{cov}(A_I,P_D)}{\sigma^2_{P_D}} = \frac{\mathrm{cov}(A_I,A_D)}{\sigma^2_{P_D}} = \frac{r_g\sigma_{A_I}\sigma_{A_D}}{\sigma_{P_D}\sigma_{P_D}} = r_g h_D \frac{\sigma_{A_I}}{\sigma_{P_D}}$$

$$\rho_{A_I,P_D} = \frac{\mathrm{cov}(A_I,P_D)}{\sigma_{A_I}\sigma_{P_D}} \cdot \frac{\sigma_{A_D}}{\sigma_{A_D}} = \frac{\mathrm{cov}(A_I,A_D)}{\sigma_{A_I}\sigma_{A_D}} \cdot \frac{\sigma_{A_D}}{\sigma_{P_D}} = r_g \cdot h_D$$

$$\sigma^2_e = \sigma^2_{P_I} \cdot h^2_I \left(1 - r^2_g h^2_D\right)$$

where r_g is the genetic correlation coefficient between the traits. The expected response is:

$$R_I = \rho_{A_I,P_D} \cdot i \cdot \sigma_{A_I} = r_g \cdot h_D \cdot i \cdot \sigma_{A_I}$$

If selection were not indirect but direct on the trait of interest, with the same intensity of selection, the response would have been:

$$R_D = h_I \cdot i \cdot \sigma_{A_I}$$

so, the relative efficiency of the methods is:

$$\frac{R_I}{R_D} = \frac{r_g \cdot h_D \cdot i \cdot \sigma_{A_I}}{h_I \cdot i \cdot \sigma_{A_I}} = r_g \frac{h_D}{h_I} = r_g \sqrt{\frac{h^2_D}{h^2_I}}$$

Note that it is not enough for indirect selection that traits are strongly correlated. If the heritability of the trait being measured, h^2_D, is lower than the heritability of the trait that is intended to be indirectly selected, h^2_I, the response to selection will be lower than if it had been selected directly.

> It is common to suggest the selection of one trait because of the beneficial effects on another trait, for example increasing sow's milk production for improving weaning weight of piglets, or increasing ovulation rate to increase litter size. The problem is that the traits have to be genetically highly correlated and the indirect trait heritability must be high enough compared with the heritability of the direct trait.

We have also assumed that the intensity of selection is the same for both traits, which might not be the case, for example if one of the traits can be measured only in one sex.

EXAMPLE 2.4

We propose to perform an indirect selection for FCR by recording only growth rate (GR). The parameters of the traits are:

$m_{FCR} = 3$ kg feed / kg liveweight	$m_{GR} = 35$ g/day	
$CV_{FCR} = 0.10$	$CV_{GR} = 0.15$	$r_g = -0.4$
$h^2_{FCR} = 0.3$	$h^2_{GR} = 0.2$	

The expected response in FCR when selecting for GR is:

$$\frac{R_I}{R_D} = -0.4 \cdot \sqrt{\frac{0.2}{0.3}} = -0.32$$

Thus, when selecting for GR, FCR would decrease by only 32% compared with what would be obtained by direct selection for FCR. The result is so poor that only the economic burden of individually recording animal consumption justifies not selecting directly by FCR. In pigs, for the same traits, the heritability of FCR and GR are approximately the same and the genetic correlation is $r_g = -0.75$, so the indirect response is 75% of what would be obtained by direct selection.

The breeder's equation and the prediction of the additive value are calculated as before, but with the new regression coefficient:

$$\sigma_{A_{FCR}} = h_{FCR} \cdot \sigma_{P_{FCR}} = h_{FCR} \cdot CV_{FCR} \cdot m_{FCR} = \sqrt{0.3} \cdot 0.10 \cdot 3 = 0.16$$

$$\sigma_{P_{GR}} = CV_{GR} \cdot m_{GR} = 0.15 \cdot 35 = 5.25$$

$$b = r_g h_{GR} \frac{\sigma_{A_{FCR}}}{\sigma_{P_{GR}}} = -0.4 \cdot \sqrt{0.2} \cdot \frac{0.16}{5.25} = -0.0054$$

The additive value for FCR of an exceptional individual having a GR of 50 g/day is:

$$\hat{A}_{FCR} = b(P_{GR} - m_{GR}) = -0.0054(50 - 35) = -0.08$$

Adding the mean, we obtain $m + \hat{A}_{FCR} = 3 - 0.08 = 2.92$, which is quite similar to the mean of the population.

The precision and reliability of the prediction are, respectively:

$$\rho_{A_{FCR},P_{GR}} = r_g \cdot h_{GR} = -0.4 \cdot \sqrt{0.2} = -0.18$$

$$\rho^2_{A_{FCR},P_{GR}} = 0.03$$

which are notoriously inferior compared with direct selection. The confidence interval length is calculated as twice the residual standard deviation:

$$\sigma_e^2 = \sigma^2_{P_{FCR}} \cdot h^2_{GR}\left(1 - r_g^2 h^2_{GR}\right) = (0.10 \cdot 3)^2 \cdot 0.3 \cdot \left(1 - (-0.4)^2 \cdot 0.2\right) = 0.026$$

$$CI_{95\%} = \pm 2\sigma_e^2 = \pm 2\sqrt{0.026} = \pm 0.32$$

Note that the additive value has a very wide interval of [–0.40, 0.24], which, adding the mean, becomes [2.60, 3.24], whose limits are below and above the population mean.

D. Selection by relatives

Sometimes, for example because the trait is expressed only in one sex (e.g. milk production or litter size), because it is too expensive to identify each individual and is more convenient to record a family mean (common in fish breeding), or because the trait cannot be measured in the candidate for selection (as in the case of carcass traits), selection is carried out by measuring the average production of a family group, in which the candidate for selection may occasionally be included.

We will look at a typical case of *selection by the mean of the offspring*, which was commonly used in cattle before use of the modern methods that we will study in Chapters 3–5. Here, the selection criterion C is the mean value of milk production of n daughters, so $C = \bar{D}$.

The parameters are (see Appendix 2.3):

$$b = \frac{\mathrm{cov}\left(A,\bar{D}\right)}{\sigma^2_{\bar{D}}} = \frac{2n}{\frac{4}{h^2} + n - 1}$$

$$\rho_{A,\bar{D}} = \frac{\mathrm{cov}\left(A,\bar{D}\right)}{\sigma_A \sigma_{\bar{D}}} = \sqrt{\frac{n}{\frac{4}{h^2} + n - 1}}$$

$$\sigma^2_e = h^2 \sigma^2_P \left(1 - \frac{1}{2}\rho^2_{A,\bar{D}}\right)$$

The prediction of the additive value of a bull is:

$$\hat{A} = b\left(\bar{D} - m\right) \pm 2 \cdot \sigma_e$$

Note that as the number of daughters of a bull increases, the precision increases and the variance of the error and the confidence interval decrease, as expected. In addition, it is interesting to note that *the prediction of the additive value, for the same mean value of the daughters, increases with increasing number of daughters* because the regression coefficient b increases when increasing the number of daughters. If we estimated the additive value simply as the mean of the daughters, $A = \bar{H}$, some bulls with few daughters would have, by chance, very high phenotypic values, and thus only bulls with little information would be selected (Fig. 2.4), as we will see in next example. Estimating the additive value by regression on the mean of the daughters, as we have done, shrinks the estimate towards the mean depending on the number of daughters of the bull. In this case, a bull with little information will be superior to a bull with a lot of information only if the mean of his daughters is extraordinarily high, which is not normally the case.

When n is very large, the denominator of b is approximately equal to n, so b is approximately equal to 2, and the additive value would be twice the mean of the phenotypic record of its daughters. This may be surprising, because we have commented that the additive value is the value transmitted to the offspring and we could think that the

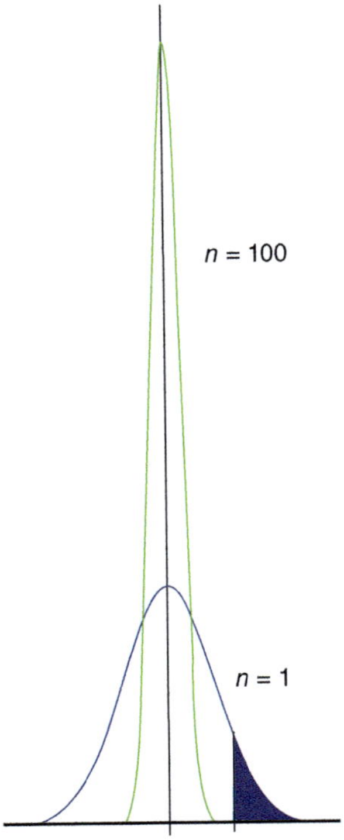

Fig. 2.4. Distribution of milk production of the daughters of a bull when the bull has 100 daughters or one daughter. Note that if we select (shaded area) only by the mean of the daughters, we only get descendants of bulls with only one daughter.

additive value of the bull would be the mean of his daughters' production, but the daughters have half the additive value of the father and half of the mother. If the cows that are inseminated by this bull are taken at random from the population, the mean value of these cows will be the mean of the population and their mean additive value will be zero; then, the mean of many daughters will be equal to half the additive value of the father. That is why the father has an additive value that is twice the mean of his daughters when n is very large.

EXAMPLE 2.5

We have three bulls, bull X with one daughter, bull Y with ten daughters, and bull Z with 100 daughters. The mean production of all daughters is exactly the same: $H_X = H_Y = H_Z = 12{,}000$ kg of milk/year. The population mean is 9000 kg/year and the coefficient of variation is 0.15. Taking a heritability value for milk production of 0.3, we are going to predict their additive values:

$$b_X = \frac{2\cdot 1}{\frac{4}{0.3}+(1-1)} = 0.15; \quad b_Y = \frac{2\cdot 10}{\frac{4}{0.3}+(10-1)} = 0.90; \quad b_Z = \frac{2\cdot 100}{\frac{4}{0.3}+(100-1)} = 1.78$$

Their predicted additive values will be:

$$A_X = 0.15(12{,}000-9{,}000) = 450 \rightarrow m + A_X = 9{,}450\text{kg}$$
$$A_Y = 0.90(12{,}000-9{,}000) = 2{,}700 \rightarrow m + A_Y = 11{,}700\text{kg}$$
$$A_Z = 1.78(12{,}000-9{,}000) = 5{,}340 \rightarrow m + A_Z = 14{,}340\text{kg}$$

Note that *the additive value of the bulls increases as the bulls have more daughters.* If we have little information, the prediction is not reliable and we 'regress' it towards the mean of the population, obtaining, in the case of bull X, a value similar to the mean because we do not have enough information. Figure 2.4 shows the means of the daughters of bulls with 100 daughters and with a single daughter. This *shrinkage* to the population mean is important, because if bulls were evaluated simply by the mean of their daughters, only bulls with little information would be selected. It is not impossible for a bull to have a single daughter with a value of 20,000 kg/year of milk, but it is practically impossible for a bull evaluated with 100 daughters to have a mean production of his daughters of 20,000 kg/year.

The accuracy, reliability and 95% confidence interval of the three bulls are also different. Applying the formulas:

$$\rho_X = 0.27 \qquad \rho_X^2 = 0.07 \qquad \sigma_e^2(X) = 506{,}892 \qquad CI_X = \pm 1{,}423$$
$$\rho_Y = 0.66 \qquad \rho_Y^2 = 0.44 \qquad \sigma_e^2(Y) = 308{,}585 \qquad CI_Y = \pm 1{,}111$$
$$\rho_Z = 0.94 \qquad \rho_Z^2 = 0.88 \qquad \sigma_e^2(Z) = 63{,}641 \qquad CI_Z = \pm 505$$

Bulls with less information have, on the one hand, worse reliabilities and higher confidence intervals and, on the other hand, lower additive values than other bulls having the same main value of their daughters.

As the amount of information increases, the precision of the additive values improves, but *the prediction of the additive value itself also increases,* because the regression coefficient used to predict the additive value increases with increasing amounts of data.

In our example, when the selection criterion is the mean of the offspring, we observe that the precision is much higher than in the previous cases of individual or indirect selection. The predicted additive values, which are the expected phenotypic mean of the offspring, are far from the population mean, which seems to contradict what is observed in Fig. 2.2, in which the phenotypic mean of the offspring is close to the population mean. Let us look at the figure in more detail. Suppose that this figure represents the weight of calves at weaning, a trait that is selected in beef cattle. Let us observe the parents that have a phenotypic value close to *S*. If we only have the mean value *S* of the parents, the *prediction* is that their offspring will have a value *R* on average. If we already have the value of the offspring, we already know that parents A in Fig. 2.2 will give rise to offspring well above *R* and that parents B will give offspring below *R*; here, the situation changes because we have the value of the offspring, so we are not making predictions about the offspring from the records of the parents. This is not to say that all offspring of parents A are going to be good (see section 1.1.1 and Fig. 1.1). There will be a variability around their mean, but it is expected that *on average* their mean will be much higher than *R*, and *on average* the offspring of B will be lower than *R* and lower than the mean.

There are other methods of selection by relatives, including: (i) *family selection*, which uses as the criterion the mean of the family (e.g. the mean of the litter); (ii) *selection by siblings* when the trait cannot be measured in the individual (e.g. carcass traits); and (iii) *within-family selection*, which uses as the criterion the difference between the value of the individual and the family mean (frequently used in fish farming, where each family is in a tank) (see Falconer and MacKay, 1996, for a full description of the methods). All are solved in the same way, applying the general formulas that we developed in section 2.3.3.A.

Appendix 2.1

$$(\bar{H}-m)=b\cdot(\bar{P}-m)$$

$$b=\frac{\operatorname{cov}(\bar{P},\bar{H})}{\sigma_{\bar{P}}^2}=\frac{\operatorname{cov}\left(\frac{P_1+P_2}{2},\frac{H_1+H_2+\cdots H_n}{n}\right)}{\operatorname{var}\left(\frac{P_1+P_2}{2}\right)}=\frac{\frac{1}{2n}\cdot\sum_{i=1}^{2}\sum_{j=1}^{n}\operatorname{cov}(P_i,H_j)}{\frac{1}{4}\left[\operatorname{var}(P_1)+\operatorname{var}(P_2)+2\operatorname{cov}(P_1,P_2)\right]}=$$

$$=\frac{\frac{1}{2n}\cdot 2n\operatorname{cov}(P_i,H_j)}{\frac{1}{4}(\sigma_P^2+\sigma_P^2)}=\frac{\frac{1}{2}\sigma_A^2}{\frac{1}{4}\cdot 2\sigma_P^2}=h^2$$

$$(\bar{H}-m)=h^2\cdot(\bar{P}-m)$$

Appendix 2.2

We will calculate the mean and variance of the selected parents. This can be applied to the phenotypic or to the breeding value distributions, and the results are general features of the normal distribution. We will assume a density $N(0, 1)$ for simplicity, but the conclusions are the same for a density $N(m, \sigma^2)$.

Mean of the selected parents

The density function is:

$$f(x)=\frac{1}{\sqrt{2\pi}}\exp\left(-\frac{x^2}{2}\right)$$

Selecting the best values from a threshold c, and calling i the *mean of the selected parents* and p the selection pressure (see Fig. 2.3), we have:

$$i=\frac{1}{p}\int_c^\infty x\cdot f(x)dx$$

$$\frac{\mathrm{d}f(x)}{\mathrm{d}x}=-x\cdot\frac{1}{\sqrt{2\pi}}\exp\left(-\frac{x^2}{2}\right)=-x\cdot f(x)\qquad\rightarrow\qquad x\cdot f(x)\mathrm{d}x=-\mathrm{d}f(x)$$

$$i=-\frac{1}{p}\int_c^\infty df(x)=-\frac{1}{p}\left[f(x)\right]_c^\infty=-\frac{1}{p}\cdot\left[\frac{1}{\sqrt{2\pi}}\exp\left(-\frac{x^2}{2}\right)\right]_c^\infty=\frac{1}{p}\cdot\frac{1}{\sqrt{2\pi}}\exp\left(-\frac{c^2}{2}\right)=\frac{1}{p}\cdot f(c)$$

Variance of the selected parents

We need some preliminary results. Consider the following expressions:

$$\frac{d}{dx}\left[x\cdot f(x)\right]=f(x)+x\frac{df(x)}{dx}=f(x)+x\left[-x\cdot f(x)\right]=f(x)-x^2f(x)$$

$$x^2 f(x) = f(x) - \frac{d}{dx}[x \cdot f(x)]$$

$$x^2 f(x)dx = f(x)dx - d[x \cdot f(x)]$$

$$\int_c^\infty x^2 f(x)\,dx = \int_c^\infty f(x)\,dx - \int_c^\infty d[x \cdot f(x)]$$

$$\frac{1}{p}\int_c^\infty x^2 f(x)\,dx = \frac{1}{p}\cdot p - \frac{1}{p}\int_c^\infty d[x \cdot f(x)] = 1 - \left[\frac{1}{p}\right][x \cdot f(x)]_c^\infty = 1 - \left[\frac{1}{p}\right][0 - c \cdot f(c)] = 1 + c \cdot i$$

The term $x \cdot f(x)$ is proportional to $x/\exp(x^2)$, which tends to zero when x tends to ∞. If V^* is the variance of the selected parents, and assuming a $N(0, 1)$ distribution, we have:

$$V^* = E(x-i)^2 = E(x^2) - i^2 = \frac{1}{p}\int_c^\infty x^2 f(x)\,dx - i^2 = 1 + c \cdot i - i^2 = 1 - i(i-c) < 1 = V$$

Applying this result to the distribution of the breeding values and where $\sigma^2_{A^*}$ is the additive variance of the selected parents, we have $\sigma^2_{A^*} < \sigma^2_A$.

Appendix 2.3

$$\mathrm{cov}(A,\bar{P}) = \mathrm{cov}\left(A, \frac{P_1 + P_2 + \cdots + P_n}{n}\right) = \frac{1}{n}\cdot n\,\mathrm{cov}(A,P_i) = \mathrm{cov}(A, A_i + E_i) = \sigma^2_A$$

$$\mathrm{var}(\bar{P}) = \sigma^2_{\bar{P}} = \mathrm{var}\left(\frac{P_1 + P_2 + \cdots + P_n}{n}\right) = \frac{1}{n^2}\cdot\left[n \cdot \sigma^2_{P_i} + \Sigma\,\mathrm{cov}(P_i,P_k)\right]$$

There are $n(n-1)$ covariances; if r_P is the phenotypic correlation between records, then:

$$\mathrm{var}(\bar{P}) = \frac{1}{n^2}\cdot\left[n \cdot \sigma^2_P + n(n-1)\mathrm{cov}(P_i,P_k)\right] = \frac{\sigma^2_P}{n} + \frac{n-1}{n} r_P \sigma_P \sigma_P = \frac{\sigma^2_P}{n}\left[1 + (n-1)r_P\right]$$

$$b = \frac{\mathrm{cov}(A,\bar{P})}{\sigma^2_{\bar{P}}} = \frac{\sigma^2_A}{\frac{\sigma^2_P}{n}\left[1+(n-1)r_P\right]} = \frac{n}{1+(n-1)r_P}h^2$$

$$\rho^2_{A,\bar{P}} = \frac{\mathrm{cov}^2(A,\bar{P})}{\sigma^2_A \sigma^2_{\bar{P}}} = b \cdot \frac{\mathrm{cov}(A,\bar{P})}{\sigma^2_A} = b \cdot \frac{\sigma^2_A}{\sigma^2_A} = b$$

$$\rho_{A,\bar{P}} = \frac{\mathrm{cov}(A,\bar{P})}{\sigma_A \sigma_{\bar{P}}} = \sqrt{b} = \sqrt{\frac{nh^2}{1+(n-1)r_P}}$$

$$\sigma^2_e = \sigma^2_A - b^2\sigma^2_{\bar{P}} = \sigma^2_A - b \cdot b \cdot \sigma^2_{\bar{P}} = \sigma^2_A - b\frac{\mathrm{cov}(A,\bar{P})}{\sigma^2_{\bar{P}}}\sigma^2_{\bar{P}} = \sigma^2_A - b \cdot \sigma^2_A = \sigma^2_A(1-b) = \sigma^2_A\left(1 - \rho^2_{A,\bar{P}}\right)$$

Applying the formulas to the case of the mean of the daughters of a sire:

$$\mathrm{cov}(A,\bar{D}) = \mathrm{cov}(A, A_{Di} + E_{Di}) = \frac{1}{2}\sigma^2_A$$

and as they are half-siblings:

$$r_P = \frac{\operatorname{cov}(D_i, D_j)}{\sigma_P \sigma_P} = \frac{\frac{1}{4}\sigma_A^2}{\sigma_P^2} = \frac{1}{4}h^2$$

$$\operatorname{var}(\bar{D}) = \frac{\sigma_P^2}{n}\left[1 + (n-1)r_P\right] = \frac{\sigma_P^2}{n}\left[1 + (n-1)\frac{1}{4}h^2\right]$$

$$b = \frac{\operatorname{cov}(A, \bar{D})}{\sigma_{\bar{D}}^2} = \frac{\frac{1}{2}\sigma_A^2}{\frac{\sigma_P^2}{n}\left[1 + (n-1)\frac{1}{4}h^2\right]} = \frac{nh^2}{2 \cdot \frac{1}{4}h^2\left[\frac{4}{h^2} + (n-1)\right]} = \frac{2n}{\frac{4}{h^2} + (n-1)}$$

$$\rho_{A,\bar{D}}^2 = b \cdot \frac{\operatorname{cov}(A, \bar{D})}{\sigma_A^2} = b \cdot \frac{\frac{1}{2}\sigma_A^2}{\sigma_A^2} = \frac{1}{2}b = \frac{n}{\frac{4}{h^2} + n - 1}$$

$$\rho_{A,\bar{D}} = \frac{\operatorname{cov}(A, \bar{D})}{\sigma_A \sigma_{\bar{D}}} = \sqrt{\frac{n}{\frac{4}{h^2} + n - 1}}$$

$$\sigma_e^2 = \sigma_A^2 - b\frac{\operatorname{cov}(A, \bar{D})}{\sigma_{\bar{D}}^2}\sigma_{\bar{D}}^2 = \sigma_A^2 - b \cdot \frac{1}{2}\sigma_A^2 = \sigma_A^2\left(1 - \frac{1}{2}b\right) = h^2\sigma_P^2\left(1 - \frac{1}{2}\rho_{A,\bar{D}}^2\right)$$

Main Concepts

- *Genotype–environment interaction:* When the best animals genetically in one environment are not the best in another environment, or their relative values change substantially, we say that there is an interaction between the genotype and the environment.
- *Genotype–environment covariance:* This is when there is an association between the genotypic and environmental values; for example, when the best farms have the best genotypes, there is a positive covariance between genotype and environment. Genotype–environment covariance has nothing to do with genotype–environment interaction; they are different concepts.
- *Infinitesimal model:* A model that explains the genetic determination of a trait by considering that it is determined by infinite independent genes each of infinitesimal effect, which implies that the trait has a normal distribution. It is a simplification of reality but is very useful, because most productive traits approximately follow the infinitesimal model.
- *Bulmer effect:* In the infinitesimal model, selection does not affect the sum of the variances of the genes, but as selection decreases the additive variance, there must necessarily be a negative covariance between genes. The appearance of this negative covariance due to selection is called the 'Bulmer effect'.
- *Genetic covariance between relatives:* Relatives look alike because they have common genes. This generates a positive covariance between the additive effects and sometimes between the dominant effects or between epistatic effects, depending on the relative.

- ***Repeatability:*** The phenotypic correlation between various measures of an individual for a trait that can be measured several times, such as litter size or traits associated with milk production.
- ***Heritability:*** The part of the phenotypic variance that is due to additive effects.
- ***Genetic correlation:*** The correlation between the additive effects of two traits. It is due to *pleiotropy* (genes acting on several traits simultaneously). It may also be due to linkage disequilibrium between genes that affect different traits.
- ***Response to selection:*** The increase (or decrease) of the mean of a trait due to selection. It is estimated as the difference between the mean in one generation and the mean in the next. A change in a trait other than the selected one is called a *correlated response*.
- ***Selection differential:*** The difference between the mean of a trait in the selected parents and the mean of the population.
- ***Selection intensity:*** The typified selection differential.
- ***Selection pressure:*** The proportion of selected individuals.
- ***Generation interval:*** The time interval between when the selected parents reproduce and when their offspring reproduce.
- ***Precision:*** The correlation coefficient between the additive value and the selection criteria.
- ***Reliability:*** The square of the precision value.
- ***Variance of the prediction error:*** The residual variance of the model that predicts the additive value from a selection criterion.

Problems

2.1. In an artificial insemination station for boars, the phenotypic values of a boar for FCR, growth rate and backfat thickness are provided. If this boar has very good phenotypic values for these traits, do these high performances guarantee that each of his offspring will be as good as he is?
2.2. In the situation provided in question 1, do the boar's high performances guarantee that the mean of his offspring will be as good as he is?
2.3. The boar in question 1 inseminates a sow with a phenotypic value of the mean of the population. Will the mean of the offspring be intermediate between the phenotypic values of the boar and the sow?
2.4. In the USA, cows are generally reared on farms and fed with concentrate. In Argentina, they are usually grazed in the field. Is there any reason to suppose that the best bulls in the USA will not be the best in Argentina?
2.5. Why is selection done using phenotypic values when genetic markers can be used instead to identify the relevant genes for a trait?
2.6. In an artificial insemination station for dairy cattle, the mean values of many daughters of each bull for milk production and percentage of fat and protein in the milk are known. Does a bull whose evaluation is good for milk production and composition, and has a reliability of 99%, guarantee that each of his daughters will be good?
2.7. In the case of question 6, will the mean of several daughters have good milk production and composition?
2.8. The additive value of a bull is 7000 kg/year of milk, predicted with five daughters. Can its additive value be increased?

2.9. Can the heritability of a trait be increased?
2.10. If two traits are non-linearly related (e.g. ovulation rate and litter size: for large ovulation rates, litter size increases less and less), will their genetic relationship also be non-linear?
2.11. Can the genetic correlation between two traits change due to selection?

Self-assessment Questions (True or False)

1. A genotype–environment interaction occurs when the best animals are in the worst environments.
2. The infinitesimal model has been overcome, because with modern genomic techniques we can detect the genes responsible for the genetic determination of a trait.
3. The additive value of an individual is the mean of the additive values of her parents.
4. Selection decreases the additive variance.
5. Repeatability is never less than the heritability.
6. A high heritability is a guarantee of a high response.
7. A clone of a very productive animal may have a low production, and its performance will generally be on average similar to that of the population mean; it will not resemble the individual from which it has been cloned.
8. Phenotypic correlation is the sum of the genetic and environmental correlations.
9. The mean growth of offspring resulting from the crossing of parents with high growth will not be as good as the mean of the parents but will be similar to the mean of the population.
10. The prediction of the additive value of a bull, for the same value of the mean of the daughters, increases with the number of daughters recorded.

References

Blasco, A. (2017) *Bayesian Data Analysis for Animal Scientists*. Springer, Cham, Switzerland.

Bulmer, M.G. (1971) The effect of selection on genetic variability. *American Naturalist* 105, 201–211. DOI: 10.1086/282718

Bulmer, M. (1985) *The Mathematical Theory of Quantitative Genetics*. Oxford University Press, Oxford, UK.

Cheverud, M. (1988) A comparison of genetic and phenotypic correlations. *Evolution* 42, 958–968. DOI: 10.1111/j.1558-5646.1988.tb02514.x

Crow, J.F. and Kimura, M. (1970). *An Introduction to Population Genetics Theory*. Harper & Row, New York.

Falconer, D.S. and MacKay, T.F.C. (1996) *Introduction to Quantitative Genetics*. Longman, Harlow, UK.

Fisher, R.A. (1918) The correlation between relatives on the supposition of Mendelian inheritance. *Transactions of the Royal Society of Edinburgh* 52, 399–433. DOI: 10.1017/S0080456800012163

Koots, K.R. and Gibson, J.P. (1996) Realized sampling variances of estimates of genetic parameters and the difference between genetic and phenotypic correlations. *Genetics* 143, 1409–1416. DOI: 10.1093/genetics/143.3.1409

Solutions

Solutions to problems

2.1. No, the offspring will be more similar to the mean of the population because the heritability values are far from unity (see Fig. 2.2).
2.2. No, because what we observe is largely due to the environment. There is no guarantee that this male is good for genetic reasons, as the heritability for productive traits is usually between 0.1 and 0.4.
2.3. No. Phenotypic values are not inherited, only additive values, and each piglet has an additive value that differs from the mean of the parent's additive values, because the gametes produce new combinations of alleles due to recombination (see Fig. 1.1).
2.4. Yes. There may be genotype–environment interactions and the best animals fed with concentrate feed may not be the best on pasture.
2.5. Because the traits are generally determined by many genes each with a very small effect, something that cannot be captured individually by markers. Large numbers of markers can be associated with phenotypic values for genomic selection (see Chapter 6) without identifying single genes.
2.6. The value of a specific individual cannot be guaranteed. By recombination and because there is a random sampling of gametes, in this specific case there may be inherited unfavourable versions of the genes (see Fig. 1.1). Moreover, a specific individual can be randomly affected by the environment, and if the environment has a negative effect, its production can be low, despite having a good genotype.
2.7. It can be guaranteed that the mean of one set of daughters of this bull will resemble the mean of the set of daughters used to evaluate his breeding value (additive value), and this guarantee is stronger if the number of daughters used to evaluate the bull is high (see section 9.1).
2.8. The additive value cannot be increased, but for the same observed mean of the daughters, the prediction of its additive value increases when there is more information (more daughters) for the prediction.
2.9. This can be done by reducing the environmental variance, or correcting the data for the environmental effects that can be estimated (e.g. seasonal effects, parity).
2.10. In the infinitesimal model, the traits are normally distributed, and therefore their genetic relationship is always linear, whatever the phenotypic relationship is.
2.11. Yes, it can. For example, if two traits are negatively related and selected in favour of both, the frequencies of genes that are non-pleiotropic may increase at the expense of those that determine both traits. In addition, there is linkage disequilibrium due to selection, which also alters the genetic correlation.

Solutions to self-assessment questions

1. F, 2. F, 3. F, 4. T, 5. T, 6. F, 7. T, 8. F, 9. T, 10. T.

3 Selection Indices

Abstract

This chapter looks at selection indices for one or more traits. The method combines information from the data of relatives with that of the individual to get the best genetic response. There is a difference between selection objectives and selection criteria (the recorded data), which may or may not be the same. Selection for several traits can also be done using selection indices. The chapter describes how to calculate economic weights for the traits to be included in the index. Indices of selection for several traits are carried out in an economically optimal way. Selection indices for several traits can be used for individual selection, but selection indices for several traits integrating information from relatives can also be proposed. Numerical examples of all types of indices are provided. The chapter ends with a series of observations regarding selection indices.

Selection indices appeared in an early article by Karl Pearson (1903) on natural selection, although it took a long time to applied them to animal breeding. First, they were applied in plants (Smith, 1936, following detailed instructions from Ronald Fisher), and Hazel (1943) was the first who applied the theory of selection indices to animal breeding. Later they were used both in laboratory experiments and in commercial selection farm nuclei, particularly in prolific species. At present, they have been displaced by best linear unbiased prediction (BLUP) (Henderson, 1963), a method that simultaneously combines familiar information with correction of the data for environmental factors, as we will see in Chapter 4. In this chapter, we will study selection indices because they are still sometimes used and because they allow a better understanding of BLUP, which is the most widely used animal breeding method today.

3.1 Selection Indices for One Trait

3.1.1 Indices with information from relatives

The basic idea of selection indices is to predict the additive value of each individual with all the information available in the population. Before the use of computers became widespread, indices were restricted to the closest relatives. As the relevant information for selection is found precisely in these relatives, the effectiveness of these indices is not very different from that of indices considering all the information, but today it is easy to handle all information computationally.

- To present the results of the indices, we will use a universal notation in animal genetic improvement texts, in which a letter in bold is a vector if it is lowercase and a matrix if it is uppercase. If the vector has a prime symbol, it is a row vector, $\mathbf{y}' = [y_1, y_2, y_3, \cdots]$ and if it does not have a prime, it is a column vector. When a matrix is symmetric, we will

DOI: 10.1079/9781836990994.0003

generally represent only the upper triangle. Transposed matrices **A** are also indicated by a prime, **A′**.
- We will also change the notation. We have represented the phenotypic values in the previous chapters as P_1, P_2, P_3...; now they will be represented as y_1, y_2, y_3, ... We have represented the additive values in the previous chapters as A_1, A_2, A_3...; now they will be represented as a_1, a_2, a_3.... Random residual values will be represented as e_1, e_2, e_3.... We will keep the notation for the variance components, σ_P^2, σ_A^2, and the correlation, $\rho_{A,I}$.
- We will assume from now that the data are already centred; that is, the mean is zero.

In practice, centring the data is not immediate because data are subject to diverse environmental effects; for example, the number of rabbits weaned from the first parity may have had, on average, 1.1 young rabbits fewer than the population mean, the second parity 0.7 more rabbits than the mean, etc. The data should be corrected in this case for a parity effect, but it could also happen that in summer the number of rabbits weaned was 0.5 times more than the average and in winter 0.5 times less, in which case it would be necessary to correct for a seasonal effect as well. When correcting for season and parity, the data are assumed to come from the same season and parity order; that is, the intention is to correct the known environmental effects that may affect the trait in a systematic way (e.g. season, parity order), so that when subtracting the general mean, the data are perfectly centred, with the mean as zero. The effective way to estimate the effects to correct the data will be discussed in detail in Chapter 4; in this chapter, we will assume that the data have a mean equal to zero.

Up to now, we have predicted the additive value using a simple regression equation on the selection criterion (see section 2.3.3.A). In this section, we will combine the information available through a multiple linear regression on the data of an animal and its relatives in order to predict the additive value of the animal. We will call the prediction of the additive value the *selection index*, *I*. We are going to assume that all individuals have the same type of data; for example, the data vector $\mathbf{y}' = [y_1, y_2, y_3, \cdots]$can be composed of one data point from the individual, another data point from their mother, another from a brother, etc., and we assume that all the individuals to be evaluated have these data. The regression equation that we will use to predict the additive value of an individual is:

$$a = \mathbf{b}'\mathbf{y} + e = [b_1, b_2, \cdots, b_n]\begin{bmatrix} y_1 \\ y_2 \\ \cdots \\ y_n \end{bmatrix} + e = b_1 y_1 + b_2 y_2 + \cdots + b_n y_n + e$$

where $\mathbf{b}' = [b_1, b_2, b_3, \cdots]$ is the vector of the regression coefficients.

As the data are centred, the variance–covariance matrix for all data is simply the expectation of the product of the data, as the mean has already been subtracted from them:

$$\mathbf{V} = E(\mathbf{y} \cdot \mathbf{y}') = E\begin{bmatrix} y_1 \\ y_2 \\ \cdots \\ y_n \end{bmatrix}[y_1, y_2, \cdots, y_n] = \begin{bmatrix} \sigma_P^2 & \text{cov}(y_1, y_2) & \cdots & \text{cov}(y_1, y_n) \\ & \sigma_P^2 & \cdots & \text{cov}(y_2, y_n) \\ & & \cdots & \cdots \\ & & & \sigma_P^2 \end{bmatrix}$$

As the data are centred, the covariance between the additive value of an individual and all the data is:

$$c' = \text{cov}(a,y') = E(a \cdot y') = E([a \cdot y_1, a \cdot y_2, \cdots, a \cdot y_n]) = [\text{cov}(a,y_1), \text{cov}(a,y_2), \cdots, \text{cov}(a,y_n)]$$

Note that $y_1 = a_1 + e_1$, $y_2 = a_2 + e_2$, ...etc., and as the additive value is independent of the environmental value:

$$c' = \text{cov}(a,y') = [\text{cov}(a,a_1), \text{cov}(a,a_2), \cdots, \text{cov}(a,a_n)]$$

The prediction of the additive value is carried out in a very similar way to what we saw in section 2.3.3.A, but taking into account that we should estimate n regression coefficients. Let us consider the multiple regression equation that we will use to predict the additive value of the individual:

$$a = b'y + e$$

Multiplying both sides by y', calculating the expectations on both sides, and assuming that the residual is not correlated with the data ($\text{cov}(e, y') = 0$), we have:

$$a \cdot y' = b'yy' + e \cdot y'$$
$$E(a \cdot y') = E(b'yy') + E(e \cdot y') = b' E(yy') + \text{cov}(e,y') = b'V + 0 = b'V$$

If $c' = \text{cov}(a,y') = E(a \cdot y')$, we have:

$$E(a \cdot y') = c' = b'V \quad \rightarrow \quad b' = c'V^{-1} \quad \rightarrow \quad \hat{a} = I = b'y = c'V^{-1}y$$

For each individual, we will have an index that is the prediction of its additive value:

$$I = c'V^{-1}y$$

Note that coefficient b has the form of a regression coefficient. In the univariate case, the regression coefficient is a covariance divided by a variance; here, it is a covariance vector multiplied by the inverse of a variance–covariance matrix.

The coefficient b should be estimated by a method ensuring independence between data and residuals ($\text{cov}(e, y') = 0$). Appendix 3.1 shows how to do this by minimizing the variance of the error, arriving at the same formula.

Note that the index, I, is the prediction of the additive value of an individual. Once we obtain this prediction, we can calculate the precision, the variance of the error and the response to the selection, as in section 2.3.3.A. To do this, we need the variance of the index and its covariance with the true additive value.

Taken into account that vector c and the matrix V are constants, we have:

$$\sigma_I^2 = \text{var}(c'V^{-1}y) = c'V^{-1} \cdot \text{var}(y) \cdot V^{-1}c = c'V^{-1}VV^{-1}c = c'V^{-1}c$$

$$\text{cov}(a,I) = \text{cov}(a, c'V^{-1}y) = c'V^{-1} \cdot \text{cov}(a,y) = c'V^{-1}c = \sigma_I^2$$

$$\rho_{A,I} = \frac{\text{cov}(a,I)}{\sigma_A \sigma_I} = \frac{\sigma_I^2}{\sigma_A \sigma_I} = \frac{\sigma_I}{\sigma_A}$$

$$R = \rho_{A,I} \cdot i \cdot \sigma_A = i \cdot \sigma_I$$

$$\sigma_e^2 = \sigma_A^2 (1 - \rho_{A,I}^2) = \sigma_A^2 - \sigma_I^2$$

Although these formulas are derived for the true values of **c** and **V**, it is obvious that we do not know them and should use estimates instead. A criticism of selection indices is that when calculating their precision and confidence interval, the error made in estimating **c** and **V** is not taken into account, so the precision of the index is actually underestimated. In Chapter 5, we will see how to include these errors.

EXAMPLE 3.1

We would like to evaluate sows by litter size with data from the first parities, one from the individual (the sow), y_I, another two data points from two half-sisters of the sow, y_{HS}, and another from her mother, y_M. The mean of the trait is ten piglets and the coefficient of variation (CV) is 0.30. The heritability of the trait is 0.10. To calculate the index, we will first calculate the phenotypic and additive variances and then the covariances of the data with the additive value of the sow:

$$\sigma_P = CV \cdot m = 0.30 \cdot 10 = 3\,\text{piglets}$$
$$\sigma_A^2 = h^2 \cdot \sigma_P^2 = 0.10 \cdot 3^2 = 0.90$$

$$\mathbf{y}' = [y_I, y_{HS}, y_{HS}, y_M]$$

$$\mathbf{c}' = \text{cov}(a_I, \mathbf{y}') = \left[\text{cov}(a_I, a_I), \text{cov}(a_I, a_{HS}), \text{cov}(a_I, a_{HS}), \text{cov}(a_I, a_M)\right] =$$
$$= \left[\sigma_A^2, \frac{1}{4}\sigma_A^2, \frac{1}{4}\sigma_A^2, \frac{1}{2}\sigma_A^2\right] = [0.90, 0.23, 0.23, 0.45]$$

For matrix **V**, we need all the variances and covariances of the vector **y**. As the data of the half-sisters and the mother are unselected samples of the population, their variance is the same and it is equal to the phenotypic variance of the trait:

$$\text{var}(y_I) = \text{var}(y_{HS}) = \text{var}(y_M) = \sigma_P^2$$

Phenotypic covariances are calculated directly from the data; however, in this case, the half-sisters must only resemble each other for genetic reasons, and the same applies to the sow and her mother, so we can calculate their covariances using the additive variance:

$$\text{cov}(y_I, y_{HS}) = \text{cov}(y_{HS}, y_{HS}) = \text{cov}(a_{HS}, a_{HS}) = \frac{1}{4}\sigma_A^2 = 0.23$$

$$\text{cov}(y_I, y_M) = \text{cov}(a_I, a_M) = \frac{1}{2}\sigma_A^2 = 0.45$$

The mother of the sow has a null covariance with the half-sisters, because she is not related to them.

$$\text{cov}(y_{HS}, y_M) = \text{cov}(a_{HS}, a_M) = 0$$

$$\mathbf{V} = \begin{bmatrix} \text{var}(y_I) & \text{cov}(y_I, y_{HS}) & \text{cov}(y_I, y_{HS}) & \text{cov}(y_I, y_M) \\ & \text{var}(y_{HS}) & \text{cov}(y_{HS}, y_{HS}) & \text{cov}(y_{HS}, y_M) \\ & & \text{var}(y_{HS}) & \text{cov}(y_{HS}, y_M) \\ & & & \text{var}(y_M) \end{bmatrix} = \begin{bmatrix} 9 & 0.23 & 0.23 & 0.45 \\ & 9 & 0.23 & 0 \\ & & 9 & 0 \\ & & & 9 \end{bmatrix}$$

Now we should invert this matrix and multiply by c' and y to obtain the index, I:

$$\hat{a} = I = c'V^{-1}y = [0.9, 0.23, 0.23, 0.45]\begin{bmatrix} 9 & 0.23 & 0.23 & 0.45 \\ & 9 & 0.23 & 0 \\ & & 9 & 0 \\ & & & 9 \end{bmatrix}^{-1}\begin{bmatrix} y_1 \\ y_{HS} \\ y_{HS} \\ y_M \end{bmatrix}$$

$I = 0.10\ y_I + 0.02y_{HS} + 0.02\ y_{HS} + 0.05\ y_M$

We will now calculate the variance of the index, the precision, the reliability and the variance of the error. In these examples, the intensity of selection when selecting only females, and using the same arguments in pigs as in rabbits in the example in section 2.3.2.A, will be $i = 0.8$:

$$\sigma_I^2 = c'V^{-1}c = [0.9, 0.23, 0.23, 0.45]\begin{bmatrix} 9 & 0.23 & 0.23 & 0.45 \\ & 9 & 0.23 & 0 \\ & & 9 & 0 \\ & & & 9 \end{bmatrix}^{-1}\begin{bmatrix} 0.90 \\ 0.23 \\ 0.23 \\ 0.45 \end{bmatrix} = 0.12$$

$$\rho_{A,I} = \frac{\sigma_I}{\sigma_A} = \sqrt{\frac{0.12}{0.90}} = 0.36 \rightarrow \rho_{A,I}^2 = 0.36^2 = 0.13$$

$$\sigma_e^2 = \sigma_A^2 - \sigma_I^2 = 0.90 - 0.12 = 0.78 \rightarrow IC_{95\%} = \pm 2\sigma_e = \pm 2\sqrt{0.78} = \pm 1.77$$

$R = i \cdot \sigma_I = 0.8 \cdot \sqrt{0.12} = 0.28$ piglets per generation

Let us estimate the additive value of a sow that had 14 piglets, her mother 15 piglets and two half-sisters 12 and 13 piglets. Because the values in the index are centred, we would have an index:

$$I = \hat{a} = 0.10(14-10) + 0.02(12-10) + 0.02(13-10) + 0.05(15-10) = 0.75$$

i.e. a prediction of the breeding value (or additive value) of 0.75 piglets above the mean. If we add the mean, we get 10.75 piglets. The confidence interval will be:

$$\hat{a} \pm CI_{95\%} = 0.75 \pm 1.77 = [-1.02,\ 2.52]$$

Performing individual selection, we obtain:

$R = h^2 \cdot i \cdot \sigma_P = 0.10 \cdot 0.8 \cdot 3 = 0.24$ piglets per generation,

$$\rho = h = \sqrt{h^2} = \sqrt{0.10} = 0.32$$

$$\rho^2 = 0.10$$

$$\sigma_e^2 = h^2\sigma_P^2\left(1 - \rho_{A,C}^2\right) = 0.10 \cdot 3^2(1 - 0.10) = 0.81$$

$$CI_{95\%} = \pm 2\sqrt{0.81} = \pm 1.80$$

$$\hat{a} \pm CI_{95\%} = 0.10(14-10) \pm 1.8 = 0.4 \pm 1.8 = [-1.4,\ 2.2]$$

The results of the index are better than the results of direct selection. The precision of the additive value is better, and note that *the additive value predicted by the index is higher* because it has more information.

So far, we have considered indices using a fixed number of relatives that we assume are available to all individuals, but the indices can use the information of all the relatives available in each individual, although it varies between individuals. The essential information is, however, in close relatives, such as parents and siblings. Individuals and relatives can also have a variable number of data values (e.g. several parities) that can also be included in the index. In Chapter 4, we will see how to include all information from all relatives.

3.1.2 Indices with information from several traits

It is possible to improve a trait by adding other traits to the selection criteria in order to use the information from their correlations with the trait of interest.

We will start with a simple example. We will select only using the records of the traits observed in the individual, without taking into account the relatives. In section 3.2.3, we will see the more general case of indices for multiple traits with information from relatives.

The selection index is built as in section 3.1.1. The data vector is the vector that collects the information on the traits measured in a single individual, $y' = [y_1,y_2,y_3,\cdots]$, where the values are *centred*, as in section 3.1.1. It may happen that the trait of interest is not recorded (e.g. if it is expensive to record), but the index is obtained in the same way:

$$V = E(y \cdot y') = E\begin{bmatrix} y_1 \\ y_2 \\ \cdots \\ y_n \end{bmatrix}[y_1,y_2,\cdots,y_n] = \begin{bmatrix} \sigma_P^2 & \text{cov}(y_1,y_2) & \cdots & \text{cov}(y_1,y_n) \\ & \sigma_P^2 & \cdots & \text{cov}(y_2,y_n) \\ & & \cdots & \cdots \\ & & & \sigma_P^2 \end{bmatrix}$$

$$c' = \text{cov}\ (a,y') = E(a \cdot y') = E([a \cdot y_1, a \cdot y_2,\cdots,a \cdot y_n]) = [\text{cov}(a,y_1),\text{cov}(a,y_2),\cdots,\text{cov}(a,y_n)]$$

Here, as in section 3.1.1, $y_1 = a_1 + e_1$, $y_2 = a_2 + e_2$, ...etc., and as the additive value is independent of the environmental value:

$$c' = \text{cov}\ (a,y') = E(a \cdot y') = [\text{cov}(a,a_1),\text{cov}(a,a_2),\cdots,\text{cov}(a,a_n)]$$

where a is the breeding value of the trait of interest, c' is the additive covariance vector of the trait of interest with the traits that are recorded, and V is the variance–covariance matrix of the traits recorded. The index is, as in section 3.1.1:

$$I = c'V^{-1}y$$

and all the formulas for the precision, response to selection, confidence interval, etc. apply here as well.

EXAMPLE 3.2

We may be interested in improving the feed conversion ratio (FCR), but we may also measure growth rate (GR) because it is a correlated trait and it is easy and cheap to record. Here, the data vector is the vector that collects the information of the traits measured in a single individual, $y' = [y_{FCR}, y_{GR}]$. The genetic parameters are calculated in the first example of section 3.2.2, and they are:

$$\mathbf{V} = \begin{bmatrix} \sigma^2_{P_{FCR}} & \operatorname{cov}(y_{FCR}, y_{GR}) \\ & \sigma^2_{P_{GR}} \end{bmatrix} = \begin{bmatrix} 0.063 & -10.8 \\ -10.8 & 5{,}184 \end{bmatrix}$$

$$c' = \operatorname{cov}(a_{FCR}, y') = [\operatorname{cov}(a_{FCR}, a_{FCR}), \operatorname{cov}(a_{FCR}, a_{GR})] = [\sigma^2_{A_{FCR}}, \operatorname{cov}(a_{FCR}, a_{GR})] = [0.013 \quad -3.1]$$

$$I = \hat{a}_{FCR} = \mathbf{c}'\mathbf{V}^{-1}\mathbf{y} = [0.013 \quad -3.1] \begin{bmatrix} 0.063 & -10.8 \\ -10.8 & 5{,}184 \end{bmatrix}^{-1} \begin{bmatrix} y_{FCR} \\ y_{GR} \end{bmatrix} = 0.16 y_{FCR} - 0.00026 y_{GR}$$

$$\sigma^2_I = \mathbf{c}'\mathbf{V}^{-1}\mathbf{c} = [0.013 \quad -3.1] \begin{bmatrix} 0.063 & -10.8 \\ -10.8 & 5{,}184 \end{bmatrix}^{-1} \begin{bmatrix} 0.013 \\ -3.1 \end{bmatrix} = 0.0029$$

For an intensity of selection as in the example of section 2.3.2.A, $i = 1.3$, the response to selection will be, taking into account that it will be negative (FCR will be reduced):

$$R = i \cdot \sigma_I = -1.3 \cdot \sqrt{0.0029} = -0.070$$

In this example, the response in FCR using the index would be –0.070, higher than the response selecting directly by FCR; applying the formula in section 2.3.2.A, the direct response is, taking the variances from $\mathbf{c}$ and $\mathbf{V}$:

$$R_{FCR} = -h^2 \cdot i \cdot \sigma_P = -\frac{\sigma^2_A}{\sigma^2_P} \cdot i \cdot \sigma_P = -\frac{0.013}{0.063} \cdot 1.3 \cdot \sqrt{0.063} = -0.067$$

which is lower than the response obtained using the index.

3.2 Selection Indices for Multiple Traits

3.2.1 Economic weights

The profit obtained from plant or animal production does not usually depend on a single trait but on a combination of them. By economically weighing the traits, we can predict the profit that an individual's descendants could produce, and define *selection objectives* based on the profit that we expect its offspring will give to us. The *profit function* is a function of traits, product prices, variable costs and fixed costs. For short periods of time (one or two generations), prices and costs are considered to be constant, regardless of the level of production, so the profit function depends only on the traits. If we have n traits, we can write the profit function, B, as:

$$B = f(y_1, y_2, \dots y_n)$$

This can be a very complex function. A simple way to work with it is to approximate it to a linear function around the means of the traits. This is done mathematically by means of the Taylor series, making the partial derivatives of the function for each of the traits, and substituting the values of the traits $y_1, y_2, \dots y_n$ for their means in the resulting expression. We will assume, as before, that the phenotypic values of the traits are *centred*, and that $\mathbf{m}' = m_1, m_2, \dots m_n$ is the value of the means of the traits before centring them:

$$B = f(y_1, y_2, \ldots y_n) \approx \left[\frac{\partial f}{\partial y_1}\right]_m y_1 + \left[\frac{\partial f}{\partial y_2}\right]_m y_2 + \cdots + \left[\frac{\partial f}{\partial y_n}\right]_m y_n$$

We will call economic weights w_i for the quantities that multiply the traits:

$B = w_1 y_1 + w_2 y_2 + \cdots + w_n y_n = \boldsymbol{w'y}$

Let us take a very simplified example. Suppose that, in rabbit production, each doe produces n kits reaching a commercial weight of p kg, after having consumed c kg of feed. If P_c is the price in € per kg of feed, the cost of feeding a young rabbit is $c \cdot P_c$. If P_p is the price in € per kg of live weight of rabbit sold, the income obtained for each rabbit sold is $p \cdot P_p$. Let us call the fixed cost of keeping a young rabbit C_I and the cost of keeping a doe C_D. The profit function will be:

$$B = \text{Returns} - \text{Costs} = p \cdot P_p - \left(c \cdot P_c + C_I + \frac{C_D}{n} \right)$$

Here, the traits are the commercial weight of the rabbit, p, the total number of kits of a doe, n, and the individual feed consumption, c. The maintenance costs of the doe throughout her life, C_D, includes the proportion of labour and amortization costs of the farm, cages and other facilities, the feeding of the female, vaccinations and all general costs related to the mother. The fixed cost of keeping an individual, C_I, includes the proportional part of labour and amortization of the fattening facilities, and the common costs per individual in the fattening facilities except the cost of feed, which varies with each individual. The economic weights are:

$$w_p = \left[\frac{\partial B}{\partial p}\right]_m = \left[P_p\right]_m = P_p;\ w_c = \left[\frac{\partial B}{\partial c}\right]_m = \left[-P_c\right]_m = -P_c;\ w_n = \left[\frac{\partial B}{\partial n}\right]_m = \left[-\frac{C_D}{n^2}\right]_m = -\frac{C_D}{\bar{n}^2}$$

where $\bar{n}$ is the average number of kits produced by a doe throughout her productive life. Suppose that in our population P_p = €2/kg live weight, P_c = €0.3/kg feed, C_H = €64 and $\bar{n}$ = 40 rabbits per female. The economic weights will be:

$$w_p = 2\frac{€}{\text{kg liveweight}};\ w_c = -0.3\ \frac{€}{\text{kg feed}};\quad w_n = -\frac{64}{40^2} = -0.04\frac{€}{\text{kits/doe}}$$

This approach implies that the relationship between profit and traits is linear; that is, if there is an increase of 10% in a trait, a profit will be obtained that is directly proportional to the increase in production, which does not always happen in the long term.

A simpler and more intuitive way to calculate the economic weights is to approximate them to the profit that would be obtained by increasing one unit of the trait:

$$w_i = \frac{\Delta B}{\Delta y_i}\ \frac{€}{\text{trait unit}}$$

In our previous example, the economic weights of live weight and feed consumption calculated as differentials or as increments are the same, because the price P_p indicates the profit when increasing 1 kg of live weight, and the price P_c indicates the increase in profit as the feed consumption trait decreases by 1 kg. Calculating the increase in profit by increasing one rabbit per female is more complicated, but if we assume that the female's costs do not vary due to having only one more kit, the profit increases because the female's costs are divided by 41 instead of 40, so the economic weight is:

$$w_n = \frac{C_H}{41} - \frac{C_H}{40} = \frac{64}{41} - \frac{64}{40} = -0.039 \frac{€}{\text{kits/doe}}$$

which is very close to what we have calculated formerly.

Linear approximations work well around the point to which they are approximated, in this case the means of the traits, but these means change with selection. Furthermore, market circumstances may change, and a valuable trait, such as the percentage of fat in milk, may fall in economic value due to changes in consumer preferences. Besides, the increase in profit is generally proportional to the increase in a trait only for small increases; there could be asymptotic profit increases (smaller as the trait increases), or there could be a production optimum in which the profit changes sign as the trait increases. As selection increments of traits are generally small, economic weights can be recalculated every two or three generations for monogastrics or every 4–5 years for cattle. They could be recalculated more frequently, but several generations are needed to confirm the results of selection, so it is not good to make decisions based on early estimates of the response to selection.

> Economic weights must be recalculated frequently, every two or three generations in the case of monogastrics, and every 4 or 5 years in programmes with long generational intervals such as beef or dairy cattle.

The behaviour of a breeding company is influenced by whether they believe the sales of their product can be affected. If a trait has a lower performance than expected and this is perceived by the industry, the breeding company will reduce its sales, so to recover their place in the market, they should increase the economic weight of this trait. It has been proposed to use the market share of the companies (the fraction of clients that buy the brand), to determine the economic weights of the traits. The economic weight of a trait could be the increase in market share that occurs as the value of the trait increases. The problem with this method is that only large companies have access to market studies that allow them to reliably calculate their increase in market share as the value of each trait increases.

As it may not be easy to determine the economic weight of some traits, sometimes the economic weights are determined by the farmers of the association that carries out the genetic programme, according to their subjective perceptions of the expected profit. Another way is to determine the *desired gain* in one or more traits relative to the other traits of the index, and find the economic weights for all traits that would lead to these relative desired gains (Yamada *et al.*, 1975), or to use *restrictions* in which one or more traits have no response to selection. The results will later depend on the true genetic parameters and the success in the process of selection. None of these methods is optimal because they do not apply the correct economic weights for the maximum economic response.

Often, the economic weights are modified according to the actual responses in the selection process. We are applying simplified models; thus, the genetic response is often not as expected; augmenting (or reducing) the weight of a trait that did not show the expected response is a common practical method used in the industry.

Sometimes in breeding schemes, there are conflicting interests. For example, farmers may care little about the carcass yield if they are not paid for it, but the slaughterhouse will care. Retail cuts of the carcass are of interest to the slaughterhouse but not necessarily to the butcher who sells meat. Meat quality is of particular interest to consumers, but the butchers may not differentiate between meats of the same type by their quality. If the economic weight is calculated by a breeding company that sells to farmers, the company will make an effort to satisfy their clients, but if the company sells to a meat company that has integrated farmers and slaughterhouses, the economic weight is more difficult to calculate due to competing interests. The same happens with national breeding programmes, which must satisfy the different players in the food chain. Some solutions have been proposed, such as calculating profits as returns divided by costs ($P = R/C$) instead of income minus costs ($P = R - C$), with the argument that the ratio is based on efficiency. It has also been proposed to include the profits in the costs and assume a zero profit, as this is the real situation for most small farmers, who earn their money basically as self-employed workers. All of these issues are discussed in detail by Weller (1994) and Blasco (1995).

3.2.2 The selection index

The selection criteria that we have seen so far consisted of several phenotypes (e.g. individual phenotype, phenotype of a correlated trait, phenotype of relatives) used to predict the additive value of a single trait. Here, we have n economically important traits, so we must predict n additive values per individual. But we need a unique index value for an animal in order to practice selection, so we will take a combination of additive values producing maximum benefit, which we will call the *economic additive value*:

$$a_E = \mathrm{w}_1 a_1 + \cdots + \mathrm{w}_n a_n = \boldsymbol{w'a}$$

where $\boldsymbol{a}' = [a_1, a_2, \ldots a_n]$ is the vector of additive values of the traits. The economic additive value is measured in monetary units. For example, if we select for milk production and milk protein in a given period, both measured in kg:

$$\text{Units}(a_E) = \text{units}(w_M a_M + w_P a_P) = \frac{€}{\text{kg milk}} \cdot \text{kg milk} + \frac{€}{\text{kg protein}} \cdot \text{kg protein} = €$$

We will select the animals whose descents produce a maximum of euros. As the traits in this example are negatively correlated, some animals will have a high economic additive value because they have genes to produce more milk and less protein, some will produce more protein and less milk, and others will be intermediate, but our objective will not be one trait or another but simply to select bulls having genes such that their daughters produce more euros, whatever the reason.

> A selection index for two traits does not necessarily improve both traits; if they are negatively correlated, one can improve and the other worsen, as long as the economic result of the combination of both is optimal.

The economic additive value is often called the *aggregate genotype*, but this name is less expressive and we will not use it. We will make the prediction of the additive values of the traits, as in the case of the indices for a single trait, from the observed phenotypic values, but here we can use the genetic correlations between traits to predict each additive value of each trait with the phenotypic values of all the traits. This adds information and therefore decreases the prediction error and increases the additive values.

The set of traits that determine the economic additive value may not be the same set of traits that are measured. For example, FCR is an economically important trait, but as it is difficult to measure, we could select it indirectly through the GR, without GR necessarily being considered an economically relevant trait to be included in the economic additive value. In this case, our *selection objective* is the conversion rate, but the *selection criterion*, the measured trait, is the GR.

> A selection index can contain data for traits that are not included in the economic additive value, and in turn the economic additive value can include traits that are not measured. The traits included in the economic additive value are called *selection objectives* and the measured traits *selection criteria*. Some traits are both objective and selection criteria.

We will start with a simple example. We will select only by the value of the trait observed in the individual, without taking into consideration the relatives. Here, the data vector is the vector that collects the information of the traits measured in a single individual; for example, in the case of the amount of milk (M) and that of protein (P), it would be $y' = [y_M, y_P]$, where the values are *centred*, as in the case of the index for a trait. The selection index is built as in the previous section:

$$\mathbf{V} = E(\mathbf{y}\cdot\mathbf{y}') = E\begin{bmatrix} y_M \\ y_P \end{bmatrix}\begin{bmatrix} y_M & y_P \end{bmatrix} = \begin{bmatrix} \sigma^2_{P_M} & \mathrm{cov}(y_M, y_P) \\ & \sigma^2_{P_P} \end{bmatrix}$$

$$c'_M = \mathrm{cov}(a_M, y') = \left[\mathrm{cov}(a_M, a_M), \mathrm{cov}(a_M, a_P)\right] = \left[\sigma^2_{A_M}, \mathrm{cov}(a_M, a_P)\right]$$

$$c'_P = \mathrm{cov}(a_P, y') = \left[\mathrm{cov}(a_P, a_M), \mathrm{cov}(a_P, a_P)\right] = \left[\mathrm{cov}(a_M, a_P), \sigma^2_{A_P}\right]$$

$$\hat{a}_M = c'_M \mathbf{V}^{-1}\mathbf{y}$$

$$\hat{a}_P = c'_P \mathbf{V}^{-1}\mathbf{y}$$

$$\hat{a} = \begin{bmatrix} \hat{a}_M \\ \hat{a}_P \end{bmatrix} = \begin{bmatrix} c'_M \\ c'_P \end{bmatrix}\mathbf{V}^{-1}\mathbf{y} = \begin{bmatrix} \sigma^2_{A_M} & \mathrm{cov}(a_M, a_P) \\ & \sigma^2_{A_P} \end{bmatrix}\mathbf{V}^{-1}\mathbf{y} = \mathbf{C}'\mathbf{V}^{-1}\mathbf{y}$$

where $\mathbf{C}'$ is the additive covariance matrix of the *selection objective* traits with the *selection criteria* (the traits that are recorded). Here $\mathbf{C}'$ is symmetric, but it may not be, as we will see in the example below. The selection index predicts the economic additive value of the traits:

$$I = \hat{a}_E = w_M\,\hat{a}_M + w_P\,\hat{a}_P = \mathbf{w}'\,\hat{\mathbf{a}} = \mathbf{w}'\mathbf{C}'\mathbf{V}^{-1}\mathbf{y}$$

Note that we could have used three or more traits and arrived at the same formula. The general formula for multi-trait selection indices is thus:

$$I = w'C'V^{-1}y$$

EXAMPLE 3.3

In pigs, the main cost of production is feeding, which can be around two-thirds of the total cost of production per piglet. The carcasses are valued according to their meat content, estimated by measuring backfat (BF) by ultrasound. The *selection objective* is to reduce the FCR and the BF of the animal, but the consumption of individual feed is expensive to measure, so GR will be measured instead; GR will be a *selection criterion*, as it is easy to measure (weighing the animal at two different ages) and is correlated with FCR. The selection objective targets are thus FCR and BF, and the selection criteria are GR and BF.

We know that a reduction of 0.1 points in FCR implies a saving of €3 per animal, and that a reduction of 1 mm of BF thickness implies a benefit of €1 per carcass sold. Table 3.1 shows the means and phenotypic deviations of the traits, as well as the genetic parameters.

Table 3.1. Means, standard deviations and genetic parameters of feed conversion ratio (FCR), growth rate (GR) and backfat (BF) in pigs. Heritabilities are on the diagonal, genetic correlations above the diagonal and phenotypic correlations below the diagonal.

	m	σ_P		FCR	GR	BF
FCR	2.5	0.25	FCR	0.2	–0.7	0.4
GR (g/day)	700	72	GR	–0.6	0.3	0.1
BF (mm)	15	3	BF	0.4	0.2	0.5

To calculate the index, we first need the variances and covariances of the traits:

$$\sigma_A^2 = h^2\sigma_P^2;\ \ \mathrm{cov}(a_1,a_2) = r_g(a_1,a_2) \cdot \sigma_{A_1}\sigma_{A_2};\ \mathrm{cov}(y_1,y_2) = r_P(y_1,y_2) \cdot \sigma_{P_1}\sigma_{P_2}$$

$$\sigma_{P_{FCR}}^2 = 0.25^2 = 0.063; \qquad \sigma_{P_{GR}}^2 = 72^2 = 5{,}184; \qquad \sigma_{P_{BF}}^2 = 3^2 = 9$$

$$\sigma_{A_{FCR}}^2 = 0.2 \cdot 0.063 = 0.013; \quad \sigma_{A_{GR}}^2 = 0.3 \cdot 5{,}184 = 1{,}555; \quad \sigma_{A_{BF}}^2 = 0.5 \cdot 9 = 4.5$$

$$\mathrm{cov}(a_{FCR}, a_{GR}) = -0.7\sqrt{0.013}\sqrt{1555} = -3.1$$
$$\mathrm{cov}(a_{FCR}, a_{BF}) = 0.4\sqrt{0.013}\sqrt{4.5} = 0.1$$
$$\mathrm{cov}(a_{GR}, a_{BF}) = 0.1\sqrt{1555}\sqrt{4.5} = 8.4$$

$$\mathrm{cov}(y_{FCR}, y_{GR}) = -0.6 \cdot 0.25 \cdot 72 = -10.8$$
$$\mathrm{cov}(y_{FCR}, y_{BF}) = 0.4 \cdot 0.25 \cdot 3 = 0.3$$
$$\mathrm{cov}(y_{GR}, y_{BF}) = 0.2 \cdot 72 \cdot 3 = 43.2$$

$$C' = \mathrm{cov}(a,y') = \mathrm{cov}\left(\begin{bmatrix} a_{FCR} \\ a_{BF} \end{bmatrix}, [y_{GR} \quad y_{BF}]\right) = \begin{bmatrix} \mathrm{cov}(a_{FCR}, a_{GR}) & \mathrm{cov}(a_{FCR}, a_{BF}) \\ \mathrm{cov}(a_{BF}, a_{GR}) & \sigma_{A_{BF}}^2 \end{bmatrix} = \begin{bmatrix} -3.1 & 0.1 \\ 8.4 & 4.5 \end{bmatrix}$$

$$\mathbf{V}=\begin{bmatrix}\sigma^2_{P_{GR}} & \text{cov}(y_{GR},y_{BF})\\ \text{cov}(y_{GR},y_{BF}) & \sigma^2_{P_{BF}}\end{bmatrix}=\begin{bmatrix}5184 & 43\\ 43 & 9\end{bmatrix}\rightarrow \mathbf{V}^{-1}=\begin{bmatrix}0.00020 & -0.00096\\ -0.00096 & 0.1157\end{bmatrix}$$

Note that the matrix **C′** is not symmetric, as the selection objective traits are not the same as the selection criteria. It does not even have to be a square matrix; for example, having another selection objective that is not a selection criterion and is not recorded, the vector of additive values would have three components, the vector of phenotypic values two components, and **C′** would be a 3 × 2 matrix.

For economic weights, profit should be expressed in the same units in which the traits are measured. For example, if a trait is measured in kg, the profit should not be expressed in €/100 g but in €/kg. In this example, the additive economic value is:

$$a_E = w_{FCR}\cdot a_{FCR} + w_{BF}\cdot a_{BF}$$

$$w_{FCR}=\frac{3€}{-0.1\ \text{kg\_feed/kg\_liveweight}}=-30\frac{€}{\text{kg\_feed/kg\_liveweight}}$$

$$w_{BF}=\frac{1€}{-1\ \text{mm}}=-1\frac{€}{\text{mm}}$$

$$\boldsymbol{w}'=[w_{FCR},w_{BF}]=[-30,-1]$$

$$I=\boldsymbol{w}'\mathbf{C}'\mathbf{V}^{-1}\mathbf{y}=[-30\ \ -1]\begin{bmatrix}-3.1 & 0.1\\ 8.4 & 4.5\end{bmatrix}\begin{bmatrix}0.00020 & -0.00096\\ -0.00096 & 0.1157\end{bmatrix}\begin{bmatrix}y_{GR}\\ y_{BF}\end{bmatrix}=0.024y_{GR}-0.95y_{BF}$$

For example, a pig having a GR of 770 g/day and a BF of 12 mm, has an economical additive value of:

$$\hat{a}_E = I = 0.024(770-700)-0.95(12-15)=+4.53€$$

The selection parameters, expected response, precision, variance of the error and confidence interval of the prediction are calculated as in the index for a single trait. When calculating the variance of the index and the covariance with the economic additive value, we take into account that the vector $\boldsymbol{w}$ and the matrices **C** and **V** are constant:

$$\sigma^2_I=\text{var}(\boldsymbol{w}'\mathbf{C}'\mathbf{V}^{-1}\mathbf{y})=\boldsymbol{w}'\mathbf{C}'\mathbf{V}^{-1}\cdot\text{var}(\mathbf{y})\cdot\mathbf{V}^{-1}\mathbf{C}\boldsymbol{w}=\boldsymbol{w}'\mathbf{C}'\mathbf{V}^{-1}\mathbf{V}\mathbf{V}^{-1}\mathbf{C}\boldsymbol{w}=\boldsymbol{w}'\mathbf{C}'\mathbf{V}^{-1}\mathbf{C}\boldsymbol{w}$$

As $a_E = \boldsymbol{w}'\boldsymbol{a} = \boldsymbol{a}'\boldsymbol{w}$ and $\mathbf{C} = [\text{cov}(\boldsymbol{a},\boldsymbol{y}')]' = \text{cov}(\boldsymbol{y},\boldsymbol{a}')$, we have:

$$\text{cov}(a_E,I)=\text{cov}(I,a_E)=\text{cov}(\boldsymbol{w}'\mathbf{C}'\mathbf{V}^{-1}\boldsymbol{y},\boldsymbol{a}'\boldsymbol{w})=\boldsymbol{w}'\mathbf{C}'\mathbf{V}^{-1}\cdot\text{cov}(\boldsymbol{y},\boldsymbol{a}')\cdot\boldsymbol{w}=\boldsymbol{w}'\mathbf{C}'\mathbf{V}^{-1}\mathbf{C}\boldsymbol{w}=\sigma^2_I$$

$$\sigma^2_{A_E}=\text{var}(\boldsymbol{w}'\boldsymbol{a})=\boldsymbol{w}'[\text{var}(\boldsymbol{a})]\boldsymbol{w}=\boldsymbol{w}'\begin{bmatrix}\sigma^2_{A_1} & \text{cov}(a_1,a_2) & \cdots & \text{cov}(a_1,a_n)\\ & \sigma^2_{A_2} & \cdots & \text{cov}(a_2,a_n)\\ & \vdots & & \vdots\\ & & & \sigma^2_{A_n}\end{bmatrix}\boldsymbol{w}$$

Note that matrix $[\text{var}(\boldsymbol{a})]$is not the same as matrix **C**, unless the selection criteria and objectives are the same.

The precision, response to selection and 95% confidence interval are:

$$\rho_{A_E,I} = \frac{\text{cov}(a_E,I)}{\sigma_{A_E}\sigma_I} = \frac{\sigma_I^2}{\sigma_{A_E}\sigma_I} = \frac{\sigma_I}{\sigma_{A_E}}$$

$$R = \rho_{A_E,I} \cdot i \cdot \sigma_{A_E} = i \cdot \sigma_I$$

$$\sigma_e^2 = \sigma_{A_E}^2\left(1-\rho_{A_E,I}^2\right) = \sigma_{A_E}^2 - \sigma_I^2 \quad \rightarrow \quad CI_{95\%} = \pm 2\sigma_e = \pm 2\sqrt{\sigma_{A_E}^2 - \sigma_I^2}$$

The response to selection for each trait, *i*, is:

$$R_i = \frac{\text{cov}(a_i,I)}{\sigma_{A_i}\sigma_I} \cdot i \cdot \sigma_{A_i} = \text{cov}\left(a_i, w'C'V^{-1}y\right)\cdot\frac{i}{\sigma_I} = w'C'V^{-1}\,\text{cov}(a_i,y)\cdot\frac{i}{\sigma_I}$$

The vector with responses to selection for all traits is:

$$\begin{bmatrix} R_1 \\ R_2 \\ \vdots \\ R_n \end{bmatrix} = w'C'V^{-1}\begin{bmatrix} \text{cov}(a_1,y) \\ \text{cov}(a_2,y) \\ \vdots \\ \text{cov}(a_n,y) \end{bmatrix}\cdot\frac{i}{\sigma_I} = w'C'V^{-1}C\cdot\frac{i}{\sigma_I}$$

EXAMPLE 3.4

We will calculate the response to selection for the former example, for an intensity of selection as in the example of section 2.3.2.A, where *i*=1.3:

$$\sigma_I^2 = w'C'V^{-1}Cw = \begin{bmatrix}-30 & -1\end{bmatrix}\begin{bmatrix}-3.1 & 0.1 \\ 8.4 & 4.5\end{bmatrix}\begin{bmatrix}0.00020 & -0.00096 \\ -0.00096 & 0.1157\end{bmatrix}\begin{bmatrix}-3.1 & 8.4 \\ 0.1 & 4.5\end{bmatrix}\begin{bmatrix}-30 \\ -1\end{bmatrix} = 9.2$$

$$\sigma_I = \sqrt{9.2} = 3.03€$$

$$\sigma_{A_E}^2 = \begin{bmatrix}w_{FCR} & w_{BF}\end{bmatrix}\begin{bmatrix}\sigma_{A_{FCR}}^2 & \text{cov}(a_{FCR},a_{BF}) \\ \text{cov}(a_{FCR},a_{BF}) & \sigma_{A_{BF}}^2\end{bmatrix}\begin{bmatrix}w_{FCR} \\ w_{BF}\end{bmatrix} = \begin{bmatrix}-30 & -1\end{bmatrix}\begin{bmatrix}0.013 & 0.1 \\ 0.1 & 4.5\end{bmatrix}\begin{bmatrix}-30 \\ -1\end{bmatrix} = 22$$

$$\rho_{A_E,I} = \frac{\sigma_I}{\sigma_{A_E}} = \sqrt{\frac{9.2}{22}} = 0.64;\ \rho_{A_E,I}^2 = 0.64^2 = 0.41$$

$$R = i \cdot \sigma_I = 1.3 \cdot 3.03 = 3.94€$$

$$\sigma_e^2 = \sigma_{A_E}^2 - \sigma_I^2 = 22 - 9.2 = 12.8 \quad \rightarrow \quad IC_{95\%} = \pm 2\sigma_e = \pm 2\sqrt{12.8} = \pm 7.2$$

$$\begin{bmatrix}R_{FCR} \\ R_{BF}\end{bmatrix} = w'C'V^{-1}C\frac{i}{\sigma_I} = \begin{bmatrix}-30 & -1\end{bmatrix}\begin{bmatrix}-3.1 & 0.1 \\ 8.4 & 4.5\end{bmatrix}\begin{bmatrix}0.00020 & -0.00096 \\ -0.00096 & 0.1157\end{bmatrix}\begin{bmatrix}-3.1 & 8.4 \\ 0.1 & 4.5\end{bmatrix}\cdot\frac{1.3}{3.03} = \begin{bmatrix}-0.072 \\ -1.7\end{bmatrix}$$

R_{FCR} = – 0.072 (kg feed/kg weight)

R_{BF} = – 1.7 mm

By individual selection, applying the formula in section 2.3.2.A, $R = h^2 \cdot i \cdot \sigma_P$, the response would have been $R_{FCR} = -0.2 \cdot 1.3 \cdot 0.25 = -0.065$ kg feed/kg live weight, and $R_{BF} = -0.5 \cdot 1.3 \cdot 3 = -1.95$ mm. BF selection pressure is lower using the index than in the case of individual selection because there are individuals who would not have been selected if the criterion had been exclusively BF but who are selected because they have a good FCR. This is an example where an index does not improve all traits but improves some and worsens others as long as the economic response is maximum. Conversely, FCR, which has a greater economic weight, has a higher response than would be obtained if selection had simply been by FCR, as it takes advantage of the extra information provided by the trait BF, which has a high heritability and a high correlation with FCR.

An error in the economic weights, such as underestimating the economic weight of the feed conversion index by 25% and overestimating BF by 25%:

$$w' = [w_{FCR}, w_{BF}] = [-22.5, -1.25]$$

would result in a response of €3.8 instead of €3.9. As can be seen, moderate errors in estimation of the economic weights do not substantially modify the results. There are several studies showing the robustness of economic weights to errors (Smith, 1983). However, a similar error in genetic parameters, in which two genetic covariances are respectively underestimated and overestimated:

$$C' = \begin{bmatrix} -3.1 & 0.2 \\ 4.2 & 4.5 \end{bmatrix}$$

would result in a response of €5.1 instead of €3.9, which is a more serious mistake.

> Selection indices for multiple traits are sensitive to estimation errors in economic weights, but they are particularly sensitive to estimation errors in genetic variances and covariances.

3.2.3 Indices for multiple traits with information from relatives

In the indices for a single trait, we have used information from relatives to predict the additive value. In the indices for several traits, we can do the same. We will calculate an index for the example above, using individual and half-sibling data as an example of using family information. We will consider the same economic weights, selection criteria and selection objectives:

$$a_E = w_{FCR} \cdot a_{FCR} + w_{BF} \cdot a_{BF}$$

The data vector will now contain the familiar information in both traits:

$$y' = \left[y_{FCR}, y_{BF}, y_{FCR}^{HS}, y_{BF}^{HS} \right]$$

where $y_{FCR}^{HS}, y_{BF}^{HS}$ are the data for GR and BF of a half-sibling. As in the previous case, the values are centred. The matrices **C** and **V** now contain the information of the individual and the relatives. The covariances between half siblings for two traits are:

$$\mathrm{cov}\left(y_1, y_2^{HS}\right) = \mathrm{cov}\left(a_1, a_2^{HS}\right) = \frac{1}{4}\mathrm{cov}\left(a_1, a_2\right)$$

$$C' = \mathrm{cov}\left(\begin{bmatrix} a_{FCR} \\ a_{BF} \end{bmatrix}, \left[y_{GR}, y_{BF}, y_{GR}^{HS}, y_{BF}^{HS}\right]\right) = \begin{bmatrix} \mathrm{cov}(a_{FCR}, a_{GR}) & \mathrm{cov}(a_{FCR}, a_{BF}) & \frac{1}{4}\mathrm{cov}(a_{FCR}, a_{GR}) & \frac{1}{4}\mathrm{cov}(a_{FCR}, a_{BF}) \\ \mathrm{cov}(a_{BF}, a_{GR}) & \sigma^2_{A_{BF}} & \frac{1}{4}\mathrm{cov}(a_{BF}, a_{GR}) & \frac{1}{4}\sigma^2_{A_{BF}} \end{bmatrix}$$

$$V = \mathrm{var}\left(\left[y_{GR}, y_{BF}, y_{GR}^{HS}, y_{BF}^{HS}\right]\right) = \begin{bmatrix} \sigma^2_{P_{GR}} & \mathrm{cov}(y_{GR}, y_{BF}) & \frac{1}{4}\sigma^2_{A_{GR}} & \frac{1}{4}\mathrm{cov}(a_{GR}, a_{BF}) \\ & \sigma^2_{P_{BF}} & \frac{1}{4}\mathrm{cov}(a_{BF}, a_{GR}) & \frac{1}{4}\sigma^2_{A_{BF}} \\ & & \sigma^2_{P_{GR}} & \mathrm{cov}(y_{GR}, y_{BF}) \\ & & & \sigma^2_{P_{BF}} \end{bmatrix}$$

Using the same data from the former example:

$$C' = \begin{bmatrix} -3.1 & 0.1 & -0.78 & 0.025 \\ 8.4 & 4.5 & 2.1 & 1.1 \end{bmatrix} \quad V = \begin{bmatrix} 5184 & 43 & 389 & 2.1 \\ & 9 & 2.1 & 1.1 \\ & & 5184 & 43 \\ & & & 9 \end{bmatrix}$$

$$\hat{a}_E = I = \mathbf{w}'\mathbf{C}'\mathbf{V}^{-1}\mathbf{y} = 0.024y_{GR} - 0.934y_{BF} + 0.004y_{GR}^{HS} - 0.114y_{BF}^{HS}$$

$$\sigma_I^2 = w'C'V^{-1}Cw = 9.31 \quad \rho_{A_E,I} = \frac{\sigma_I}{\sigma_{A_E}} = \sqrt{\frac{9.31}{22}} = 0.65 \quad \rho^2_{A_E,I} = 0.65^2 = 0.42$$

$$\sigma_e^2 = \sigma^2_{A_E} - \sigma_I^2 = 22 - 9.31 = 12.7 \quad \rightarrow \quad CI_{95\%} = \pm 2\sigma_e = \pm 2\sqrt{12.7} = \pm 7.1$$

$$R = i \cdot \sigma_I = 1.3 \cdot \sqrt{9.31} = 3.97€$$

Note that, compared with the example of indices for a single trait, the data for the half-sibling produce a much lower weight than that for the individual, because this weight depends not only on the relationship but also on the genetic parameters. When the heritabilities are high, the individual becomes important with respect to the data from the relatives. The response, precision and confidence interval here are very similar to those of the selection with only individual information, and adding the data of a half-brother contributes little to the index in this case.

3.3 Some Important Features of Selection Indices

Selection indices have good properties if we assume that the variance and covariance components are approximately true and that the pre-correction and centring of the data had no error. In Chapter 4, we will discuss how to correct data and predict additive values simultaneously.

The most important properties of selection indices are:

- Selection indices are unbiased and minimize the variance of the prediction error σ_e^2, which in statistics is known as the prediction risk (see Appendix 3.1).

- Selection indices maximize the precision, $\rho_{A,I}$ (see Appendix 3.2), and therefore the reliability of the prediction $\rho^2_{A,I}$.
- Selection indices maximize the response to selection, which is a direct consequence of maximizing precision.
- The advantages of the indices are higher when the heritabilities are low. If the heritability of a trait is very high, the individual value is already a good predictor of its additive value.

However, when considering family information, selection indices give a certain weight to complete families, which tends to increase consanguinity, particularly when heritabilities are low and the weight of individual information is small. We can moderate this effect by computing the index using an artificially high value of heritability, as this will decrease the weight of family information, but this has to be done with caution, because the real response will be lower; as the heritability is not the right one, it is not an optimal index.

In general, indices have great flexibility to adapt to the characteristics of an animal breeding programme.

- *It is possible to select indices with restrictions, so that a trait is forced to remain unchanged.* This is useful in species such as beef cattle, in which it is intended to increase GR but not adult weight, because they are positively correlated. We have not covered this technique in this chapter, because it is generally not recommended. As both traits are positively correlated, preventing the change in one of them markedly decreases the response to selection in the other trait, because many genes that could increase their frequencies by selecting one trait do not do that in order to avoid the increment of the other trait. This means that the response is not economically optimal, compared with an index that considers the economic weight of the trait whose change is to be prevented. A negative economic weight for adult weight can be more efficient than restricting its change.
- *The index can predict the additive values of individuals with no data.* An example is in the case of litter size or milk production, where males have no data. In this case, the prediction of the additive value of these individuals is made through relatives.
- *Some traits can have repeated measures, and the number of relatives considered can be variable according to each individual.* This is the case, for example, for litter size or milk production. In these cases, the construction of simple indices such as those we have described in this chapter is more complex. Moreover, inverting the variance–covariance matrix V is difficult if the number of individuals in the index is large. In these cases, linear models are used, as we will see in subsequent chapters.
- *Selection indices are the most efficient method for improving two traits compared with other methods* such as tandem selection (selecting one trait in one generation and the other trait in the next generation), or selection by independent levels (selecting first by a trait and selecting by the other trait on the previously selected individuals). But this is always true *only when the genetic parameters of the index are the true ones*, and we have already commented before that the indices are sensitive to estimation errors of the genetic parameters. The estimation of genetic correlations between traits requires a large amount of data to be estimated with a certain precision, in general at least around

2000 data points per trait. Furthermore, genetic parameters must be re-estimated after a certain number of generations, because they are affected by selection.

- *The estimation error of the genetic parameters is not taken into account when the confidence interval of an index is calculated*, so the variance of the error and the confidence intervals are underestimated. This difficulty can only be solved by applying Bayesian statistics, which we will see in Chapter 5.
- *In practice, selection indices do not have many traits, generally no more than six or seven.* There are two reasons for this, the first is that the probability of making serious errors in the estimation of genetic parameters increases with the number of traits, and the second is that, when selecting several traits, the responses in each of them are relatively small compared with selecting by one trait, making it difficult to check the actual responses.
- *Selection indices modify genetic parameters.* We know that additive variances decrease by the Bulmer effect, as we saw in section 2.1.2.B2, but genetic correlations also change. For example, if two traits are negatively correlated, there will be a tendency to increase the frequencies of the non-pleiotropic alleles; that is, the alleles that influence each trait separately, or that have a greater influence on one trait than the other, particularly when the heritabilities and economic weights of both traits are similar. These changes are slow, but genetic parameters would have to be estimated again after several generations of selection.
- *The data may or may not be normally distributed.* Note that the normality of the data has not been invoked at any time. If the data have a normal distribution, we know from the statistical properties of the multivariate normal distribution that the relationship with the additive values (also normal) is necessarily linear. If the data do not have a normal distribution, the indices are the best linear approximation to the prediction of the additive values. In both cases, the index is the best linear prediction (BLP).
- *In practice, the indices are modified by increasing or decreasing the pressure on the traits, depending on their actual responses to selection.* This is done in a somewhat artisanal way, but it is necessary to do it because the indices are based on simplified models, and the distance to the true models can be considerable. First, we have assumed that data were centred; that is, we know the means of the traits without error. We have also assumed that the economic weights and genetic parameters are well estimated and without error, the model is additive (no dominance or epistasis), there are no genotype–environment interactions, etc. Therefore, it is convenient to modify the values of the indices according to the responses obtained.

There is a wide range of free-access educational software that includes the construction of selection indices. There is a popular program, GENUP, by Brian Kinghorn, which can be found at https://bkinghor.une.edu.au/genup.htm (accessed 4 July 2025).

Appendix 3.1

In statistics, the quadratic *risk* of an estimate is defined as the expectation of the squared error. Errors may not have a zero mean; in this case, the mean of the errors, m_e, is called *bias*:

$$Risk = E(e^2) = E(e - m_e + m_e)^2 = E(e - m_e)^2 + m_e^2 = \sigma_e^2 + (bias)^2$$

There are infinite estimators with the same *risk*, for different values of the *bias* and the *variance of the error*. In the case of the selection indices, as the data are centred, the errors have a zero mean, and the risk coincides with the variance of the error, so that a selection index is *unbiased* and of *minimal risk*. This is obviously an approximation, because pre-correcting and centring the data produce prediction errors that have not been considered.

We will estimate the regression coefficient by minimizing the variance of the error:

$$\hat{a} = I = b'y$$
$$a = b'y + e$$
$$e = a - \hat{a} = a - b'y$$

Taking into account that $b'c = c'b$:

$$Risk(I) = \sigma_e^2 = \mathrm{var}(a - b'y) = \sigma_A^2 + b'\mathrm{var}(y)b - 2b'\mathrm{cov}(a,y) = \sigma_A^2 + b'Vb - 2c'b$$

$$\frac{\partial \sigma_e^2}{\partial b} = 2b'V - 2c' = 0 \quad \rightarrow \quad \hat{b}' = c'V^{-1}$$

Data and residuals are uncorrelated:

$$\mathrm{cov}(e,y') = \mathrm{cov}(a - b'y,y') = \mathrm{cov}(a,y') - \mathrm{cov}(b'y,y') = c' - b'V = c' - c'V^{-1}V = 0$$

Appendix 3.2

To find the value of b that maximizes the precision, we derive the precision and set it equal to zero. As the value of the maximum of a function is the same as its logarithm, we will take logarithms because it is operationally simpler:

$$I = b'y$$
$$\mathrm{var}(I) = b'\mathrm{var}(y)b = b'Vb$$
$$\mathrm{cov}(a,I) = \mathrm{cov}(a,b'y) = b'\mathrm{cov}(a,y) = b'c$$

$$\rho_{A,I} = \frac{\mathrm{cov}(a,I)}{\sigma_A \sigma_I} = \frac{b'c}{\sigma_A \sqrt{b'Vb}}$$

$$\log(\rho_{A,I}) = \log(b'c) - \frac{1}{2}\log(b'Vb) - \log(\sigma_A)$$

$$\frac{\partial \log(\rho_{A,I})}{\partial b'} = \frac{c}{b'c} - \frac{1}{2} \cdot \frac{2Vb}{b'Vb}$$

The estimate $\hat{b} = V^{-1}c \quad \rightarrow \quad \hat{b}' = c'V^{-1}$ is the value that makes the derivative of the logarithm of the precision zero and corresponds to the maximum of the precision:

$$\frac{\partial \log(\rho_{A,I})}{\partial b'} = \frac{c}{c'V^{-1}c} - \frac{VV^{-1}c}{c'V^{-1}VV^{-1}c} = 0$$

Moreover, $\hat{b}$ has been estimated to minimize σ_e^2, thus as $\sigma_A^2 = \sigma_I^2 + \sigma_e^2$ and σ_A^2 is a constant, σ_I^2 should be maximum and precision is maximum as well:

$$\rho_{A,I} = \frac{\sigma_I}{\sigma_A}$$

Main Concepts

- *Selection index for a trait with familiar information:* A prediction of the additive value of an individual for a trait that combines individual and familiar information in an optimal way.
- *Individual selection index for several traits:* A prediction of the economic additive value of an individual that combines the additive values of several traits in an economically optimal way, using information from the individual but not from their relatives.
- *Selection index for several traits with familiar information:* A prediction of the economic additive value of an individual that combines the additive values of several traits in an economically optimal way, using individual (when available) and family information also optimally combined.
- *Economic additive value* or economic breeding value: The combination of additive values of several traits of an individual that indicate the expected economic gains of their offspring.
- *Economic weight of a trait:* The increase in economic benefit that occurs when increasing a unit of a trait. If the relationship between profit and trait increase is not linear, the economic weight is considered as a linear approximation.
- *Selection objectives and selection criteria:* In an index for several traits, the selection objectives are the traits that determine the economic additive value, which are not necessarily the same as the measured traits (selection criteria). Some selection objectives are not measured due to various difficulties, and they are selected indirectly through the selection criteria.

Problems

3.1. Calculate a selection index for poultry layers, using the following information from a hen and its mother: mean = 300 eggs, CV = 0.15, h^2 = 0.1.
3.2. Calculate a selection index for milk production and the percentage of protein in milk from the following information. Milk: average 9000 kg, CV = 0.15, h^2 = 0.4. An increase of 1 kg of milk produces a profit of €0.3. Protein: mean 3%, CV = 0.05, h^2 = 0.5. An increase of 0.1% protein produces a profit of €0.3 per kg of milk. $r_g = -0.5$, $r_P = -0.25$.
3.3. Calculate a selection index for two genetically and phenotypically uncorrelated traits, with economic weights of 2 and 1, and heritabilities of 0.1 and 0.4, respectively.

Self-assessment Questions (True or False)

1. The amount of milk and the percentage of protein in the milk are negatively correlated. A selection index for these traits should improve both.
2. The variance of a selection index is smaller the more family information is available to build the index.
3. The residual variance is smaller the more family information is available to build the index.
4. Indices with family information are useful especially when the heritability of the trait is low.
5. Selection indices for several traits are constructed by weighing the observed values with economic weights.

6. Selection indices for several traits take into account the genetic and phenotypic relationships between traits.
7. There can be traits in an index that are selected but not recorded, and traits that are recorded but not selected.
8. If you want a trait to remain unchanged, you can build a selection index with restrictions on that trait.
9. It is recommended to use selection indices for the largest number of traits possible, in order to maximize the benefit.
10. Selection indices are robust to the errors made when calculating the economic weights; the response to the selection that is obtained is similar, although the economic weights are not calculated exactly.

References

Blasco, A. (1995) Los pesos económicos en mejora genética animal. *ITEA* 91A, 59–79.

Hazel, L.N. (1943) The genetic basis for constructing selection indexes. *Genetics* 28, 476–490. DOI: 10.1093/genetics/28.6.476

Henderson, C.R. (1963) Selection index and expected genetic advance. In: Hanson, W.D. and Robinson, H.F. (eds) *Statistical Genetics and Plant Breeding*. National Academy of Sciences, Washington, DC, pp. 141163.

Pearson, K. (1903) Mathematical contributions to the theory of evolution. XI. On the influence of natural selection on the variability and correlation of organs. *Philosophical Transactions of the Royal Society A* 200, 1–66. DOI: 10.1098/rsta.1903.0001

Smith, C. (1983) Effects of changes in economic weights on the efficiency of index selection. *Journal of Animal Science* 56, 1057–1064. DOI: 10.2527/jas1983.5651057x

Smith, F. (1936) A discriminant function for plant selection. *Annals of Eugenics* 7, 240–250. DOI: 10.1111/j.1469-1809.1936.tb02143.x

Weller, J.I. (1994) *Economic Aspects of Animal Breeding*. Chapman & Hall, London.

Yamada Y., Yokouchi K. and Nishida A. (1975) Selection index when genetic gains of individual traits are of primary concern. *Japanese Journal of Genetics* 50, 33–41. DOI: 10.1266/jjg.50.33

Solutions

Solutions to the problems

3.1. $\sigma_P^2 = 2025$, $\sigma_A^2 = 202$, $\text{cov}_P(\text{Hen, Mother}) = \text{cov}_A(\text{Hen, Mother}) = 101$.
3.2. $I = 0.098$ (Hen's eggs − 300) + 0.045 (Mother's eggs −300).
$\sigma_M^2 = 1{,}822{,}500; \sigma_M^2 = 729{,}000; \sigma_{P_P}^2 = 0.0225; \sigma_{A_P}^2 = 0.0112$; $\text{cov}_A(\text{Milk,Protein}) = -45$; $\text{cov}_P(\text{Milk,Protein}) = -50$; $w_{Milk} = 0.3$€/kg, $w_{Protein} = 3$€/%.
$I = 0.11$ (Milk − 9,000) + 354 (% Protein − 3).
3.3. $I = w_1 h_1^2 y_1 + w_2 h_2^2 y_2 = 0.2y_1 + 0.4y_2$

Solutions to the self-assessment questions

1. F, 2. F, 3. T, 4. T, 5. F, 6. T, 7. T, 8. T, 9. F, 10. T.

4 The Linear Model

Abstract

Linear models are the most commonly used models in selection. This chapter starts with a description of classical fixed effects models and moves on to a description of the mixed model (with fixed and random effects). This is followed by a discussion of the differences between fixed and random effects and when to use one or the other. The most widely used selection method in current programmes, best linear unbiased prediction (BLUP), is then presented with a numerical example, and different models for different circumstances are described (e.g. multi-trait BLUP, BLUP with maternal or dominance effects, BLUP with genetic groups). The most relevant properties of BLUP are discussed. The chapter ends with a description of the maximum likelihood and restricted maximum likelihood (REML) methods for variance component estimation, the latter being the most widely used method at present. The calculation of standard errors from variance components is also discussed, giving an example with heritability.

4.1 The Linear Model

4.1.1 The fixed effects model

When deducing the selection indices in section 3.1, we said that the data should be *centred*. If the data are not centred, we cannot deduce the selection indices as we have done, because then $E(y \cdot y')$ is no longer the phenotypic variance–covariance matrix, nor is $E(a \cdot y')$ a covariance vector of the additive values. Subtracting the means is a more complicated procedure than it seems, because the animals have different means depending on whether they are measured in winter or summer, the data are taken from one farm or another, the data comes from the first parity or subsequent parities, etc.

Linear models are used to estimate the different means of the data. Take, for example, litter size in pigs, in which we consider litter size to be affected by season and parity effects. We can use the model:

$$y = \mu + E + P + e$$

which indicates that litter size y depends on an effect of season E and one of parity P, referred to a general mean μ. For example, if we knew the true values of the effects, litters from the first parity could have 1.1 piglets fewer than the average, so the first parity has a negative effect of $P_1 = -1.1$ piglets. Litters born in summer could have a positive effect of 0.5 piglets, and if the global mean is $\mu = 10$ piglets, the mean of those born in summer and first parity would be $10 + 0.5 - 1.1 = 9.4$ piglets. We will see below how to estimate these effects from our data.

Not all females having their first litter in summer have the same number of piglets, which is why the residual term e is added, which is also called the 'error'. Thus, if we knew the true values, a female from the first parity and in summer having 12 piglets has a residual of $12 - 9.4 = 2.6$, and a female from the same season and parity having eight

DOI: 10.1079/9781836990994.0004

piglets has a residual of 8 − 9.4 = −1.4 piglets. The residual is the data y once its mean has been subtracted; that is, once the general mean μ and the effect of season and parity are discounted. The mean of the data from season 1 (e.g. summer) and parity 1 would be:

$$E(y_{11}) = \mu + E_1 + P_1$$

We say the data are *centred* as in selection indices (see section 3.1) when the general mean μ and the effects corresponding to the season and parity in which the litter took place have been subtracted. Centred data behave as if all the data were measured in the same season and came from the same parity. This correction is approximate, because season and parity affect each female differently, and here we assign all females in season 1 and parity 1 the same mean $E(y_{11})$; we are acting as if summer 1 and parity 1 affect all females exactly the same, which is not the case. *We are not removing the effect of parity and season completely*, but we do hope to correct for the most important part of these effects.

The model can include variables measured for the individual, for example when the objective is to analyse litter size at constant female weight, or at constant ovulation rate. If a variable measured on the animal has a linear relationship with the data, a linear regression coefficient can be introduced in the model, as if all females weighed the same or had the same ovulation rate. The model would then be:

$$y = \mu + E + P + \beta x + e$$

where x is the so-called covariate, which in this example could be the weight of the female, and β is the regression coefficient on the covariate. It is convenient that the covariate is centred; thus, the interpretation of the season and parity effects is that of these effects when the covariate is zero. Again, *this correction is approximate* for the above reasons (not all animals are affected equally by the covariate) and because it is also possible that the relationship between the data and the covariate is not exactly linear.

As we do not know the true values of the effects and the covariates, we should estimate them. We will now give an example where litter size is corrected for season (with two levels) and parity (with three levels), and is also corrected for the covariate 'ovulation rate'. Note that when correcting for ovulation rate, we intend litter size to be expressed as if all females had the same ovulation rate. The corrected litter size is no longer the same litter size as before but rather is a way of expressing prenatal survival; at a constant ovulation rate, individuals with the largest litter size are those with a higher prenatal survival.

If E_1 and E_2 are the season effects, P_1, P_2, P_3 the parity effects, and β the regression coefficient of the covariate, we will represent the data as follows:

$$\begin{aligned}
12 &= \mu + E_1 + P_1 + \beta \cdot 20 + e_1 \\
7 &= \mu + E_2 + P_2 + \beta \cdot 9 + e_2 \\
9 &= \mu + E_1 + P_1 + \beta \cdot 15 + e_3 \\
11 &= \mu + E_2 + P_2 + \beta \cdot 16 + e_4 \\
8 &= \mu + E_2 + P_3 + \beta \cdot 14 + e_5 \\
&\dots \dots \dots \dots \dots \dots \dots \dots \dots \dots \\
&\dots \dots \dots \dots \dots \dots \dots \dots \dots \dots
\end{aligned}$$

which in matrix form, $\boldsymbol{y} = \boldsymbol{Xb} + \boldsymbol{e}$, is:

$$\begin{bmatrix} 12 \\ 7 \\ 9 \\ 11 \\ 8 \\ \vdots \end{bmatrix} = \begin{bmatrix} 1 & 1 & 0 & 1 & 0 & 0 & 20 \\ 1 & 0 & 1 & 0 & 1 & 0 & 9 \\ 1 & 1 & 0 & 1 & 0 & 0 & 15 \\ 1 & 0 & 1 & 0 & 1 & 0 & 16 \\ 1 & 0 & 1 & 0 & 0 & 1 & 14 \\ \vdots & \vdots & \vdots & \vdots & \vdots & \vdots & \vdots \end{bmatrix} \cdot \begin{bmatrix} \mu \\ E_1 \\ E_2 \\ P_1 \\ P_2 \\ P_3 \\ \beta \end{bmatrix} + \begin{bmatrix} e_1 \\ e_2 \\ e_3 \\ e_4 \\ e_5 \\ \vdots \end{bmatrix}$$

$$\boldsymbol{y} = \boldsymbol{X} \quad \boldsymbol{b} + \boldsymbol{e}$$

where $\boldsymbol{b}$ contains the effects and covariates to be estimated (not to be confused with vector $\boldsymbol{b}$ of regression coefficients of the selection indices in Chapter 3), and $\boldsymbol{X}$ is an incidence matrix; that is, a matrix composed of ones and zeros indicating the presence or absence of an effect for a specific individual. In the case that there are covariates, a column of $\boldsymbol{X}$ contains the values of the covariate for each individual (there can be several covariates in several columns of $\boldsymbol{X}$).

The model assumes that the mean of the data is calculated with the effects contained in $\boldsymbol{b}$; that is, the effects contained in $\boldsymbol{b}$ are constant and, therefore:

$$E(\boldsymbol{y}) = \boldsymbol{Xb}$$

$$E(\boldsymbol{e}') = E[e_1, e_2, \ldots, e_n] = [0, 0, \ldots, 0] = \boldsymbol{0}$$

In this model, it is generally assumed that the residuals are uncorrelated and all have the same variance:

$$var(\boldsymbol{e}) = \begin{bmatrix} \sigma_e^2 & 0 & \cdots & 0 \\ & \sigma_e^2 & \cdots & 0 \\ & & \ddots & \vdots \\ & & & \sigma_e^2 \end{bmatrix} = \boldsymbol{I}\sigma_e^2$$

Usually, the data will be unbalanced; for example, we will have more data from the first parity than from the second and third parities, and the amount of data in winter will be different from in summer, so one effect cannot be estimated separately from another. If in winter we have much more data from the first parity than in summer, and we compare winter with summer without taking this into account, we would attribute to winter an effect that should actually be attributable to the negative effect of the first parity. The way to estimate the effects $\boldsymbol{b}$ correcting the unbalance of the data is the *estimation by least squares*, which is a classic result of statistics, the derivation of which can be found in Appendix 4.1:

$$\hat{\boldsymbol{b}} = (\boldsymbol{X'X})^{-1}\boldsymbol{X'y}$$

However, note that in matrix $\boldsymbol{X}$, the first column is the sum of the two other columns. This means that in the $\boldsymbol{X'X}$ matrix, the first column is also the sum of the other two, but when in a matrix one column is a linear combination of two other columns, the matrix has no inverse and the former formula for $\hat{\boldsymbol{b}}$ cannot be applied. Taking a simpler example, suppose that we have only the first five data points of our example and that the model has only seasonal effects; in this case the matrix **X'X** is:

$$\begin{bmatrix}12\\7\\9\\11\\8\end{bmatrix}=\begin{bmatrix}1&1&0\\1&0&1\\1&1&0\\1&0&1\\1&0&1\end{bmatrix}\begin{bmatrix}\mu\\E_1\\E_2\end{bmatrix}+\begin{bmatrix}e_1\\e_2\\e_3\\e_4\\e_5\end{bmatrix}\quad\rightarrow\quad X'X=\begin{bmatrix}1&1&1&1&1\\1&0&1&0&0\\0&1&0&1&1\end{bmatrix}\cdot\begin{bmatrix}1&1&0\\1&0&1\\1&1&0\\1&0&1\\1&0&1\end{bmatrix}=\begin{bmatrix}5&2&3\\2&2&0\\3&0&3\end{bmatrix}$$

Note that the diagonal of **X'X** is the number of individuals that each effect has: μ is in all of them, E_1 in two data points and E_2 in three data points. Numbers off the diagonal indicate when two effects coincide: μ matches E_1 and E_2, and E_1 and E_2 never match.

To solve the problem of the lack of inverse of **X'X**, some restrictions are needed; for example, the effects can be estimated referred to the mean and make $E_1 + E_2 = 0$, or the estimates can be referred to the first value by making $E_1 = 0$, or to the smallest of the values of E, or any other restriction, and the same can be done with the P parity effect. In our example we will introduce the most frequent restriction, $E_1 + E_2 = 0$, and a more infrequent one, $P_1 = 0$. This is easily done by adding two rows to the end of the **X** matrix and two zeros to the end of the data. In the first added row, ones are included in order to multiply by E_1 and by E_2, and zeros in the rest, and in the case of $P_1 = 0$, a single 1 is included in order to multiply by P_1 and zeros to multiply by the rest of components of $\boldsymbol{b}$:

$$\begin{bmatrix}12\\7\\9\\11\\8\\\vdots\\0\\0\end{bmatrix}=\begin{bmatrix}1&1&0&1&0&0&20\\1&0&1&0&1&0&9\\1&1&0&1&0&0&15\\1&0&1&0&1&0&16\\1&0&1&0&0&1&14\\\vdots&\vdots&&&&&\vdots\\0&1&1&0&0&0&0\\0&0&0&1&0&0&0\end{bmatrix}\cdot\begin{bmatrix}\mu\\E_1\\E_2\\P_1\\P_2\\P_3\\\beta\end{bmatrix}+\begin{bmatrix}e_1\\e_2\\e_3\\e_4\\e_5\\\vdots\end{bmatrix}$$

$$\boldsymbol{y} = \boldsymbol{X}\quad\boldsymbol{b} + \boldsymbol{e}$$

Instead of adding restrictions, it is also possible to estimate combinations of the effects. Returning to the simple example we used before, with only five data points and season effect, if we jointly estimate $\mu + E_1$ and $\mu + E_2$, we would have:

$$\begin{bmatrix}12\\7\\9\\11\\8\end{bmatrix}=\begin{bmatrix}1&0\\0&1\\1&0\\0&1\\0&1\end{bmatrix}\begin{bmatrix}\mu+E_1\\\mu+E_2\end{bmatrix}+\begin{bmatrix}e_1\\e_2\\e_3\\e_4\\e_5\end{bmatrix}\quad\rightarrow\quad X'X=\begin{bmatrix}1&0&1&0&0\\0&1&0&1&1\end{bmatrix}\cdot\begin{bmatrix}1&0\\0&1\\1&0\\0&1\\0&1\end{bmatrix}=\begin{bmatrix}2&0\\0&3\end{bmatrix}$$

where **X'X** has an inverse because the rows of **X** are no longer a linear combination of each other and the same happens with the columns. These combinations of effects allowing us to invert $X'X$ are called *estimable functions*. Therefore, it is not possible to estimate the effects of season, parity and the general mean separately; it is only possible to estimate 'estimable functions', which in our example is combinations of season and parity effects with the general mean.

4.1.2 The mixed model

The fixed effects model is incomplete, as the individuals are related and there is a genetic effect common to the relatives. Suppose in our previous example that we only have one litter size per female and that the data are only genetically correlated, so there are no environmental correlations between relatives. The model will be:

$$y = \mu + E + P + \beta x + a + e$$

where a is the additive value of the animal having the data y. If we have more than one litter per female, the litters of the same female share a common environmental effect and they are more correlated than they would be only by genetic effects. This environmental effect is called the *permanent effect* of the female on her litters. It is common to have a permanent environmental effect when having several data points for a trait; for example when there is a permanent environmental effect on milk production over several lactations of a cow. In this case, the model is:

$$y = \mu + E + P + \beta x + a + p + e$$

where p is the permanent environmental effect common to the data from the same female. In the example above:

$$\begin{aligned}
12 &= \mu + E_1 + P_1 + \beta\cdot 20 + a_1 + p_1 + e_1 \\
7 &= \mu + E_2 + P_2 + \beta\cdot\ 9 + a_1 + p_1 + e_2 \\
9 &= \mu + E_1 + P_1 + \beta\cdot 15 + a_2 + p_2 + e_3 \\
11 &= \mu + E_2 + P_2 + \beta\cdot 16 + a_2 + p_2 + e_4 \\
8 &= \mu + E_2 + P_3 + \beta\cdot 14 + a_2 + p_2 + e_5 \\
\dots &\quad \dots\dots\dots\dots\dots\dots\dots\dots\dots \\
\dots &\quad \dots\dots\dots\dots\dots\dots\dots\dots\dots
\end{aligned}$$

Note that the first two data points belong to female 1, who has two parities, and the next three to female 2 with three parities. In matrix form:

$$\mathbf{y} = \mathbf{Xb} + \mathbf{Za} + \mathbf{Wp} + \mathbf{e}$$

$$\begin{bmatrix} 12 \\ 7 \\ 9 \\ 11 \\ 8 \\ \vdots \end{bmatrix} = \begin{bmatrix} 1 & 1 & 0 & 1 & 0 & 0 & 20 \\ 1 & 0 & 1 & 0 & 1 & 0 & 9 \\ 1 & 1 & 0 & 1 & 0 & 0 & 15 \\ 1 & 0 & 1 & 0 & 1 & 0 & 16 \\ 1 & 0 & 1 & 0 & 0 & 1 & 14 \\ \vdots & \vdots & \vdots & \vdots & \vdots & \vdots & \vdots \end{bmatrix} \cdot \begin{bmatrix} \mu \\ E_1 \\ E_2 \\ P_1 \\ P_2 \\ P_3 \\ \beta \end{bmatrix} + \begin{bmatrix} 1 & 0 & \cdots & 0 \\ 1 & 0 & \cdots & 0 \\ 0 & 1 & \cdots & 0 \\ 0 & 1 & \cdots & 0 \\ 0 & 1 & \cdots & 0 \\ \vdots & \vdots & \vdots & \vdots \end{bmatrix} \begin{bmatrix} a_1 \\ a_2 \\ \vdots \end{bmatrix} + \begin{bmatrix} 1 & 0 & \cdots & 0 \\ 1 & 0 & \cdots & 0 \\ 0 & 1 & \cdots & 0 \\ 0 & 1 & \cdots & 0 \\ 0 & 1 & \cdots & 0 \\ \vdots & \vdots & \vdots & \vdots \end{bmatrix} \begin{bmatrix} p_1 \\ p_2 \\ \vdots \end{bmatrix} + \begin{bmatrix} e_1 \\ e_2 \\ e_3 \\ e_4 \\ e_5 \\ \vdots \end{bmatrix}$$

$$y \quad = \quad X \quad b \ + \quad Z \quad a \ + \quad W \quad p \ + \ e$$

Note that the data vector has n components corresponding to the n data, but if we have m animals (females and males), the vector of additive values has m components, and that of permanent environmental effects has as many components as females with data. If an animal has no data, for example a male for the litter size trait, columns of zeros are introduced in the $\mathbf{Z}$ matrix; in our example, if the first individual were a male, the first column of $\mathbf{Z}$ would be a column of zeros, and thus the equations in matrix form would remain the same; in section 4.3.1 we give an example.

The model assumes that the mean of the additive and permanent effects is zero, as well as the mean of the residuals:

$$E(\boldsymbol{a}) = \boldsymbol{0};\ E(\boldsymbol{p}) = \boldsymbol{0};\ E(\boldsymbol{e}) = \boldsymbol{0};\ E(\boldsymbol{b}) = \boldsymbol{b};\ E(\boldsymbol{y}) = \boldsymbol{Xb}$$

where $\boldsymbol{0}$ is a vector of zeros.

When data come from a selection programme, the additive values no longer have a zero mean from the first generation of selection if genetic progress took place. However, with certain conditions that we will see in section 4.3.2.B, this selection effect can be ignored and the model assuming $E(\boldsymbol{a}) = \boldsymbol{0}$ can be used for estimating the breeding values as if selection had been taken into account.

In this model, additive and permanent effects are not 'fixed' as in the former section but rather are random variables, normally distributed. As this model includes 'fixed' and 'random' effects, it is called a *mixed model*. We will see the differences between fixed and random effects in section 4.2.

It is assumed, as before, that all residuals have the same variance and they are not correlated, and it is also assumed that the permanent environmental effects have the same variance and are also unrelated but that the additive effects are correlated because the individuals are relatives. If we take into account that, as we saw in section 2.1.3.B4, the covariance between two individuals X and Y is:

$$cov(a_X, a_Y) = 2r_{XY}\sigma_A^2$$

then we can write the variance–covariance matrixes of the random effects as:

$$var(\boldsymbol{e}) = \boldsymbol{I}\sigma_e^2$$

$$var(\boldsymbol{p}) = \boldsymbol{I}\sigma_p^2 \text{ (not to be confused with the phenotypic variance } \sigma_P^2\text{)}$$

$$var(\boldsymbol{a}) = \begin{bmatrix} \sigma_A^2 & cov(a_1,a_2) & \cdots & cov(a_1,a_m) \\ & \sigma_A^2 & \cdots & cov(a_2,a_m) \\ & & \ddots & \vdots \\ & & & \sigma_A^2 \end{bmatrix} = \begin{bmatrix} 1 & 2r_{12} & \cdots & 2r_{1m} \\ & 1 & \cdots & 2r_{2m} \\ & & \ddots & \vdots \\ & & & 1 \end{bmatrix} \sigma_A^2 = \boldsymbol{A}\sigma_A^2$$

where matrix $\boldsymbol{A}$ contains the relationship coefficients of all the individuals in the population, which is why it is known as the *relationship matrix*. On the diagonal should go twice the relationship coefficient of each individual with themself; that is, $1 + F_i$, where F_i is the inbreeding coefficient of each individual (see section 1.2.1). This makes it more difficult to calculate $\boldsymbol{A}$ and its inverse, but as F_i is generally small, it is often ignored; in general, the results taking into account or not the consanguinity are similar (Aguilar *et al.*, 2020). When there is selection, the vector $\boldsymbol{a}$ and the relationship matrix usually include the individuals of the initial generation (also called generation zero or the *base population*), whose animals we assume are unrelated, to fulfil the conditions for ignoring selection that we will see in section 4.3.2.B.

The random effects $\boldsymbol{a}$ and $\boldsymbol{p}$ are normally distributed, as is the residual random effect $\boldsymbol{e}$, with mean zero:

$$\boldsymbol{a} \sim N(0, \boldsymbol{A}\sigma_A^2) \quad \boldsymbol{p} \sim N(0, \boldsymbol{I}\sigma_p^2) \quad \boldsymbol{e} \sim N(0, \boldsymbol{I}\sigma_e^2)$$

We assume that the random effects $\boldsymbol{a}$ and $\boldsymbol{p}$ are unrelated to each other and unrelated to the residual random effect $\boldsymbol{e}$:

$$cov(\boldsymbol{a},\boldsymbol{p}) = 0;\ cov(\boldsymbol{a},\boldsymbol{e}) = 0;\ cov(\boldsymbol{p},\boldsymbol{e}) = 0$$

As $\boldsymbol{b}$ is constant, var($\boldsymbol{b}$) = 0, and the covariances of $\boldsymbol{b}$ with $\boldsymbol{a}$, $\boldsymbol{p}$ and $\boldsymbol{e}$ are also zero.

These models can become more complex by adding new effects, such as, in this case, the common environmental effect between full-siblings of the same litter that we talked about in section 2.1.1, or the genetic effect of dominance, but this model is the most used in animal breeding.

This model is used for traits related to milk production, where the fixed effects include, among others, the herd–year–season effect (one level for each herd in each year and each season), the parity effect and covariates to standardize the period of milk production.

In growth characteristics, siblings from the same litter have common environmental effects attributable to being born and lactating with the same mother and in the same litter; in these cases, there is a random common environmental effect *c*, which is called the *litter effect*, and in the matrix $\boldsymbol{W}$, the values of '1' indicate being siblings of the same litter. A permanent effect *p* common to all litters of the same female could also be included in the model.

Note that the model has a parity effect, *P*, and a permanent environmental effect of the female *p* throughout her litters. This does not indicate that we are correcting twice for the same effect. The P_1 parity effect is common to all females in their first parity (e.g. they have 1.2 piglets fewer than in other parities, on average), but the permanent environmental effect *p* is specific to each female; in our example, female 2 has three parity effects, P_1, P_2 and P_3, but only one permanent effect, p_2.

This model makes it possible to estimate the additive effects of individuals, including those without data. The model is called the *animal model*, because the additive value of each animal in the population is predicted using all animals. In the example we have given, the model is called the *repeatability animal model*, because it takes into account several data points from the same animal and therefore the permanent environmental effects. In section 4.3.2.C, we will look at other models.

4.2 Fixed and Random Effects

In the examples of mixed models shown in section 4.1.2, the environmental effects contained in vector $\boldsymbol{b}$ are constant; all data from the first parity have the same P_1 effect and all winter data have their winter effect, which is the same for all litters born in winter. If we repeat the experiment, we expect to find the same effects; we can find different estimates, but we assume that the true effect of winter and first parity are the same. The environmental effects contained in vector $\boldsymbol{b}$ are what we call *fixed effects*. The same does not happen with the effects of the residual, the permanent environmental effect and the additive genetic effect. If we repeat the experiment, we expect a constant negative effect of the first parity, but we do not expect a constant effect of the residual; the residual is a value drawn at random from a distribution of a trait, so the residual is a random value. The same happens with the permanent environmental effect of the female. If we repeat the experiment, we will have another female that will not have the same permanent effect; the female has been drawn at random from our population, so the permanent effect is what we call a *random effect*. The additive genetic effect of each individual is, for the same reason, a random effect; if we repeat the experiment, we expect the individual to change and so the additive effect also changes.

According to the nomenclature that we have used so far, we have talked about *estimation* for fixed effects and *prediction* for random effects because this is the common use in animal breeding, although some authors use 'estimation' in both cases (e.g. Lynch and Walsh, 1998, pp. 745–778). When we estimate the fixed effects or predict the random ones, an error occurs, which is in both cases the difference between the true value and the estimated or predicted value.

If we call an effect u, whether it is fixed or random, and call its estimate or prediction $\hat{u}$, the error will be, in both cases:

$$e = u - \hat{u}$$

In the case of fixed effects, an estimator is called *unbiased* when its values are distributed around the true value, but this way of defining unbiased estimators cannot be applied to random effects. In Fig. 4.1, we can see how the estimates or predictions of the effects are distributed when repeating the experiment. In the case of the fixed effect, the true value of the fixed effect remains constant and all its estimates are distributed around this single true value. In the case of the random effect, each repetition of the experiment leads to a new true value, so the true value is not constant; it varies and is distributed around its own mean.

Note that the errors are in both cases the difference between the true values and those estimated or predicted, but in the case of random effects, they are not the distance between the prediction and the mean of the predictions, because the true value changes in

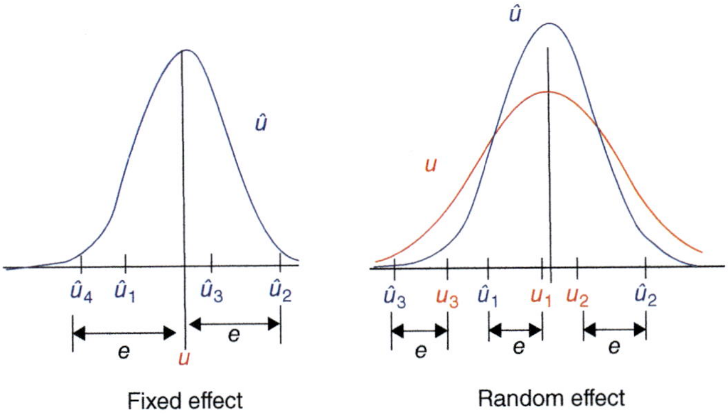

Fig. 4.1. Distribution of the estimators of a fixed effect, and distribution of the predictors and the true values of a random effect, when repeating the experiment.

each case when repeating the experiment. One way to define an *unbiased estimator or predictor* for both types of effects is to define it as one whose *mean of errors is zero*. In this case, taking into account that, when repeating the experiment, for a fixed effect the true value u is constant and for a random effect the true value u is not constant, then:

Fixed effect $\quad E(e) = E\,(u - \hat{u}) = E(u) - E(\hat{u}) = u - E(\hat{u}) = 0 \;\rightarrow\; E(\hat{u}) = u$

Random effect $\quad E(e) = E(u - \hat{u}) = E\,(u) - E(\hat{u}) = 0 \;\rightarrow\; E(\hat{u}) = E(u)$

In the case of fixed effects, the concept of unbiasedness is the usual one: the mean of the estimates coincides with the true value. *In the case of random effects, unbiasedness is much less attractive*, because only the mean of the true values coincides with the mean of the estimates.

In section 2.3.3.D and Fig. 2.4, we saw how the predictions of breeding values of dairy cattle bulls decreased if they were calculated with little information. For the same daughters' milk production average of 12,000 kg, depending on whether the bull had 100, ten or one daughters, the predictions of the bull's breeding values became 5340, 2700 or 450 kg of milk above the population mean. This regression to the mean (called sometimes *shrinkage*) is common in all random effects, whose characteristic is that not only does the prediction error increase when the amount of information decreases but the prediction itself also decreases in this case.

A common mistake is to consider that an effect is fixed or random due to its nature. This is true for additive effects because we know the relationships between relatives and we want to use them in selection; if they were fixed, they would be constant and their covariances would be zero. For environmental purposes, however, this is not always true; *the effects can be taken as fixed or random depending on our interests*. Suppose we have a model for dairy cattle with a fixed effect of the farm. If most farms have very few cows, the estimation of the farm effect for each farm would be very unreliable because they would be done with few data points. We would be in the same case as when estimating the value of a bull only by the average production of his daughters when he has few daughters. In this case, we can decide to consider the farm effect as random, so that when we have few data points, the prediction of the farm effect would return towards the mean having a lower value, depending on the amount of data used to predict it. In this case, the consequences of a bad estimation would be mitigated.

In general, it is recommended to consider the environmental effects as fixed when having many data points per level, and random if not. Nevertheless, it is often preferred to increase the number of data points to estimate a fixed effect based on combining levels, such as combining farms in the same area, or joining season levels or parities (e.g. using three levels for parity: first, second and third onwards). The disadvantage of this procedure is that the correction lacks precision when several levels are joined, as all environmental incidents of the period are summarized in a single mean. If the period for correction if short (e.g. a month instead of a season), fewer environmental incidences would have occurred, and the correction for the effect of this period would be more accurate. The usual recommendation is to try several options and observe the results and their accuracy before making a decision.

One aspect that is often overlooked is that *random effects corrections are generally weak*. An effect is usually considered random when it has many levels and few data points within each level, as we have said, so when being *shrunk*, the effect values are lower and the correction less important than when it is corrected as a fixed effect. Therefore, caution must be exercised when correcting for a random effect if it is important, because the correction made on the data is weak.

Ignoring random effects has two consequences. One is error in predictions; for example, ignoring permanent effects leads to predicting the additive value of the cow i as a_i + p_i, higher or lower than the correct estimate depending on the sign of p_i. Another consequence is error when calculating confidence intervals. For example, if we apply the fixed effects model $y = Xb + e$, we are assuming that the correlation between the data is null, which is not true because animals are related. The variance of the error is the sum of the variances for each data point plus the covariances, but if we ignore these covariances, we will find that our estimate of the variance of the error will be lower and so the confidence intervals are shorter, which means that we are offering estimation errors smaller than the real ones, so we are underestimating errors and confidence intervals.

4.3 Best Linear Unbiased Prediction (BLUP)

4.3.1 The mixed model equations

We will estimate and predict the fixed and random effects of the previous models. We will start with a simple model with one data point per individual. The model has a component with *fixed effects* $\boldsymbol{b}$, and another component with *random effects* $\boldsymbol{a}$, so having components of both types is known as a *mixed model*:

$$y = Xb + Za + e$$

If we know the fixed effects $\boldsymbol{b}$, we could centre the data and predict $\boldsymbol{a}$ in the model:

$$y - Xb = Za + e$$

which would be equivalent to calculating a selection index like the one we used in section 3.1. Prediction of additive genetic effects using centred data is:

$$\tilde{a} = C'V^{-1}(y - Xb)$$

where $\boldsymbol{V}$ is the covariance variance matrix of the data and $\boldsymbol{C'}$ the covariance matrix between the additive values and the data. As $\boldsymbol{b}$ is not known, it must be estimated. It can be done in several ways; Appendix 4.2 shows that:

$$\hat{b} = \left(X'V^{-1}X\right)^{-1} X'V^{-1}y$$

$$\hat{a} = C'V^{-1}(y - X\,\hat{b})$$

Note that $\tilde{a}$ and $\hat{a}$ are different predictions, because $\hat{b}$ is an estimate and $\boldsymbol{b}$ is the supposedly known true value.

Matrix $\boldsymbol{V}$ is very large (as many rows and columns as data points, which in the case of dairy cattle can be millions), and therefore it is difficult to invert. Henderson (1963) proposed equations that greatly simplify the estimation of $\boldsymbol{b}$ and $\boldsymbol{a}$, mainly because the

V matrix does not have to be inverted. Henderson's equations are known as *mixed model equations*. These equations are:

$$\begin{bmatrix} X'X & X'Z \\ Z'X & Z'Z + A^{-1}\dfrac{\sigma_e^2}{\sigma_A^2} \end{bmatrix} \begin{bmatrix} \hat{b} \\ \hat{a} \end{bmatrix} = \begin{bmatrix} X'y \\ Z'y \end{bmatrix}$$

It can be shown that the result of these equations is the same as that of the estimates $\hat{a}$ and $\hat{b}$ we have shown above. The derivation of these equations can be done in different ways; the simplest is by using Bayesian statistics, and we will see this in Appendix 5.1.

The problem of inverting the relationship matrix A remains, because it is also a very large matrix with as many rows and columns as individuals whose breeding values we want to predict. However, Henderson (1976) found some simple rules for calculating A^{-1} directly, and today there is standard software using these rules for constructing A^{-1}. This has made it possible to use the equations of the mixed model with millions of animals without entailing a prohibitive computational cost, and nowadays the mixed model equations are the standard procedure for predicting additive values in animal breeding.

These equations can be generalized to the model with permanent environmental effects and to other models with other effects. For example, for the model with permanent effects, $y = Xb + Za + Wp + e$ (see section 4.1.2), the equations are:

$$\begin{bmatrix} X'X & X'Z & X'W \\ Z'X & Z'Z + A^{-1}\dfrac{\sigma_e^2}{\sigma_A^2} & Z'W \\ W'X & W'Z & W'W + I\dfrac{\sigma_e^2}{\sigma_p^2} \end{bmatrix} \begin{bmatrix} \hat{b} \\ \hat{a} \\ \hat{p} \end{bmatrix} = \begin{bmatrix} X'y \\ Z'y \\ W'y \end{bmatrix}$$

EXAMPLE 4.1

We will write the mixed model equations for a simple case in which only the season fixed effect is taken into account for the trait litter size in pigs. We have two sows with additive values a_1 and a_2 that are half-siblings and a male with additive value a_3 that is a full sibling of the first female and a half-sibling of the second. The heritability of litter size is 0.1. The equations of the linear model are:

$$\begin{aligned} 12 &= \mu + E_1 + a_1 + e_1 \\ 7 &= \mu + E_2 + a_1 + e_2 \\ 9 &= \mu + E_1 + a_2 + e_3 \\ 11 &= \mu + E_2 + a_2 + e_4 \\ 8 &= \mu + E_2 + a_2 + e_5 \end{aligned}$$

Note that the male does not appear here, because he does not have his own data. If we put the equations in matrix form, we can include the male in the vector of additive values to be predicted, adding a column of zeros in the Z matrix:

$$\begin{bmatrix}12\\7\\9\\11\\8\end{bmatrix}=\begin{bmatrix}1&0\\0&1\\1&0\\0&1\\0&1\end{bmatrix}\cdot\begin{bmatrix}\mu+E_1\\\mu+E_2\end{bmatrix}+\begin{bmatrix}1&0&0\\1&0&0\\0&1&0\\0&1&0\\0&1&0\end{bmatrix}\cdot\begin{bmatrix}a_1\\a_2\\a_3\end{bmatrix}+\begin{bmatrix}e_1\\e_2\\e_3\\e_4\\e_5\end{bmatrix}$$

$$y \;=\; X \quad b \;+\; Z \quad a \;+\; e$$

Note that we are estimating the season effects adding the overall mean μ, so we are using *estimable functions*. This is one of the ways to prevent matrix ***X*** from having rows that are linear combinations of each other, as we saw in section 4.1.1.

In this simple example, we will consider that there is no selection and we will not consider the 'zero generation' of the parents of these three animals. We will use the first data points from the example in section 4.1.2. The relationship matrix is, taking into account that the sows are half-siblings, and the male is a full sibling of the first and a half-sibling of the second:

$$A=\begin{bmatrix}1&0.25&0.5\\&1&0.25\\&&1\end{bmatrix}\quad\rightarrow\quad A^{-1}=\begin{bmatrix}1.36&-0.18&-0.64\\&1.09&-0.18\\&&1.36\end{bmatrix}$$

The ratio between the error variance and the additive variance is:

$$\frac{\sigma_e^2}{\sigma_A^2}=\frac{\sigma_P^2-\sigma_P^2h^2}{\sigma_P^2h^2}=\frac{1-h^2}{h^2}=9\quad\rightarrow\quad A^{-1}\cdot 9=\begin{bmatrix}12&-1.6&-5.7\\&9.8&-1.6\\&&12\end{bmatrix}$$

The mixed model equations are:

$$\begin{pmatrix}\begin{matrix}2&0\\0&3\end{matrix}&\begin{matrix}1&1&0\\1&2&0\end{matrix}\\ \begin{matrix}1&1\\1&2\\0&0\end{matrix}&\begin{pmatrix}\begin{matrix}2&0&0\\0&3&0\\0&0&0\end{matrix}+\begin{matrix}12&-1.6&-5.7\\-1.6&9.8&1.6\\-5.7&-1.6&12\end{matrix}\end{pmatrix}\end{pmatrix}\begin{bmatrix}\mu+E_1\\\mu+E_2\\a_1\\a_2\\a_3\end{bmatrix}=\begin{bmatrix}21\\26\\19\\28\\0\end{bmatrix}$$

$$\begin{pmatrix}X'X & X'Z\\ Z'X & \left(Z'Z+A^{-1}\dfrac{\sigma_e^2}{\sigma_A^2}\right)\end{pmatrix}\quad\begin{pmatrix}\hat{b}\\\hat{a}\end{pmatrix}\quad\begin{pmatrix}X'y\\Z'y\end{pmatrix}$$

leading to the system of equations:

$$\begin{bmatrix}2&0&1&1&0\\0&3&1&2&0\\1&1&12&-1.6&-5.7\\1&2&-1.6&13&-1.6\\0&0&-5.7&-1.6&12\end{bmatrix}\begin{bmatrix}\mu+E_1\\\mu+E_2\\a_1\\a_2\\a_3\end{bmatrix}=\begin{bmatrix}21\\26\\19\\28\\0\end{bmatrix}\quad\rightarrow\quad\begin{bmatrix}\mu+E_1\\\mu+E_2\\a_1\\a_2\\a_3\end{bmatrix}=\begin{bmatrix}10.5\\8.7\\-0.012\\0.012\\-0.004\end{bmatrix}$$

Note that there is an individual without data, so we had to include a column of zeros in the Z matrix, which leads to a row and a column of zeros in $Z'Z$. This would produce an indeterminate system, but when the matrix A^{-1} is added to $Z'Z$, the resulting matrix no longer has a row and column of zeros, and the system can be solved.

This system often has many equations, hundreds of thousands, or even millions in the case of dairy cattle, so it is solved by standard numerical methods commonly used to solve systems of equations. There are also approximate methods to calculate confidence intervals. In Chapter 5, we will see how to estimate these intervals using Bayesian statistics.

4.3.2 Remarks about BLUP

A. Fixed or random?

Note that, when solving the system of equations, the estimates of the seasonal fixed effects are involved in predicting the random additive effects. For example, solving from the first equation, in the example above:

$$2(\mu + E_1) + a_1 + a_2 = 21 \quad \rightarrow \quad a_1 = 21 - 2(\mu + E_1) - a_2$$

which implies that if the values of the fixed effects have few data points and they are not well estimated, this will affect the predictions of the random values. We have said in section 4.2 that season effect could be considered random when there are few data points to estimate it. This problem was discussed for the first time in dairy cattle when considering the herd fixed effect. Through artificial insemination, bulls have daughters in many herds. Some herds are well managed but in others herds the care of the animals is of poorer quality. The herd effect was considered fixed from its inception because considering it as random led to several problems. The first is that the variance of the error was not the same in all herds, as there are herds that are very constant in their feeding and management and others that are not, so that all the cows in a poorly managed herd had error variances much larger than cows from well-managed herds. The second problem was that well-managed herds buy more expensive semen from genetically better bulls, so the covariances between additive and residual environmental values were no longer zero. When estimating the herd effect as fixed, these problems disappear because, as the fixed effect is constant, the variance of a fixed effect is null and its covariances are also null. Deciding whether an effect is fixed or random must be done on a case-by-case basis.

B. Genetic trends

A problem in estimating the response to selection is that, when comparing the mean of the offspring with the mean of the parents, we do not know whether the differences are due to genetic or environmental causes. It may happen that the environment has improved in the generation of the offspring, and the higher value of the offspring is due only to this improvement in the environment. However, as BLUP can separate environmental from genetic causes (at least the most relevant environmental causes), the response to selection can be estimated by calculating the mean of the predictions of additive values at the end of the process (at the beginning of the process, the mean of the additive values is zero).

We can also estimate the average of the additive values of the individuals in each generation, and observe the evolution of the genetic progress, as shown in Fig. 4.2; this is known as the *genetic trend*.

However, this presents a problem: we have assumed when deducing BLUP that the means of all additive values are zero. This means that, if we repeat the experiment infinite times, the sum of the additive values for an animal would be zero, and this applies to each of the individuals, whether selected or not. However, if the additive values progress from generation to generation, it seems that the deduction of the BLUP formulas may be incorrect, because, due to selection, when repeating the experiment infinite times, the mean of the additive values of the selected individuals must be greater than zero.

Indeed, the additive values increase with selection and their predictions should also increase, but under certain conditions BLUP can ignore the fact that data are selected. This means that we can apply the model we have used so far (where the means of all the additive values are assumed to be zero) and yet obtain correct predictions of the additive values. There are several ways to prove that selection can be ignored; the simplest one is using Bayesian inference, which we will see in the next chapter (Appendix 5.2, Chapter 5).

The conditions for this analysis to be correct are as follows:

- The complete relationship matrix $\mathbf{A}$ from generation zero should be used.
- All the data used in selection should be analysed together.
- The true genetic parameters must be used. As the true values are not known, good estimates of the parameters *in the base generation* should be used, because genetic trends are very sensitive to the values of the parameters used to estimate them.

Using the correct genetic parameters is particularly important. A geneticist who did not obtain results and obtained a null phenotypic tendency could camouflage his failure by

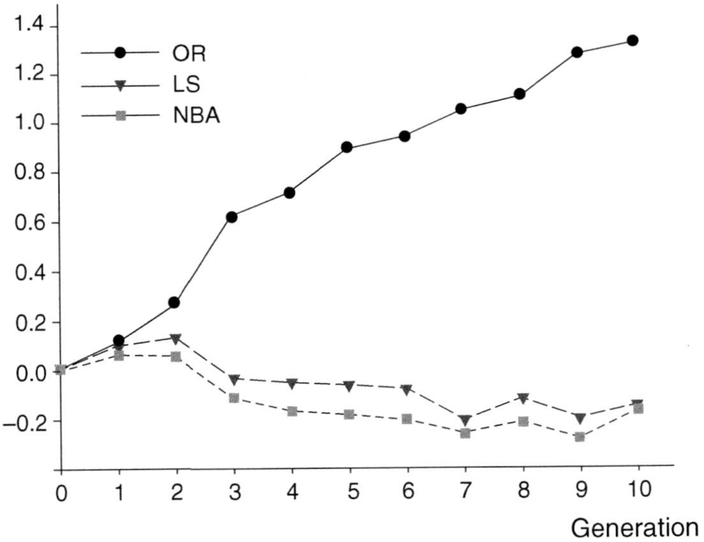

Fig. 4.2. Genetic trends for ovulation rate (OR), total litter size (LS) and number born alive (NBA) of a line of rabbits selected by OR. Note that selection for OR has been successful but has not produced correlated responses in LS or NBA. From Laborda *et al.* (2011). Figure used with permission from Oxford University Press.

using genetic parameters such that they produced positive genetic trends and negative environmental trends. In order to avoid overestimating genetic results, it is prudent to use genetic parameters that do not give rise to a negative environmental trend, unless there is some reason to believe that the environment has indeed deteriorated. However, genetic parameters change with selection (reduction of additive variance by the Bulmer effect; see section 2.1.2.B2; changes in genetic correlations discussed in section 3.3), but this is not a problem, as the parameters used in BLUP are *the genetic parameters in generation zero* or the base population.

C. Other models

BLUP is, like selection indices, flexible, and the mixed model equations are applicable to several mixed models. Some examples of these models are as follows.

C1. MULTI-TRAIT BLUP As in the case of indices, BLUP can be applied to multiple traits to optimize profit. In Chapter 5, we will see an example of how to predict additive values for multiple traits in a multi-trait model.

C2. MATERNAL EFFECTS MODEL As we have seen in section 2.1.3.A, some traits, such as weaning weight in beef cattle, are strongly influenced by the maternal environment. A dam producing more milk than the average will have calves with high weaning weight, and this is an environmental effect for the calf. However, a dam produces more milk for genetic and environmental reasons, so we could separate the part of the maternal environment that is due to genes. Using a slightly different nomenclature than in section 2.1.3.A, representing fixed effects with capital letters and random effects with lowercase letters, the model is:

$$y = \mu + F + a + m_a + m_p + e$$

where F is the fixed effects, a is the additive value of the calf, m_a is the environmental effect on the calf's weaning weight due to the mother's genes for milk, and m_p is the environmental effect on the calf's weaning weight due to the environment of the mother that causes her to produce more milk, which leads to a positive effect on the growth of the calf. Actually, m_a and m_p do not refer only to milk production of the mother but to all kinds of maternal care. Notice that m_a and m_p are both *environmental effects for the weaning weight of the calf*, and that they are measured in grams of weaning weight of the calf. The sum of m_a and m_p is the *maternal effect*.

As the mothers are related, there are correlations between the m_a effects of different mothers, so that the relationship matrix for m_a is the same as that for a, but multiplying by the respective additive variances for a and m_a. In a more complex model, it can be assumed that the effects a and m_a can be correlated, as some of the genes that make a mother produce more milk can be the same as those that lead to the calf having a higher weaning weight; for example, we can imagine that genes acting on growth hormone act both on weaning weight and milk production. In this case, genetic covariances appear between a and m_a, and both effects can be considered as if they were two different traits. We will see the multi-character model in Chapter 5. It is interesting to observe that these genetic correlations between direct and maternal effects are in many cases negative.

The maternal effects model is highly parameterized and needs large samples to be estimated with precision.

C3. SIRE MODEL In the origins of BLUP, when computers were in their infancy, the model that was used to explain the lactation data of a cow was:

$$y = \mu + F + s + e$$

where s is the genetic effect of the cow's father. The relationship matrix only contains the relationship of the sires, and is a diagonal matrix if they are not related, so its inverse is immediate. Furthermore, the number of sires is much smaller than the number of individuals, so the size of the matrix is much smaller. The variance of the random effect, the sire effect, is calculated as the covariance between the daughters of the sire; as they are half-sisters, it is ¼ σ_A^2, as we saw in section 2.1.3.B2. This model is out of date for common dairy traits but is still useful for characteristics whose phenotype is very uninformative such as fertility. Threshold models are commonly used to analyse dichotomous traits (e.g. Falconer and MacKay, 1996), but as the value of y is only 1 or 0, sometimes the fixed effects contain only ones or contain only zeros, and estimation problems arise. Moreover, sometimes the trait has little incidence and there are a large number of zeros, making it less informative. In these cases, it is difficult to apply the threshold models and it is more convenient to use the sire model, because instead of predicting the additive value of the cows, we predict the additive value of sires having several daughters, so that each sire has different values, not just ones and zeros. This model can be made more complex, for example by adding the effect of the dam, the father of the dam, etc.

C4. MODEL WITH DOMINANCE Litter size and other traits showing heterosis present genetic dominance effects, although these are usually less important than the additive effects. The model used for including dominance effects is:

$$y = \mu + F + a + d + e$$

where d is the dominance effects and they have a variance–covariance matrix similar to the relationship matrix, in which the components are the identity coefficients that we saw in section 2.1.3.B4. The *identity coefficient* is the probability that in two individuals, both alleles of a locus are identical by descent. This matrix cannot be inverted directly like the relationship matrix, and for this reason dominance models are more complex to calculate. As dominance is often a much smaller effect than additivity, it is generally ignored and these models are rarely used.

C5. MODEL WITH GENETIC GROUPS In intensive animal production, particularly in dairy cattle and pigs, it is common for a breeding programme to introduce genes from another programme. For example, in pigs it is common to import males from other companies into a selection farm nucleus, in order to take advantage of genes that may have been left by chance in low frequency in the nucleus but that are in higher frequencies in the animals that are introduced (obviously the performance of both companies must be similar; if the imported animal has unfavourable genes in high frequencies, the results of the programme may be worse than before). In cattle, it is common to import semen from animals for which the genealogy is not available because it is not provided by the seller or for other reasons. In these cases, the relationship matrix is incomplete and this affects the evaluations. A model with a group effect can then be used:

$$y = \mu + F + G + a + e$$

where G is the group fixed effect, an effect including in a group all the animals introduced from other programmes. In matrix form, the model is:

$$y = Xb + ZQg + Za + e$$

where $\mathbf{g}$ is a fixed group effect and $\mathbf{Q}$ is the incidence matrix that relates each individual to each genetic group. The additive value $\mathbf{a}$ refers to the group mean $\mathbf{Qg}$, so that its additive value would be $\mathbf{Qg} + \mathbf{a}$. A variant of this model consists of assigning 'ghost parents' to those individuals whose father or mother is unknown because they come from imported semen or from migration. Details of this model can be found in Westell *et al.* (1988).

C6. REDUCED ANIMAL MODEL In cases such as poultry or pigs, in which most of the animals produced go to slaughter, only a few animals leave descendants. In this model, the animals that leave offspring are separated from those that do not, and the procedure used greatly simplifies the calculations, but the solution is exactly the same as if the complete animal model had been applied. Given the current developments in computing, this model is rarely necessary, and has fallen into disuse. The interested reader can consult Blair and Pollack (1984).

D. BLUP properties

BLUP is a selection index in which the data are centred by simultaneously estimating their means, and therefore the properties of selection indices also apply to BLUP.

- *BLUP maximizes the response to selection* if the genetic parameters used to calculate it are the true ones (otherwise, the predictions are not strictly BLUP and are only approximations).
- Individuals with little information have a prediction with lower reliability and precision, so their variance of the prediction error is higher. This implies greater uncertainty about the breeding value of individuals with little information, but it should be ignored when selecting. As in the case of the indices and as we saw in Fig. 2.4 of section 2.3.3.D, the additive value predicted by BLUP for individuals with less information is penalized, it is *regressed* towards the mean and its lower reliability is already taken into account.
- *BLUP maximizes the correlation between the predicted and the true value.*
- In BLUP, the values of $\mathbf{y}$ are not centred, so their mean must be estimated. The risk of a prediction is usually taken as the mean square error and is equal to the squared bias plus the variance of the prediction error (see Appendix 3.1). The problem is that there are infinite minimum risk predictions, combining the bias and the variance of the prediction error; thus, the system is indeterminate. A solution is to restrict the search to a subset of the predictors. In our case, a reasonable subset may be that of unbiased predictors, even though we already saw in section 4.2 that unbiasedness in the case of random effects is a rather unattractive property. BLUP is the best of the predictions within the subset of unbiased predictions.
- *In the infinitesimal model, there is no genetic drift due to the sampling effect*, because there are infinite genes with infinitesimally changing frequencies. However, the same phenomenon occurs that when there is genetic drift due to sampling, the prediction errors increase with the number of generations; this is due to the increase in relationships, which add covariances to the relationship matrix.

Continued

- BLUP predictions are not affected by the reduction in variance due to selection (the Bulmer effect; see section 2.1.2.B2), as *they only use the variances of the base population*. This is not necessarily the case for selection indices, which typically only consider one or two generations of relatives.
- *Data should not necessarily be normally distributed to apply BLUP*. We know that, in the multivariate normal, the relationships between the variables are always linear, and thus if the data are not normally distributed, BLUP is the best linear approximation to the prediction of the breeding value.

The efficiency of BLUP with regard to selection indices has often been exaggerated. Actually, BLUP is only slightly more efficient than an index applied to data pre-corrected by well-estimated means, either by least-squares or more approximate methods such as comparison between contemporaries (Hagger and Dempfle, 1983, in cattle; Baselga *et al.*, 1985, in rabbits; Sorensen, 1988, in pigs). The advantages of BLUP come from the enormous development of computation, which now allows solving of large systems of equations quickly and at low cost. The ease with which these systems are solved makes BLUP a convenient instrument to use. BLUP results also allow better monitoring of the response to selection (genetic trends) and the evolution of environmental effects, as long as the genetic parameters that are introduced are reliable.

BLUP is easy to program, and there are accessible versions in R and other languages. Some very popular programs for various mixed models are those of Ignacy Mizstal, which are freely accessible and can be found at http://nce.ads.uga.edu/software (accessed 4 July 2025).

4.4 Estimation of Genetic Parameters

The estimation of heritabilities and genetic correlations can be done by simple regression methods and analysis of variance. The problem is that to estimate the genetic parameters with enough precision, many data points are needed. Consider that for a heritability of 0.20, the correlation between half-siblings is one-quarter of this (0.05). This implies that we must estimate correlations very close to zero, which are associated with large standard errors unless large samples are used. Another problem is that data are often under a selection process, so it is necessary to have the complete relationship matrix and use all the data that were used in the selection process in order to ignore selection; otherwise, the estimates are biased (see section 4.4.2). For this last reason, these simple methods have fallen into disuse, and genetic parameters are estimated at present using maximum likelihood or the Bayesian estimators that we will see in the next chapter; both procedures can be used ignoring that data are selected (see Appendix 5.2).

4.4.1 The maximum likelihood method

Maximum likelihood was proposed by several statisticians in the 18th century (Hald, 1999), but its modern formulation is due to Ronald Fisher (1922). Suppose we are interested in knowing the mean weight at 8 weeks of age of a population of rabbits and we take a sample to estimate it. To facilitate understanding, in this example the sample will be from a single rabbit weighing y_0 = 1.6 kg, but the same example can be used for the mean of n rabbits. Three possible populations are presented in Fig. 4.3; in the first one the mean is 1.5 kg, in the second 1.6 kg and in the third 1.8 kg.

Weights are normally distributed. To define a normal distribution, it is enough to give its mean and its variance, which is usually represented as $f(y|\mu,\sigma^2)$, and is read as the 'distribution of data y given the mean μ and variance σ^2'. Suppose, for simplicity, that all distributions have a known variance, for example $\sigma^2 = 0.01$ kg^2, and we are only interested in knowing the population mean, μ. If the true value of the mean weight at 8 weeks of age were $\mu = 1.8$ kg (Fig. 4.3C), it would be very unlikely to extract a rabbit with a value of $y_0 = 1.6$ kg as the probability density for the sample $f(y_0|\mu = 1.8)$ is very small. If the true value were $\mu = 1.5$ kg (Fig. 4.3A), drawing a sample of value $y_0 = 1.6$ would be more likely; here $f(y_0|\mu = 1.5)$ is greater. But the sample would have the maximum probability of being extracted if it came from the population in which the true value is exactly that of the sample, $\mu = 1.6$ kg, as in Fig. 4.3B; in this case, the probability density $f(y_0|\mu = 1.6)$ is the highest of all. If we do not have more information, it seems intuitively reasonable to estimate the value of the mean weight at 8 weeks of age as $\hat{\mu} = 1.6$ kg.

The function $f(y_0|\mu)$ is called the *likelihood function* and contains for the sample y_0 the probability densities of each distribution as the mean changes. $f(y_0|\mu)$ is a function of the mean μ but not of the sample y_0. As the variable is μ and the fixed quantity is y_0, the likelihoods are the values of a *family of density functions* for a single point y_0:

$$[f(y_0|\mu)]_{\mu\,\text{variable}} = [f(y_0|\mu_1), f(y_0|\mu_2), f(y_0|\mu_3), \ldots]$$

The likelihood function is usually represented as $L(\mu \mid y_0)$, which is somewhat confusing, because the representation of the family of densities is lost with this notation. The *maximum likelihood method of estimation* (ML) consists of taking the maximum value of the likelihood function of a parameter (μ in our case) as an estimate of this parameter. This method, unlike the least squares method, needs to know the distribution of the data, which in livestock will be a normal distribution in most cases.

> Note that likelihood is not a probability. Each likelihood value comes from a different probability density distribution. A likelihood four times greater than another does not give four times more evidence in favour of this value (see section 5.1.2).

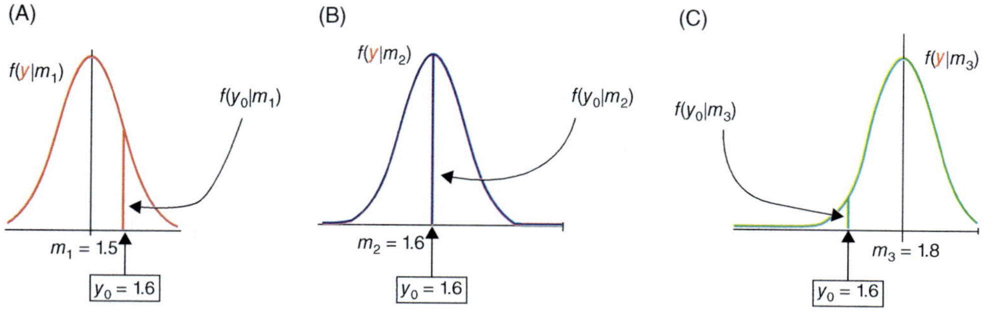

Fig. 4.3. Three likelihoods for the sample y_0 = 1.6. (A) Likelihood if the true mean of the population is 1.5. (B) Likelihood if the true mean of the population is 1.6. (C) Likelihood if the true mean of the population is 1.8.

It is often said that the ML method obtains the value that maximizes the probability of the sample. This is strictly incorrect, because the sample has the probability that it has, and we do not know it; we do not know what the true value of the mean is, so we cannot guarantee that we have drawn a sample with maximum probability. In our example, the ML method consists of obtaining the value whereby, *if the mean actually had that value*, then the sample would have maximum probability.

The ML method became popular when it was found to have optimal properties if the sample was large enough. The method is asymptotically unbiased, and the distribution of the estimates in asymptotically normal, which means that the variance of the ML estimator and its standard error can be approximately calculated.

This method is applied to any parameter of a distribution, and we will apply it to the estimation of genetic variances and covariances.

4.4.2 Restricted maximum likelihood

Apart from some early attempts, restricted (or residual) maximum likelihood (REML) was proposed in its current formulation by Patterson and Thompson (1971). Suppose the data are normally distributed and we want to estimate the mean and variance of the distribution. The likelihood function is the multivariate normal, with mean $\boldsymbol{Xb}$ and variance $\mathbf{V}$:

$$f(y|b,\mathbf{V}) = \frac{1}{(2\pi)^{\frac{n}{2}}|\mathbf{V}|^{\frac{n}{2}}} exp\left[-\frac{1}{2}(y-\boldsymbol{X}b)'\,\mathbf{V}^{-1}(y-\boldsymbol{X}b)\right]$$

The simplest linear model is:

$$y = \mu + e$$

$$\begin{bmatrix} y_1 \\ \vdots \\ y_n \end{bmatrix} = \begin{bmatrix} \mu \\ \vdots \\ \mu \end{bmatrix} + \begin{bmatrix} e_1 \\ \vdots \\ e_n \end{bmatrix} = \begin{bmatrix} 1 \\ \vdots \\ 1 \end{bmatrix}\mu + \begin{bmatrix} e_1 \\ \vdots \\ e_n \end{bmatrix};\quad y = \boldsymbol{1}\mu + e;\quad \mathbf{V} = var(e) = \begin{bmatrix} \sigma_e^2 & 0 & 0 \\ & \ddots & \vdots \\ & & \sigma_e^2 \end{bmatrix} = \boldsymbol{I}\sigma_e^2$$

The likelihood is:

$$f(y|\mu,\sigma_e^2) = \frac{1}{(2\pi)^{\frac{n}{2}}(\sigma_e^2)^{\frac{n}{2}}} exp\left[-\frac{1}{2\sigma_e^2}(y-\boldsymbol{1}\mu)'(y-\boldsymbol{1}\mu)\right]$$

which is a function of μ and σ_e^2, but not a function of *y*, because the data are fixed, as for y_0 in section 4.4.1. The maximum of the likelihood function (see Appendix 4.3) is:

$$\hat{\sigma}_e^2 = \frac{1}{n}(y-\boldsymbol{1}\mu)'(y-\boldsymbol{1}\mu) = \frac{1}{n}\sum_1^n (y_i-\mu)^2$$

$$\hat{\mu} = \frac{1}{n}\sum_1^n y_i$$

Substituting the true value of μ by $\hat{\mu}$, we have the ML estimate of the variance:

$$\hat{\sigma}^2_{ML} = \frac{1}{n}\left(y - 1\hat{\mu}\right)'\left(y - 1\hat{\mu}\right) = \frac{1}{n}\sum_1^n\left(y_i - \hat{\mu}\right)^2$$

Note that the error of estimation of $\hat{\mu}$ is not taken into account, as the formula for estimating the variance does not change by putting $\hat{\mu}$ in place of the true μ.

When there are many fixed effects, as usually happens in animal breeding programmes, the mean of the data is $\boldsymbol{Xb}$, and the estimate of the variance is:

$$\hat{\sigma}^2_{ML} = \frac{1}{n}\left(y - X\hat{b}\right)'\left(y - X\hat{b}\right)$$

But as there are many effects to be estimated in $\boldsymbol{b}$, it is somewhat confusing to use the estimates of $\boldsymbol{b}$ as if they were their true value, and it is also confusing to ignore the estimation errors. One solution may be to eliminate the fixed effects from the model using linear combinations of the data (what in statistics are called *contrasts*). For example, in the simple model $y = \mu + e$, we can use differences between data:

$$y_i - y_k = \mu + e_i - (\mu + e_k) = e_i - e_k$$

which is equivalent to multiplying the data by a suitable matrix, for example:

$$K'y = \begin{bmatrix} 1 & -1 & 0 & 0 & \dots & \dots & 0 \\ 1 & 0 & -1 & 0 & \dots & \dots & 0 \\ 1 & 0 & 0 & -1 & 0 & \dots & 0 \\ \dots & \dots & \dots & \dots & \dots & \dots & \dots \\ 1 & 0 & 0 & 0 & \dots & \dots & -1 \end{bmatrix} \begin{bmatrix} y_1 \\ y_2 \\ y_3 \\ \vdots \\ y_n \end{bmatrix} = \begin{bmatrix} y_1 - y_2 \\ y_1 - y_3 \\ y_1 - y_4 \\ \vdots \\ y_1 - y_n \end{bmatrix} = \begin{bmatrix} y_1^* \\ y_2^* \\ y_3^* \\ \vdots \\ y_{n-1}^* \end{bmatrix} = y^*$$

The estimation of the variance is made on the new data vector $\boldsymbol{y}^*$ whose mean is zero. For n data, there are only $n - 1$ differences between data, so the new vector $\boldsymbol{y}^*$ has one less element than the original data vector $\boldsymbol{y}$. In the case where there are many fixed effects, it is necessary to make several contrasts to eliminate the fixed effects, as we have done with the mean, so the new data vector $\boldsymbol{y}^*$ is smaller than the original. In our simple example, if we were to make a geometric representation of our sample $\boldsymbol{y}$, we would need an n-dimensional space; however, $\boldsymbol{y}^*$ is represented in a space of $n - 1$ dimension; it is a more restricted space, because we have lost one degree of freedom. When there are many fixed effects, the space is even more restricted, so this method is called restricted or residual maximum likelihood (REML), because it works with contrasts of residuals.

The variance of $\boldsymbol{y}^*$ is:

$$var(K'y) = K'var(y)K = K'VK = K'\left(I\sigma_e^2\right)K = K'K\sigma_e^2$$

and the likelihood is:

$$f\left(y^* | K, \sigma_e^2\right) = \frac{1}{(2\pi)^{\frac{n}{2}}\left|K'K\sigma_e^2\right|^{\frac{n}{2}}} exp\left[\sum_1^n -\frac{1}{2}\frac{(K'y)'(K'K)^{-1}(K'y)}{\sigma_e^2}\right]$$

The maximum of the likelihood function can be shown to be:

$$\hat{\sigma}^2_{REML} = \frac{1}{n-1}\left(y - 1\frac{1}{n}\sum_1^n y_i\right)'\left(y - 1\frac{1}{n}\sum_1^n y_i\right)$$

Note that the mean μ no longer appears, because we have analysed the data vector y^* whose mean is zero. We can also write the formula in a simpler way as:

$$\hat{\sigma}^2_{REML} = \frac{1}{n-1}\left(y - 1\hat{\mu}\right)'\left(y - 1\hat{\mu}\right)$$

but here we have not substituted the true value of μ by its estimate $\hat{\mu}$, because we have no μ in the data y^* used for estimating the variance, so the mean of y^* is zero. What happens is that when obtaining the maximum value of the likelihood, the following term appears:

$$\frac{1}{n}\sum_1^n y_i$$

Note that the variance estimate is divided by $n - 1$, indicating the loss of one degree of freedom, which means that we are restricting the estimation space to the $n - 1$ dimensions defined by the data of y^* instead of the n dimensions defined by the data of y.

In the case of the linear model with many fixed effects, the estimate is:

$$\hat{\sigma}^2_{REML} = \frac{1}{n - rg(X)}\left(y - X\hat{b}\right)'\left(y - X\hat{b}\right)$$

where $rg(X)$ is the rank of the incidence matrix X (the number of rows or columns that are not linear combinations of each other).

In the case of the mixed model, the variance has two components, σ^2_A for the additive values and σ^2_e for the error, and both components should be estimated:

$$y = Xb + Za + e\ ; \quad E(y) = Xb\ ; \quad V = Z'var(a)Z + var(e) = Z'AZ\sigma^2_A + I\sigma^2_e$$

The likelihood is written as at the beginning of this section, but the matrix V is more complex:

$$f\left(y|b, A, \sigma^2_A, \sigma^2_e\right) = \frac{1}{(2\pi)^{\frac{n}{2}}\left|Z'AZ\sigma^2_A + I\sigma^2_e\right|^{\frac{n}{2}}} exp\left[-\frac{1}{2}(y - Xb)'\left(Z'AZ\sigma^2_A + I\sigma^2_e\right)^{-1}(y - Xb)\right]$$

It is difficult to invert V; moreover, when calculating the derivatives to find the maximum, multiplications of large matrixes are needed in the case of animal breeding, as large data bases are used. Therefore, the maximum of the function is not calculated by differentiating and equating to zero; instead, numerical analysis methods are used to find approximately the maximum of the function. For the same reasons, the exact values of the standard errors are not calculated but rather are approximated using numerical analysis methods, which are often underestimating the actual standard errors.

As ML and REML are asymptotically normal, the standard error of the estimates of the variance components can be approximately calculated. It can be shown (e.g. Blasco, 2017, p. 63) that the variance and the standard error (*s.e.*) of a parameter θ of a likelihood function $f(y|\theta)$ (remember this is a function of θ and not of y) is:

$$Var\left(\theta^2\right)=\frac{1}{\left(\frac{\partial^2 logf\left(y|\theta^2\right)}{\partial\theta^2}\right)^2_{\theta=\hat{\theta}_{ML}}}\ ;\ s.e.\left(\theta^2\right)=\sqrt{Var\left(\theta^2|y\right)}$$

For example, on the simplest model $y = \mu + e$, the variances and standard error of the ML estimators of μ and σ^2 are obtained from the likelihood function $f=f\left(y|\mu,\sigma_e^2\right)$ shown at the beginning of this section. We are not detailing the development here, but the final result is:

$$var(\mu)=\frac{1}{\left[-\frac{\partial^2 logf}{(\partial\mu)^2}\right]_{\substack{\mu=\hat{\mu}\\ \sigma_e^2=\hat{\sigma}_e^2}}}=\frac{\hat{\sigma}^2}{n}\quad ;\quad var\left(\sigma^2\right)=\frac{1}{\left[-\frac{\partial^2 logf}{\left(\partial\sigma^2\right)^2}\right]_{\substack{\mu=\hat{\mu}\\ \sigma_e^2=\hat{\sigma}_e^2}}}=\frac{2\left(\hat{\sigma}^2\right)^2}{n^2}$$

4.4.3 Genetic parameters estimation

Once $\hat{\sigma}_A^2$ and $\hat{\sigma}_e^2$ are available, to estimate the heritability and its standard error we will use a linear approximation like the one we used when calculating the economic weights in section 3.2.1. Here the heritability is approached as a linear function of its components, around their estimates:

$$h^2=\frac{\sigma_A^2}{\sigma_A^2+\sigma_e^2}\approx\left[h^2\right]_{\substack{\sigma_A^2=\hat{\sigma}_A^2\\ \sigma_e^2=\hat{\sigma}_e^2}}+\left[\frac{\partial h^2}{\partial\sigma_A^2}\right]_{\substack{\sigma_A^2=\hat{\sigma}_A^2\\ \sigma_e^2=\hat{\sigma}_e^2}}\left(\sigma_A^2-\hat{\sigma}_A^2\right)+\left[\frac{\partial h^2}{\partial\sigma_e^2}\right]_{\substack{\sigma_A^2=\hat{\sigma}_A^2\\ \sigma_e^2=\hat{\sigma}_e^2}}\left(\sigma_e^2-\hat{\sigma}_e^2\right)$$

$$=\frac{\hat{\sigma}_A^2}{\hat{\sigma}_A^2+\hat{\sigma}_e^2}+k_1\left(\sigma_A^2-\hat{\sigma}_A^2\right)+k_2\left(\sigma_e^2-\hat{\sigma}_e^2\right)$$

where k_1 and k_2 are known, because after differentiating, we substitute the values of the variances σ_A^2 and σ_e^2 by their known $\hat{\sigma}_A^2$ and $\hat{\sigma}_e^2$ estimates. ML and REML estimates are asymptotically unbiased, so that $E\left(\hat{\sigma}_A^2\right)=\sigma_A^2$ and $E\left(\hat{\sigma}_e^2\right)=\sigma_e^2$, and therefore we can estimate the heritability from the former expression as:

$$\hat{h}^2=E\left(h^2\right)\approx\left[h^2\right]_{\substack{\sigma_A^2=\hat{\sigma}_A^2\\ \sigma_e^2=\hat{\sigma}_e^2}}+k_1E\left(\sigma_A^2-\hat{\sigma}_A^2\right)+k_2E\left(\sigma_e^2-\hat{\sigma}_e^2\right)=\left[h^2\right]_{\substack{\sigma_A^2=\hat{\sigma}_A^2\\ \sigma_e^2=\hat{\sigma}_e^2}}+0+0=\frac{\hat{\sigma}_A^2}{\hat{\sigma}_A^2+\hat{\sigma}_e^2}$$

Now we can obtain the standard error of the heritability. First, we obtain the variance of the heritability; taking into account that the covariance between the additive and residual variance is null:

$$var\left(\hat{h}^2\right) = E\left[h^2 - E\left(h^2\right)\right]^2 \approx E\left[k_1\left(\sigma_A^2 - \hat{\sigma}_A^2\right) + k_2\left(\sigma_e^2 - \hat{\sigma}_e^2\right)\right]^2 = k_1^2 var\left(\hat{\sigma}_A^2\right) + k_2^2 var\left(\hat{\sigma}_e^2\right)$$

$$s.e.\left(\hat{h}^2\right) = \sqrt{var\left(\hat{h}^2\right)}$$

where k_1 and k_2 are known, and the variances of the additive and residual variances are obtained from the REML estimation analysis as we did at the end of the previous section for μ and σ^2 in the simplest model. In Chapter 5, we will see a more precise way to determine the uncertainty of the genetic parameters using Bayesian techniques.

Genetic correlations are estimated in the same way, using multivariate linear mixed models to obtain estimates of additive variances and covariances, which we will see in Chapter 5. The standard errors of the correlations can also be approximated in the same way that we used for the standard errors of the heritability. The interested reader can consult Bulmer (1985, p. 82).

Today, there are several computer programs that allows estimation of genetic parameters and their standard errors in linear models such as those we have seen in this chapter, and also in more complex models. Some examples are (all accessed 4 July 2025):

- ASREML: www.vsni.co.uk/es/software/asreml/
- VCE: www.openagrar.de/servlets/MCRFileNodeServlet/openagrar_derivate_00022208/
- WOMBAT: http://didgeridoo.une.edu.au/km/wombat.php
- BLUPF90: http://nce.ads.uga.edu/software/

Appendix 4.1

If $\boldsymbol{y}$ is normally distributed, we can estimate $\boldsymbol{b}$ by ML (see sections 4.4.1 and 4.4.2):

$$f(y) = \frac{1}{(2\pi)^{\frac{n}{2}}|\mathbf{V}|^{\frac{1}{2}}} exp\left[-\frac{1}{2}(y - \mathbf{X}b)'\,\mathbf{V}^{-1}(y - \mathbf{X}b)\right]$$

$$logf(y) = -log\left((2\pi)^{\frac{n}{2}}|\mathbf{V}|^{\frac{1}{2}}\right) - \frac{1}{2}(y - \mathbf{X}b)'\,\mathbf{V}^{-1}(y - \mathbf{X}b)$$

$$\frac{\partial logf(y)}{\partial b} = -\frac{1}{2}2\mathbf{X}'\mathbf{V}^{-1}(y - \mathbf{X}b) = 0 \rightarrow \hat{b} = \left(\mathbf{X}'\mathbf{V}^{-1}\mathbf{X}\right)^{-1}\mathbf{X}'\mathbf{V}^{-1}y$$

$$\text{If } \mathbf{V} = \mathbf{I}\sigma_e^2 \rightarrow \hat{b} = (\mathbf{X}'\mathbf{X})^{-1}\mathbf{X}'y$$

Appendix 4.2

Here we will change the notation used in Chapter 3 for the index, and call $\boldsymbol{d}' = [d_1, d_2, d_3, \cdots]$ the regression coefficients, to avoid confusion with the fixed effects vector $\boldsymbol{b}$. The selection index in section 3.1 would be $I = \hat{a} = \boldsymbol{d'y}$, but this is a centred index in which $E(\boldsymbol{y}) = 0$.

In our case, the data are not centred, $E(y) = Xb$, and we also know that $E(a) = E(\hat{a}) = 0$ (section 4.2). Thus:

$$E(a) = E(\hat{a}) = E(d'y) = d' \cdot E(y) = d'Xb = 0 \quad \rightarrow \quad d'X = 0$$

as b cannot be zero because if so, the mean of y would be zero and the values would be centred, which is not our case.

Our objective is to find the value of d that minimizes the variance of the error σ_e^2 with the condition that $d'X = 0$. To do this, we will add a null quantity $2d'Xq$ to the expression of the residual variance (this quantity is null because $d'X = 0$), so the residual variance will not change. Taking into account that $d'c = c'd$, from Appendix 3.1, we have:

$$\sigma_e^2 = \sigma_A^2 + d'Vd - 2d'c + 2d'Xq$$

Differentiating and equating to zero to find the minimum:

$$\frac{\partial \sigma_e^2}{\partial d'} = 2Vd - 2c + 2Xq = 0 \rightarrow \hat{d} = V^{-1}(c - Xq)$$

$$\frac{\partial \sigma_e^2}{\partial q} = 2d'X = 0 \rightarrow X'd = X'V^{-1}(c - Xq) = 0 \rightarrow \hat{q} = \left(X'V^{-1}X\right)^{-1} X'V^{-1}c$$

$$\hat{a} = \hat{d}'y = \left(c - X\hat{q}\right)' V^{-1}y = c'V^{-1}y - \hat{q}'X'V^{-1}y = c'V^{-1}y - c'V^{-1}X\left(X'V^{-1}X\right)^{-1} X'V^{-1}y = c'V^{-1}\left(y - X\hat{b}\right)$$

where $\hat{b} = \left(X'V^{-1}X\right)^{-1} X'V^{-1}y$

Appendix 4.3

The maximum of the normal distribution function is at the same point as its logarithm, so to find the maximum we will differentiate the logarithm because it is easier:

$$f\left(y|\mu,\sigma^2\right) = \frac{1}{(2\pi)^{\frac{n}{2}}\left(\sigma^2\right)^{\frac{n}{2}}} exp\left[\sum_1^n -\frac{(y_i - \mu)^2}{2\sigma^2}\right]$$

$$log f = -\frac{n}{2}log(2\pi) - \frac{n}{2}log\left(\sigma^2\right) - \frac{1}{2\sigma^2}\left[\sum_1^n (y_i - \mu)^2\right]$$

$$\frac{\partial}{\partial \sigma^2} log f = -\frac{n}{2\sigma^2} + \frac{1}{2\sigma^4}\left[\sum_1^n (y_i - \mu)^2\right] = 0 \quad \rightarrow \quad \hat{\sigma}^2 = \frac{1}{n}\sum_1^n (y_i - \mu)^2$$

$$\frac{\partial}{\partial \mu} log f = \frac{1}{2\sigma^2}\left[2\sum_1^n (y_i - \mu)\right] = \sum_1^n (y_i) + n\mu = 0 \quad \rightarrow \quad \hat{\mu} = \frac{1}{n}\sum_1^n (y_i)$$

Main Concepts

- *Fixed effects:* The effects of a model that remain constant in the conceptual repetitions of the experiment. They are usually environmental effects, although sometimes they can be genetic (e.g. effect of a major gene).
- *Random effects*: The effects of a model that change with each conceptual repetition of the experiment. Typically, they are effects associated with the animal, such as its additive value, its common environmental value of several parities or the environmental effect of belonging to a common litter. The residual effect is another example of a random effect.
- *BLUP (best linear unbiased prediction):* A selection index that predicts the additive value of individuals and simultaneously corrects the data for fixed effects and random environmental effects. Like selection indices, it can be for a single trait or multi-trait.
- *Genetic trend:* The evolution of the mean of the predictions of the breeding values provided by BLUP, calculated by generation or by year or by another time period.
- *Maximum likelihood:* An estimation method, usually applied in animal breeding to the variance components, which provides the value with which the observed sample has the maximum probability if this value were the true one. Obviously, we do not know what the true value is, nor does our sample have a maximum probability; the method is used because it has good properties (it leads asymptotically to unbiased estimates with minimal error as the sample size increases).
- *REML (restricted maximum likelihood):* A maximum likelihood estimation method applied to linear combinations of the data to avoid the fixed effects when estimating the variance components.

Problems

4.1. Formulate the equations from five data points from two cows with a mixed model with effects of farm and season for milk production, and three lactations per cow at most. Include a male, the father of the two cows. The heritability of milk production is 0.4.
4.2. State the equations of the mixed model in the previous problem.

Self-assessment Questions (True or False)

1. The true value of a fixed effect does not change in the conceptual repetitions of an experiment, but the true value of a random effect is different in each repetition.
2. BLUP predictions are not affected by using data under a selection process, only if they are calculated with all the data used in the selection and with the complete relationship matrix.
3. The ML method is the one that maximizes the probability of obtaining the sample.
4. The ML method is the one that obtains a value that, if it were the true value, would result in the sample being the most probable.
5. A genetic trend estimates the response to selection.

6. To calculate genetic trends and estimate responses to selection, a control population is required.
7. BLUP is the best of all selection methods.
8. BLUP is the best among the linear unbiased predictions.
9. To apply BLUP, the data must be normally distributed.
10. BLUP maximizes the response to selection, but only if the genetic parameters used are the true ones.

References

Aguilar, I., Fernandez, E.N., Blasco, A., Ravagnolo, O. and Legarra, A. (2020) Effects of ignoring inbreeding in model-based accuracy for BLUP and SSGBLUP. *Journal of Animal Breeding and Genetics* 137, 356–364. DOI: 10.1111/jbg.12470

Baselga, M., Blasco, A. and Estany, J. (1985) Evaluación de reproductores. II. Posibilidades de simplificación de los métodos de selección. In: *I Jornadas sobre Producción Animal, Zaragoza*. ITEA, Vol. supplement 5, pp. 41–52.

Blair, H.T. and Pollak, E.J. (1984) Comparison of an animal model and an equivalent reduced animal model for computational efficiency using mixed model methodology. *Journal of Animal Science* 58, 1090–1096. DOI: 10.2527/jas1984.5851090x

Blasco, A. (2017) *Bayesian Data Analysis for Animal Scientists*. Springer, Cham, Switzerland.

Bulmer, M. (1985) *The Mathematical Theory of Quantitative Genetics*. Oxford University Press, Oxford, UK.

Falconer, D.S. and MacKay, T.F.C. (1996) *Introduction to Quantitative Genetics*. Longman, Harlow, UK.

Fisher, R.A. (1922) On the mathematical foundations of theoretical statistics. *Philosophical Transactions of the Royal Society A* 222, 309–368. DOI: 10.1098/rsta.1922.0009

Hagger, C. and Dempfle, L. (1983) Comparison of the efficiency of BLUP with other estimation procedures in dairy sire evaluation. II. Empirical investigations. *Zeitschrift für Tierzüchtung und Züchtungsbiologie* 100, 266–270. DOI: 10.1111/j.1439-0388.1983.tb00732.x

Hald, A. (1999) On the history of maximum likelihood in relation to inverse probability and least squares. *Statistical Science* 14, 214–222. DOI: 10.1214/ss/1009212248

Henderson, C.R. (1963) Selection index and expected genetic advance. In: Hanson, W.D. and Robinson, H.F. (eds) *Statistical Genetics and Plant Breeding*. National Academy of Sciences, Washington, DC.

Henderson, C.R. (1976) A simple method for computing the inverse of a numerator relationship matrix used in prediction of breeding values. *Biometrics* 32, 69–83. DOI: 10.2307/2529339

Laborda, P., Mocé, L., Santacreu, M.A. and Blasco A. (2011) Selection for ovulation rate in rabbits: genetic parameters, direct response and correlated response on litter size. *Journal of Animal Science* 89, 2981–2987. DOI: 10.2527/jas.2011-3906

Lynch, M. and Walsh, B. (1998) *Genetics and Analysis of Quantitative Traits*. Sinauer, Sunderland, Massachusetts.

Paterson, H.D. and Thompson, R. (1971) Recovery of inter-block information when block sizes are unequal. *Biometrika* 58, 545–554. DOI: 10.1093/biomet/58.3.545

Sorensen D. (1988) Effect of selection index versus mixed model methods of prediction of breeding value on response to selection in a simulated pig population. *Livestock Production Science* 20, 135–148. DOI: 10.1016/0301-6226(88)90058-9

Westell, R.A., Quaas R.L. and van Vleck, L.D. (1988) Genetic groups in an animal model. *Journal of Dairy Science* 71, 1310–1318. DOI : 10.3168/jds.S0022-0302(88)79688-5

Solutions

Solutions to the problems

4.1. This is the same example as in section 4.1.2. The data vector of the same example can be considered, expressing milk production as thousands of kilograms of milk at 305 days. The fixed effect here would be the herd effect instead of the season effect. In this case, the relationship matrix would change, and in the second row and third column, the value would be 0.5. The ratio between additive and error variance would also change because the heritability of milk is around 0.4, so the value of this ratio is around 1.5. The calculations would be the same.
4.2. This is the same example as in section 4.1.2.

Solutions to self-assessment questions

1. T, 2. T, 3. F, 4. T, 5. T, 6. F, 7. F, 8. T, 9. F, 10. T.

5 Bayesian Methods

Abstract

In this chapter, we examine Bayesian inference techniques used in animal breeding. First, a brief overview of the basics of Bayesian inference is given, underlining the special interest of this type of inference, as it can work with probabilities. Bayesian inference is then applied to the mixed model for one trait, and the Markov chain Monte Carlo numerical method is presented, which avoids the need to solve the integrals necessary in this type of inference. It is shown that best linear unbiased prediction (BLUP) and restricted maximum likelihood estimate (REML) have a Bayesian interpretation. The multivariate model is then discussed, describing the 'data augmentation' technique, which allows us to solve multi-character analyses in which the characters may be expressed under different models. An appendix discusses when and how selected databases can be analysed without taking selection into account.

5.1 Bayesian Inference

Bayesian inference was proposed separately by the Reverend Thomas Bayes in 1763 and by the Marquis de Laplace in 1774, and was developed by Pierre-Simon Laplace in the late 18th and early 19th centuries (Hald, 1998). Certain philosophical doubts about its use, as well as serious computational difficulties, caused it to be replaced in the 1930s by the classical statistics that are commonly applied today. However, the development of computers has made it possible to solve the computational difficulties of this school of inference, and since the 1990s, the Bayesian school of thought has been increasingly applied, being used nowadays in virtually all scientific fields, including animal science, especially in animal breeding. The application of Bayesian techniques in animal science textbook is explained in more detail in Blasco (2017).

5.1.1 Bayes' theorem

The way of working in the Bayesian model of thinking consists of describing the uncertainty around the value of an unknown parameter or quantity using probability as a measure of the uncertainty, basing the inference on the observed data and, if possible, on prior knowledge about the unknowns. For example, in a Bayesian context, we determine the probability of a heritability being lower than 0.10, or the probability of a difference between two treatments being higher than a given relevant value. In classical statistics, this is not possible. If a difference between two treatments has a P value of 3%, this does not indicate that the probability that the treatments are different is 3%; what it indicates is that if we repeat the experiment, the probability that a new sample would have values equal to or greater than those obtained would be 3%, which is not the same. In fact, it has been shown that P values of 0.05, commonly used to establish significant differences,

DOI: 10.1079/9781836990994.0005

do not indicate 95% evidence that the treatments are different, but rather that this evidence is around 75%; that is, we will find 'false positives' one out of every four times that significant differences are found. This is why it is so common to state that two treatments are different without sufficient evidence (Johnson, 2013). These problems of interpretation of the *P* values and the very limited usefulness of these values inspired a recent letter to the journal *Nature* signed by 800 scientists (Amrhein *et al.*, 2019) recommending avoiding its use. Moreover, in a special issue on *P* values by the *Journal of the American Statistical Association*, removing *P* values from scientific papers is proposed as a conclusion of the whole issue (Wasserstein *et al.*, 2019). A detailed discussion on the meaning and utility of *P* values can be found in Blasco (2017).

In contrast, Bayesian inference shows, for example, that the difference between two treatments being greater than zero has a probability of 97%. This is the actual probability of this difference given the only sample we have, and it is not related to repeated sampling, *P* values or significance. This result is easier to understand and is also more precise, and is one of the reasons why Bayesian inference is increasingly popular.

The second reason for the popularity of Bayesian inference, particularly in the area of animal breeding, is that it allows analyses that are not possible in classical statistics. For example, we have seen in Chapters 3 and 4 that both selection indices and best linear unbiased prediction (BLUP) use genetic parameters without taking into account the estimation error of these parameters. Taking into account the error of estimation of the genetic parameters is something that cannot be done in classical statistics but can be done using Bayesian techniques. Moreover, Bayesian statistics transform multivariate problems in univariate ones using marginalization, as we will see later, a technique that is not possible to apply in classical statistics. We will also see that genomic selection leads to equations with a large number of unknowns, much more than the available data; these equations have no solution in classical statistics but have a straightforward solution using Bayesian inference.

Bayesian statistics is based on the application of Bayes' theorem, which was not actually proposed by Bayes but by Laplace. To understand the theorem, we first need to understand the concept of conditional probability. We will give an example. Let us define two events:

A: Living in America
B: Being British

Suppose we have *N* individuals of which N_A is the number of people living in America, N_B the number of British (living in America or not), and N_{AB} the number of those who live in America and are British. The probability of being British is:

$$P(B) = \frac{N_B}{N}$$

and the probability of living in America and being British is:

$$P(A,B) = \frac{N_{AB}}{N}$$

Taking into account only British people, the probability of a person living in America given that this person is British is:

$$P(A|B) = \frac{N_{AB}}{N_B}$$

where '|' means 'given'. This is called the *conditional probability* of A given B. We find that:

$$P(A, B) = \frac{N_{AB}}{N} = \frac{N_{AB}}{N_B} \cdot \frac{N_B}{N} = P(A|B) \cdot P(B)$$

which can also be written as:

$$P(A, B) = P(B|A) \cdot P(A) = P(A|B) \cdot P(B)$$

leading to the Bayes' theorem:

$$P(A|B) = \frac{P(B|A) \cdot P(A)}{P(B)}$$

The theorem is not controversial, as it is a result of the laws of probability.

5.1.2 Prior probability

This theorem can be applied to the data of an experiment. For example, of we are estimating the heritability of a trait and we decide not to select unless the heritability is greater than 0.10, Bayes' theorem allows us to calculate the probability that the heritability is greater than 0.10, based on the data from our experiment:

$$P(h^2 > 0.1|y) = \frac{P(y|h^2 > 0.1) \cdot P(h^2 > 0.1)}{P(y)}$$

To calculate the probability of the heritability being greater than 0.10 *given our data* y, we have to know the probability of our data y when the heritability is greater than 0.10, but this is not a problem because the distribution of the data y is usually known, and is often the normal distribution, so that this probability can be calculated, as we will see later. Besides, the probability of the data $P(y)$ is a constant, so our data have a probability, but it is not necessary to calculate it, as we will see later. The problem lies in $P(h^2 > 0.1)$, which is the probability that the heritability of the trait is greater than 0.10 regardless of our data. This probability is based on the information we have before performing the experiment and does not depend on our data; this is known as *prior probability*.

Let us give an example of *prior probability* in an experiment. In mice, the colour black depends on a dominant gene. If the population is randomly mated and the allele frequencies are the same ($p = q = 0.5$), the black mouse can be black for three reasons: for having two A alleles, for having an A allele from the father and an a allele from the mother, or for having an a allele from the father and an A allele from the mother. We will try to find out if a black mouse is homozygous AA or heterozygous Aa by crossing it with a brown mouse (Fig. 5.1).

Before conducting the experiment, we already know that the black mouse has a prior probability of being homozygous of 1/3. After conducting the experiment, we have both our prior information and the information derived from our data. For example, if the black mouse has a brown offspring, we would know with certainty that the individual is heterozygous, as only a heterozygous individual could have provided the recessive a allele necessary for the mouse to be brown. However, if the offspring are composed of several black mice, we are still in doubt as to whether it was homozygous or a heterozygote that

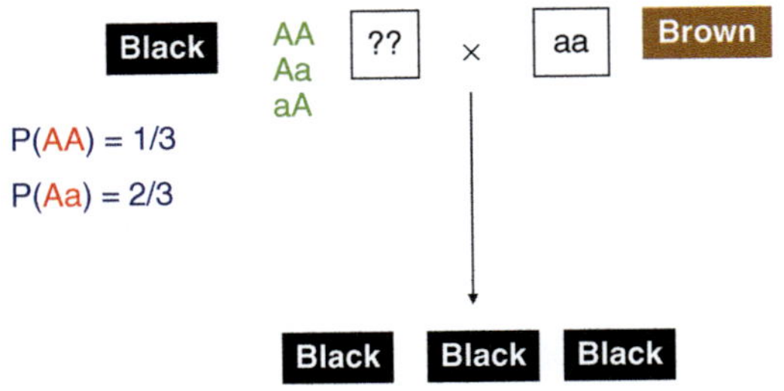

Fig. 5.1. Experiment to find out whether a black mouse is homozygous or heterozygous. Before conducting the experiment, we know that it has a 1/3 probability of being homozygous.

passed the A allele to its offspring. To calculate the probability that the mouse is homozygous *given that it had three black offspring*, we will use both the a priori information and the information from the data, applying Bayes' theorem:

$$P(\text{AA}|y = 3\,\text{black}) = \frac{P(y = 3\,\text{black}|\text{AA}) \cdot P(\text{AA})}{P(y = 3\,\text{black})}$$

where the prior probabilities are (see Fig. 5.1):

$$P(\text{AA}) = \frac{1}{3} \quad ; \quad P(\text{Aa}) = \frac{2}{3}$$

The probability of having three black offspring *given* that the black parent is AA, will be:

$$P(y = 3\,\text{black}|\text{AA}) = 1$$

because if the black parent is AA, necessarily the offspring are Aa and therefore all of them will be black. This will not happen if the black parent is Aa because half of its offspring will inherit the A allele and will be Aa (black) and half will inherit the a allele and will be aa (brown); in this case, the probability of having three black offspring *given* that the black parent is Aa, will be:

$$P(y = 3\,\text{black}|\text{Aa}) = \frac{1}{2} \cdot \frac{1}{2} \cdot \frac{1}{2} = \frac{1}{8}$$

To calculate the probability of our data, $P(y = 3\text{ black})$, we have to consider two possible cases, either the black parent is AA and the offspring are necessarily black, or it is Aa and it has had three black offspring by chance. As these are two distinct events, the probability is the sum of their probabilities:

$$P(y = 3\,\text{black}) = P(y = 3\,\text{black}, \text{AA}) + P(y = 3\,\text{black}, \text{Aa})$$

We saw in section 5.1.1 that $P(A,B) = P(A|B) \cdot P(B)$, so we can apply this now:

$$P(y = 3\,\text{black}, \text{AA}) = P(y = 3\,\text{black}|\text{AA})P(\text{AA}) = 1 \cdot \frac{1}{3} = \frac{1}{3}$$

$$P(y = 3\,\text{black}, \text{Aa}) = P(y = 3\,\text{black}|\text{Aa})P(\text{Aa}) = \frac{1}{8} \cdot \frac{2}{3} = \frac{1}{12}$$

$$P(y = 3\,\text{black}) = P(y = 3\,\text{black}, \text{AA}) + P(y = 3\,\text{black}, \text{Aa}) = \frac{1}{3} + \frac{1}{12} = \frac{5}{12}$$

Therefore, applying Bayes' theorem:

$$P(\text{AA}|y=3\,\text{black}) = \frac{P(y=3\,\text{black}|\text{AA})\cdot P(\text{AA})}{P(y=3\,\text{black})} = \frac{1\cdot\frac{1}{3}}{\frac{5}{12}} = 0.80$$

$$P(\text{Aa}|y=3\,\text{black}) = \frac{P(y=3\,\text{black}|\text{Aa})\cdot P(\text{Aa})}{P(y=3\,\text{black})} = \frac{\frac{1}{8}\cdot\frac{2}{3}}{\frac{5}{12}} = 0.20$$

Before performing the experiment, we had more evidence that the black mouse could be heterozygous, because the *prior probability* for Aa is higher, but after the experiment, the probability of Aa given the data or *posterior probability* is lower.

Note that $P(y = 3\,\text{black}|\,\text{AA})$ and $P(y = 3\,\text{black}|\,\text{Aa})$ are the likelihoods that we defined in section 4.4.1. If AA were the true genotype of the black mouse, the probability of our sample would be 1, and if Aa were the true genotype of this mouse, the probability of our sample would be 0.125.

> Using the maximum likelihood method, we would decide that AA is more likely, but we do not have a *measure of the uncertainty* for this case. The sum of the likelihoods is 1.125 and not 1 because the likelihoods are not probabilities, and therefore the evidence they provide is not proportional to their value; we cannot say how much evidence we have for AA, we can only choose AA. However, in a Bayesian context, we can say that it is four times more probable that the black mouse is AA than Aa, so we have a measure of our evidence and a measure of our uncertainty.

Note that if the sample were larger, for example y = 7 black mice, the probability of AA given the data is much higher; recalculating, we can see that $P(y = 7\,\text{black}|\,\text{Aa}) = 0.99$.

> Prior probabilities become less important as the number of data points increases. In practice, for almost all the problems that we will deal with in animal breeding, prior probabilities do not influence the results in practical terms.

It is not frequent, however, to find exact prior information as in our example. Usually, there is prior information, but it is not clear how to formalize it. For example, if we are going to estimate the heritability of litter size of a pig population, we might know that this heritability has also been estimated in other populations and has often been given values between 0.05 and 0.15. However, in our example, the estimate was 0.30, but the standard error was high. We also have a high heritability from this experiment, but there are reasons not to believe the results, such as because we believe that there are maternal effects that have not been taken into account. In addition, some of the experiments have been performed in situations that are not similar to our own experiment. Therefore, we have prior information, but it is not obvious how to integrate it. For these situations, it has been proposed that probability could be considered as a *state of belief*. This does not mean that this state of belief is arbitrary; several experts may agree that it is probable that litter size has a heritability around 0.10, it is less probable that heritability has higher values and it is very improbable that it has values greater than 0.30. However, this way of defining the probability is considered to be inappropriate by many statisticians. Moreover, it is difficult or impossible to define probabilities as states of belief in a multivariate case; for example, you have to decide what is your belief about the probability of the first trait having a heritability of 0.1 when the second has 0.2, while its genetic correlation is 0.5, and so on for each of the possible cases. In these cases, it is practically impossible to construct a probability function representing beliefs, because beliefs are difficult to determine. Another difficulty is *how to represent ignorance*. Although it is very rare to have complete ignorance about the possible value of an unknown before performing an experiment, we would prefer to avoid the use of prior information when estimating its value. Representing ignorance is not possible, as we will see later.

This situation has led modern Bayesians to ignore actual prior probabilities. They use priors that do not usually represent prior belief but introduce little prior information, for example proposing that all events have the same prior probability. If priors are barely informative and they do not affect the results in practical terms, then posterior probabilities depend essentially only on the data. In our previous example, we can consider that the prior probabilities are $P(\text{AA}) = P(\text{Aa}) = 0.5$, and we will obtain, with seven black offspring, almost the same posterior probabilities that we obtained with the true prior probabilities.

> In modern Bayesian inference, the prior probabilities are often mere mathematical artefacts whose information is irrelevant, so that the posterior probabilities are practically determined solely by the data.

We have an exception in animal breeding, where we have precise prior information from the relationship matrix, which allows us to propose an objective prior structure of the variance–covariance matrix of the additive values.

5.1.3 Probability density functions

To describe our uncertainty, we will use probability density functions. In these functions, the area between two points represents the probability that the parameter we estimate is between those two points. This allows the creation of any confidence interval to be useful for us. In Fig. 5.2, two confidence intervals for the heritability of a trait are represented using the same density function. If we make the decision not to select the trait unless its heritability is higher than a *relevant value*, such as 0.10, we can see in Fig. 5.2A the probability of the heritability being higher than 0.10, in this case 92%, which can give us enough guarantees for selection. We can also determine the minimum value that can be guaranteed with a certain probability, for example 80%. Fig. 5.2B shows that the *minimum guaranteed value* of the heritability is 0.2 with a probability of 80%. The probabilities that will allow us to take a decision depend on the trait and on the problem; for human health treatments, we would normally ask for high probabilities because we need to be sure that the treatment is going to be effective; however, for some animal traits (e.g. meat-quality traits), we can often accept lower probabilities for taking a decision.

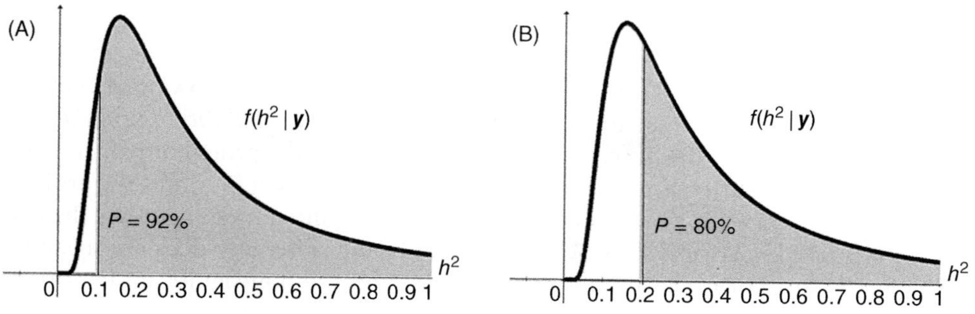

Fig. 5.2. Confidence intervals for the probability density function of the heritability of a trait given the data $f(h^2|y)$. (A) Probability of $h^2 > 0.10$ (shaded area). (B) Guaranteed minimum value ($h^2 = 0.2$) with a probability of 80% (shaded area).

Relevant value

The relevant value is the minimum value of a parameter or the minimum difference between treatments from which we take a decision. For example, when comparing the effect of several feeds for the trait feed conversion ratio in order to choose the best feed, the relevant value is the minimum difference between the feeds from which we decide to choose. If the value of the difference between two treatments is less than the relevant value, both treatments are considered to be the same in practice. They can never be completely equal, as the difference between treatments cannot be exactly zero (i.e. 0.0000...), but for practical purposes we can say that there are no differences between treatments if the difference is less than the relevant value. The relevant value should be decided before performing the experiment, and is in fact independent of the experiment; a difference between treatments or a correlation between traits may be relevant even if no experiment is performed.

> The *relevant value* is the minimum value of an unknown (e.g. a parameter or a difference between effects) from which a decision is taken.

A breeder or a scientist performing an experiment should know which values of the traits are relevant to make decisions. However, sometimes it is difficult to decide what is relevant, particularly in certain experiments. For example, in meat-quality experiments, when analysing organoleptic traits from a panel test, it is difficult to decide which score is relevant for traits such as anise odour, anise flavour, liver flavour, etc. Another example is deciding which enzyme activity is relevant when comparing treatments. In these cases, a fraction of the standard deviation of the trait can be chosen as the relevant value. Most productive traits have relevant economic values between 1/2 and 1/3 of their standard deviation; we can choose the same range when we do not have economic or biological information for deciding the actual relevant value of the trait.

Defining the relevant value allows us to define the *probability of relevance* as the probability of finding values equal or higher than the relevant value , and the *probability of similarity* as the probability of finding values lower than the relevant value. For example, we can consider that all correlations between −0.1 and 0.1 are irrelevant for an experiment.

Figure 5.3A shows that, although there is a high probability that a correlation is positive, its probability of being relevant is not so high (82%). For another experiment, considering the same relevant values (−0.1 and 0.1) for correlations, Fig. 5.3B shows that the probability that a correlation is null for practical purposes (the *probability of similarity*) is high (94%). In the case of the difference between treatments, the probability of similarity indicates whether this difference is likely to be irrelevant; that is, it indicates the probability that the treatments are similar and that there are no differences between them for practical purposes. In classical statistics, a 'non-significant' difference does not mean that the difference is small or irrelevant; it can be relevant and still non-significant because the sample is small – large samples give significant differences that may be irrelevant, small samples give often non-significant differences that can be important. The advantage of having a probability of similarity is that it can be allow us to state that some differences are irrelevant, measuring our uncertainty about this statement.

Confidence intervals

The concept of the confidence interval is different in the case of Bayesian statistics. Here, it indicates *the limits between which the true value lies with a probability of 95%* (or another probability decided by the researcher), but in classical statistics this is not the case. In classical statistics, each time the experiment is repeated, a different confidence interval is obtained, and 95% of these intervals contain the true value. For example, in classical

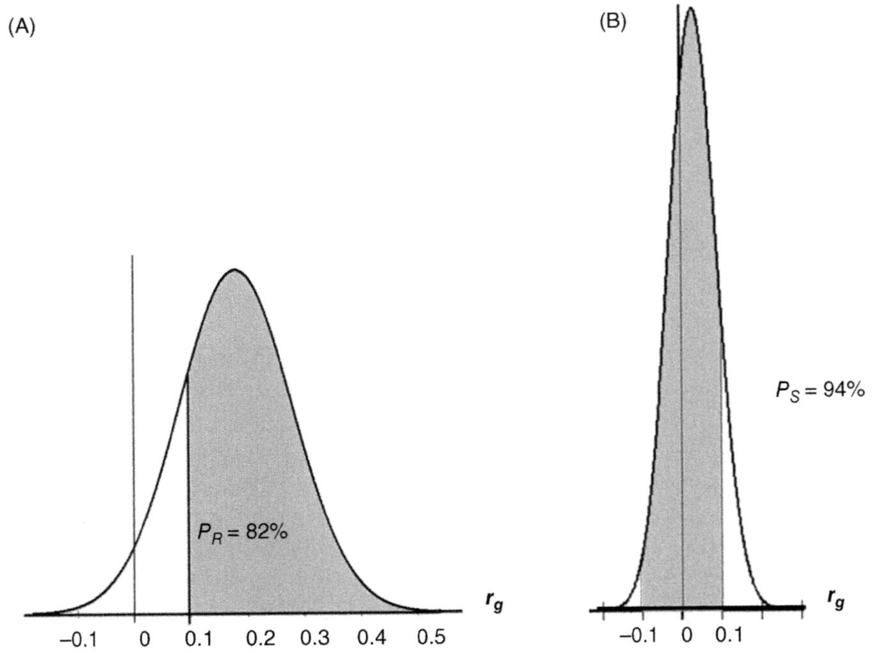

Fig. 5.3. Probability density function $f(r_g|y)$ for the genetic correlation between two traits, for two different experiments. Relevant value = 0.1, which is the absolute value for both experiments. (A) Probability of relevance, P_R, for the first experiment. (B) Probability of similarity, P_S, for the second experiment.

statistics the confidence interval of our experiment 0.3 ± 0.1 is only one of the possible intervals, and in each conceptual repetition we would obtain different results, such as 0.2 ± 0.1, 0.4 ± 0.1, etc. We maintain that 95% of these intervals would contain the true value and that the interval of our experiment is one of those, expecting to be wrong throughout our career 5% of the time when making a similar statement. In Bayesian statistics, a confidence interval indicates that the heritability is between 0.2 and 0.4 with 95% probability. The frequentist confidence interval determines uncertainty in a less clear and less intuitive way than the Bayesian confidence interval. As they are different concepts, in a Bayesian context we usually speak of *credibility intervals* instead of confidence intervals, but we will continue to use the nomenclature 'confidence intervals' to facilitate understanding when being introduced to Bayesian inference.

Note that the true value is not represented in Bayesian inference, as we do not know where it is. In the function $f(r_g|y)$ in Fig. 5.3, the parameter r_g *is not the true value of the genetic correlation*, which is unknown, but an auxiliary variable that we call r_g, allowing us to express the uncertainty about the true value r_{gTRUE} of the genetic correlation. This generates some confusion, because in frequentist statistics we also call the true value of the genetic correlation r_g, but this is not the case in a Bayesian context.

We can give a point value for the parameter that we are estimating. We can take the mode (the most probable value of the parameter), or the mean of the distribution, or the median (the value with a 50% probability of being greater or lesser than the true value). The properties of these estimators are different: the mean minimizes the expectation of the squared error of estimation, and the median minimizes the expectation of the absolute value of the error. The mode has a peculiar loss function that makes it less interesting, but it still has the interesting property of being the most probable value (see Blasco, 2001, 2017, for details).

The point value can be accompanied by a confidence interval indicating the uncertainty around the true value of the unknown being estimated. This interval usually contains 95% of the probability that the value true is within that range. There are many confidence intervals containing 95% probability. The most common is the shortest possible confidence interval, because the precision is better; it is not the same to say that the heritability is 0.40 with an interval of 0.10–0.60 as it is to say that it is 0.40 with an interval of 0.35–0.65, both with a probability of 95%, as the second interval gives a higher precision. The shortest interval with 95% probability is often referred to as the 95% highest posterior density (95% HPD).

> Note that the shortest confidence interval is estimated independently of the point value, which can lead to asymmetric intervals around the point value.

In Fig. 5.4, two point estimates of the heritability are given, based on the mode and the mean of the posterior density, with an asymmetric interval around each of the two estimates. Confidence intervals can also be given without giving point estimates.

A common confidence interval is that representing $P > 0$. For example, in Fig. 5.3, the probability of $r_g > 0$ is 96%. However, the shortest confidence interval is [−0.02, 0.38] and contains zero. In a frequentist context, when zero is within the confidence interval, generally a parameter or a difference between treatments (A − B) is non-significant, but in this

Bayesian context, we do not apply a hypothesis test; instead, we estimate $P(r_g > 0)$ or P(A – B) > 0 instead (Fig. 5.5). The concept of 'significance' is meaningless in a Bayesian context, as we do not consider what will happen when repeating the experiment many times.

> Note that a confidence interval containing zero does not mean that a difference between treatments is null or that we cannot say that it is positive. It can be that the shortest confidence interval contains zero and the probability *P* of the difference between treatments can still be *P* > 95%.

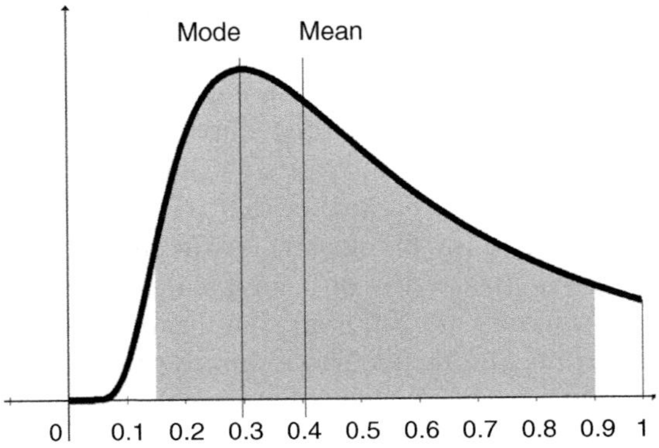

Fig. 5.4. Posterior density function of the heritability of a trait $f(h^2|y)$. Point estimates (mode and mean) with the shortest confidence interval that includes the 95% probability (shaded area) (95% highest posterior density).

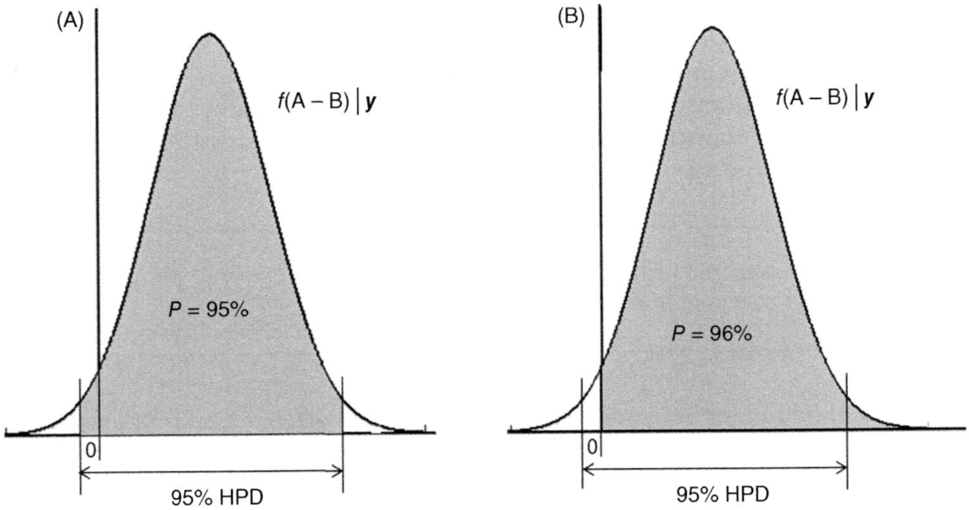

Fig. 5.5. Posterior density function *f*(A – B|**y**) of the difference between treatments A and B on a trait. (A) Shortest confidence interval at 95% probability (shaded area). (B) Probability of the difference between treatments being higher than zero (shaded area).

Note that the posterior density function is not represented as $f(h^2)$ but as $f(h^2|\mathbf{y})$; that is, the probability density function of the heritability, *given the data*. This indicates that with other data we would have another situation of uncertainty and therefore another probability density function. In Fig. 5.6, two posterior density functions are represented for two different samples. In the second sample, there are much more data points and therefore the uncertainty is lower and the confidence intervals are smaller.

Application of Bayes' theorem to density functions

Bayes' theorem can be applied not only to probabilities but also to density functions. These functions are obtained by applying Bayes' theorem. To see that Bayes' theorem can be applied not only to probabilities but also to density functions, let us look at Fig. 5.7, in which a probability density function is broken down into small rectangles. The base of the rectangle at one point, x, is a small increment from one value of x to the next (Δx), and the height is the value of $f(x)$ at that point. As the areas in the figure are probabilities, the area of each rectangle, $f(x)\cdot\Delta x$, is approximately the probability that x lies between x and $x + \Delta x$. The same happens with $f(x|y)\cdot\Delta x$ and $f(y|x)\cdot\Delta y$, they are also probabilities (approximately). Applying Bayes' Theorem, we have:

$$f(x|y)\Delta x = \frac{f(y|x)\Delta y \cdot f(x)\Delta x}{f(y)\Delta y} \quad \rightarrow \quad f(x|y) = \frac{f(y|x)\cdot f(x)}{f(y)}$$

where y is the data, x is the unknown to be estimated, $f(x)$ is the a priori *probability density* of the unknown, $f(x|y)$ is the *posterior probability density* given the data, $f(y|x)$ is the probability density of the data assuming x is known, and finally $f(y)$ is the density of the data (i.e. taking into account all possible values of x), which is constant (notice that y is 'given' in the formula: y is the set of data).

Note that if we add the rectangles between two points A and B, we obtain approximately the area of the curve between A and B. If the rectangles have an infinitesimal small base dx, the area of the rectangle would be an infinitesimal probability $f(x)\cdot dx$, and the probability

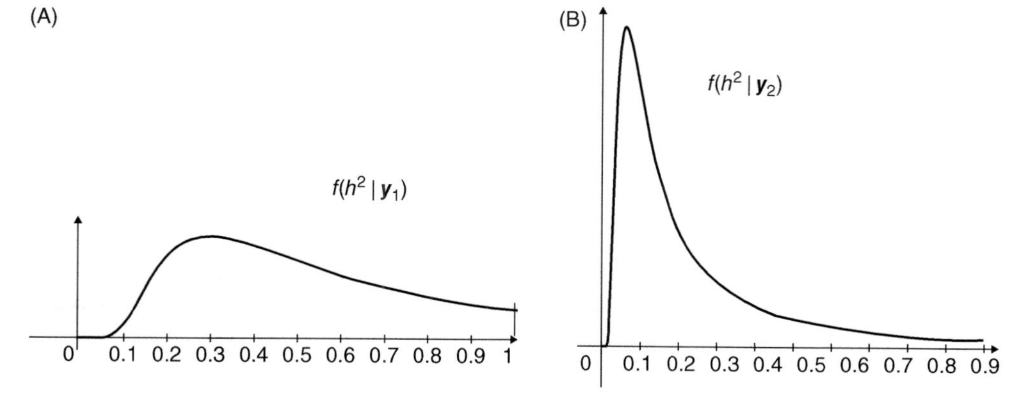

Fig. 5.6. Posterior densities of the heritability of a trait for two different samples. (A) A small sample, $\mathbf{y}_1$. (B) A large sample, $\mathbf{y}_2$. The second sample $\mathbf{y}_2$ has more data than the first and its posterior density reflects less uncertainty; the confidence intervals with this density are shorter.

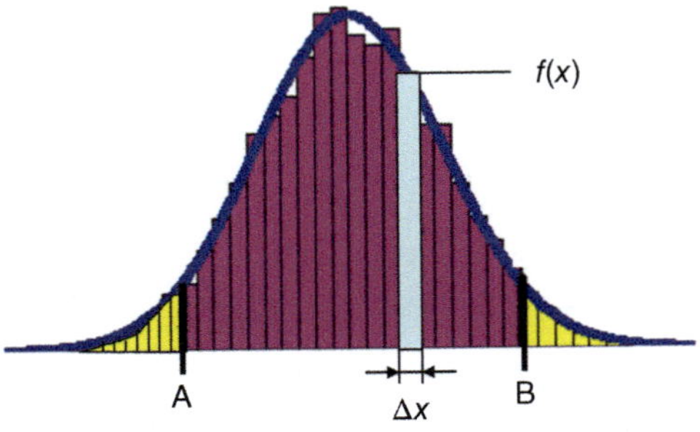

Fig. 5.7. Area of the probability density approximated by the addition of small rectangles.

that the true value lies between A and B would be the area between A and B, and would be calculated by adding all the infinitesimal rectangles through an integral:

$$P(\mathrm{A} \le x \le \mathrm{B}) = \text{area between A and B} = \int_{\mathrm{A}}^{\mathrm{B}} f(x)dx$$

For example, in the case of the heritability:

$$f(h^2|\mathrm{y}) = \frac{f(\mathrm{y}|h^2)\cdot f(h^2)}{f(\mathrm{y})} \propto f(\mathrm{y}|h^2)\cdot f(h^2)$$

where the sign ∝ means 'proportional to'. We do not need to know the value of $f(\mathbf{y})$, as we can make the same inferences proportionally, as can be seen in Fig. 5.8. The inferences are the same when using a function proportional to the posterior density.

The function $f(\mathrm{y}|h^2)$ is known, as we will see in the next section, because the distribution of the data is usually known. The function $f(h^2)$ represents the *prior information* on the heritability, and it is not clear how it can be determined, because the prior information on the heritability usually comes from several heterogeneous experiments. Interested readers can consult Blasco *et al.* (1998) for an example of how prior information can be entered to estimate the heritability, but this is rarely used in modern Bayesianism.

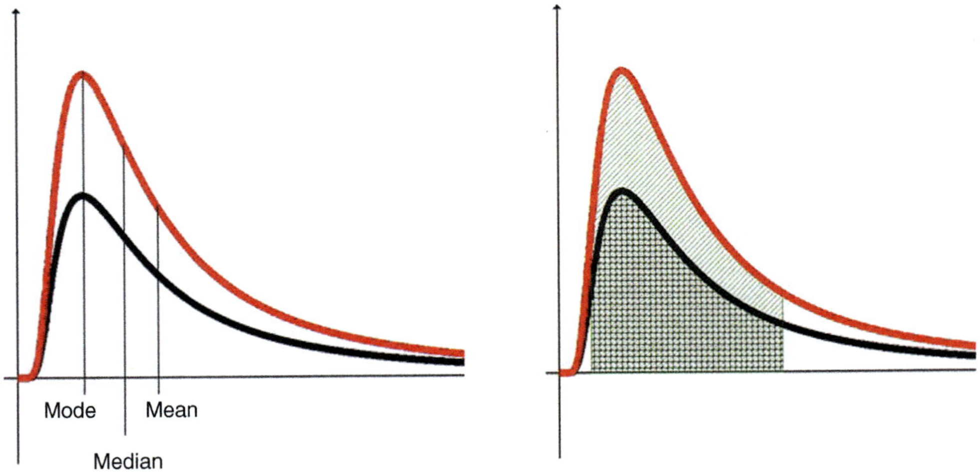

Fig. 5.8. Point estimates and credibility intervals in density functions and in functions proportional to density functions are the same.

Flat priors

If we do not have appropriate prior information about the heritability, we can assume that we have the same ignorance about the heritability from zero to 1; that is, all confidence intervals of heritability with the same length are equally probable, so that $f(h^2)$ = constant (Fig. 5.9).

This prior density function is known as a *flat prior* or non-informative prior. However, it is not true that it is not informative; actually it gives very vague information and should not affect the result of the heritability estimation, as heritabilities are estimated with a large number of data points and prior information loses importance when this is the case. If $f(h^2)$ = constant, the posterior density function is:

$$f\left(h^2|\mathbf{y}\right) \propto f\left(\mathbf{y}|h^2\right)$$

or, as a function of the variances:

$$f\left(\sigma_A^2, \sigma_e^2|\mathbf{y}\right) \propto f\left(\mathbf{y}|\sigma_A^2, \sigma_e^2\right)$$

Note that the function $f(\mathbf{y}|h^2)$ is coincident with the function that we defined as *likelihood* in section 4.4.1; that is, the probability density of $\mathbf{y}$ for each possible value of h^2. At first sight, it seems that Bayesian methods should give the same results as those of maximum likelihood. However, this is not so. Likelihoods are not probabilities, as we said in section 4.4.1, and therefore probability confidence intervals cannot be established with them, nor can other operations such as marginalization, which we will see later. Our inferences do not come from the likelihood but from the posterior density $f(h^2|\mathbf{y})$, which, although it has the same form as the likelihood, is nevertheless a true probability density function that allows establishment of confidence intervals and assignment of probabilities to events. It is true that in this case Bayesian inference has a disadvantage in not being able to clearly integrate prior information, but nevertheless, even in this case, Bayesian inference allows us to work with probabilities, which is a natural way of understanding uncertainty about a parameter and also has many operational advantages, as we will see later.

5.2 The Mixed Model in Bayesian Inference

We saw in section 4.3 that a model with fixed effects $\boldsymbol{b}$ and with random effects $\boldsymbol{a}$ is known as a mixed model. In a Bayesian context, there is no distinction between *fixed* and *random* effects, they are all random as the uncertainty is described using probabilities. However, we will continue using the terms 'fixed' and 'random' to facilitate understanding, and we will see that the effects considered as 'fixed' are, in a Bayesian context, random effects with flat prior density functions. The same happens with the parameters (variances,

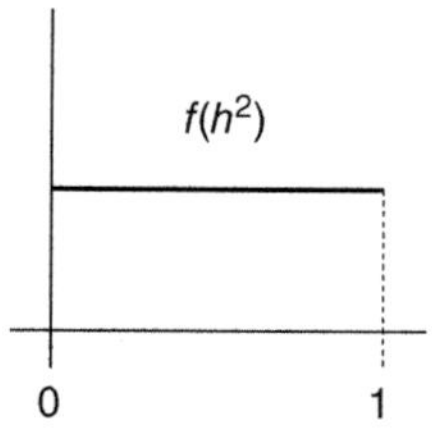

Fig. 5.9. Flat prior density for the heritability.

covariances and correlations) – in a Bayesian context they are random as their uncertainty is also described using probabilities, but we will still use the word *estimation* for 'fixed effects' and parameters, and *prediction* for additive values and other 'random effects', as this is the common use in animal breeding.

5.2.1 Predicting additive values and estimating genetic parameters

All of the models in Chapter 4 can be used in Bayesian inference. We will now look at the model in section 4.3 used for predicting additive values and genetic parameters:

$$y = Xb + Za + e$$

where $\boldsymbol{b}$ is a vector with fixed effects, also called systematic effects (e.g. season, calving order) and $\boldsymbol{a}$ is a vector with random effects, in this case with the additive values of the animals. The vector $\boldsymbol{y}$ contains the data and $\boldsymbol{e}$ is the random vector of the residuals. $\boldsymbol{X}$ and $\boldsymbol{Z}$ are design matrices that contain ones and zeros indicating the presence or absence of the effect, as we saw in section 4.1.2. The structure of the variance–covariance matrices of the random effects is the one we saw in section 4.1.2 when defining the mixed model.

If we had repeated data, as in the case of several lactations in a cow or litter size in several parities of a female, we would use a model with permanent effects like the one we saw in section 4.1.2., or other possible models, as explained in section 4.3.2.C.

To solve the model and predict the additive values of the individuals, we need to know the prior variances of the random effects $\boldsymbol{a}$ and $\boldsymbol{e}$. Therefore, the unknown parameters that should be estimated are $\boldsymbol{b}$, $\boldsymbol{a}$, σ^2_A and σ^2_e. In a Bayesian context, to predict the unknowns and express our uncertainty about them, we should calculate their posterior probability density functions given the data:

$$f(b|y), f(a|y), f(\sigma^2_A|y), f(\sigma^2_e|y)$$

To do this, we will apply Bayes' theorem, first to the joint density function of all unknown parameters:

$$f(b, a, \sigma^2_A, \sigma^2_e, |y) \propto f(y|b, a, \sigma^2_A, \sigma^2_e) \cdot f(b, a, \sigma^2_A, \sigma^2_e)$$

where, if the residuals are normally distributed (as usual), the probability density function of the data given the unknowns $f(y|b, a, \sigma^2_A, \sigma^2_e)$ is the same as we saw in section 4.1.2; but in this case, because not only $\boldsymbol{b}$ but also $\boldsymbol{a}$ is given, $\boldsymbol{a}$ is also considered constant, as if it were a fixed effect, so that the mean and variance of $\boldsymbol{y}$ are:

$$E(y) = Xb + Za\,; \quad V = \text{var}(e) = I\sigma^2_e$$

and if the residuals are normally distributed, the multivariate normal density of the data, given the unknowns, is:

$$f(y|b, a, \sigma^2_A, \sigma^2_e) = \frac{1}{(2\pi)^{\frac{n}{2}} (\sigma^2_e)^{\frac{n}{2}}} exp\left[-\frac{1}{2\sigma^2_e}(y - Xb - Za)'(y - Xb - Za)\right]$$

For example, if we repeat the experiment as in a frequentist context, the values of the residuals would vary but those of $\boldsymbol{a}$ and $\boldsymbol{b}$ will remain constant.

We have to propose now the a priori density function $f\left(b,a,\sigma_A^2,\sigma_e^2\right)$. We will assume that the distributions of b and a are independent a priori. As we discussed in section 4.4.1, this assumption is not always reasonable; for example, in dairy cattle, good herds with a good environment buy semen from genetically better bulls, so there is a positive covariance between additive values a and farm environmental effects contained in b. This does not usually have important consequences, because the amount of data used in dairy cattle and in animal breeding in general is large and the prior densities for b become irrelevant. For the same reason, we will usually assume flat priors for b and for the variances:

$$f(b)=f\left(\sigma_e^2\right)=f\left(\sigma_A^2\right)=\text{constant}$$

We will also assume that the residual variance is independent of the other parameters. We can assume that the additive variance is a priori independent of b, and of σ_e^2, but not of a, because the additive values depend on their variance, as we saw in section 4.1.2. Therefore:

$$f\left(b,a,\sigma_A^2,\sigma_e^2\right)\propto f(b)f\left(a,\sigma_A^2\right)f\left(\sigma_e^2\right)=f\left(a,\sigma_A^2\right)$$

We can apply the rules of probability to density functions as we have seen when applying Bayes' theorem. We know that $P(\text{A, B}) = P(\text{A}|\text{B})\ P(\text{B})$, thus:

$$f\left(a,\sigma_A^2\right)=f\left(a|\sigma_A^2\right)f\left(\sigma_A^2\right)$$

Regarding the prior density of a, we can assume that most traits are controlled by many genes with very small effect, so that the distribution of their additive values is approximately normal with a zero mean, as in the previous chapter. If the data are selected, it is not expected that the mean of all additive values will be zero, but as we said in section 4.3.2.B, this can be ignored under certain conditions. Appendix 5.2 shows why selection can be ignored and we can apply a zero-mean model for all additive values even if the data are selected.

For m additive values:

$$f\left(a|\sigma_A^2\right)=\frac{1}{(2\pi)^{\frac{m}{2}}\left(\sigma_A^2\right)^{\frac{m}{2}}}\exp\left(-\frac{1}{2\sigma_A^2}a'A^{-1}a\right)$$

Here, we also assume that A is given, as it includes the relationships between the animals of the experiment, and sometimes instead of using the expression $f\left(a|\sigma_A^2\right)$ for the prior density of the additive values, we can use $f\left(a|A,\sigma_A^2\right)$. Note that in a Bayesian context, this density expresses the prior uncertainty about the additive values *before taking the data*. With no data from any animal, all of them have a priori an additive value of zero, with an uncertainty around this prediction expressed by the density function $f\left(a|A,\sigma_A^2\right)$. This density does not depend on the data, and contains objective prior information, that coming from the kinship relationships contained in matrix A.

Finally, the *joint posterior density* will be:

$$f\left(b,a,\sigma_A^2,\sigma_e^2\mid y\right)\propto f\left(y\mid b,a,\sigma_A^2,\sigma_e^2\right)\cdot f\left(a\mid A,\sigma_A^2\right)$$

$$\propto\frac{1}{\left(\sigma_e^2\right)^{\frac{n}{2}}}\exp\left[-\frac{1}{2\sigma_e^2}(y-Xb-Za)'(y-Xb-Za)\right]\cdot\frac{1}{\left(\sigma_A^2\right)^{\frac{m}{2}}}\exp\left(-\frac{1}{2\sigma_A^2}a'A^{-1}a\right)$$

Note that if the population is being selected, the additive variance is reduced over time by the Bulmer effect, which we saw in section 2.1.2.B2. As we are under the conditions of ignorable selection (section 4.3.2.B and Appendix 5.2), *the variances that are estimated are those of the base population.*

Thus, if we calculate the variances with the first five generations of selection and then, for example, recalculate them with the first ten generations, the estimates will differ because we have more data in the second case, but in any case, we will not see how the variance decreases due to selection and the Bulmer effect, because we will be estimating in both cases the variances of the base population. To examine the evolution of variances with selection, we should use other methodology (see Sorensen *et al.*, 2001).

To calculate the individual posterior densities from the joint density, an operation called marginalization should be performed, which we will discuss next.

5.2.2 Marginalization

One of the main advantages of Bayesian inference is the possibility of marginalization. This implies that in a model with several variables and parameters, we can examine the uncertainty of each variable and parameter one by one, taking into account the estimation and prediction errors of all the other unknowns. We will see how it works in a simple example. Suppose we have a set of men and women, British and Spanish, who earn a certain amount per year. Table 5.1 shows the proportion of each sex and nationality, and their annual income.

Table 5.1. Annual income (thousands of euros) and proportion of individuals in two groups of Spanish and British people, separating men and women.

	British (*B*)	Spanish (*S*)
Men (*M*)	36 (40%)	26 (30%)
Women (*W*)	30 (20%)	20 (10%)

In order to find the income of British and Spanish people, regardless of whether they are men or women, we need to *marginalize* our data. The probability of being British is 0.4 + 0.2 = 0.6. The probability of being a man, given that he is British is:

$$f(M,B) = f(M|B) \cdot f(B) \rightarrow f(M|B) = \frac{f(M,B)}{f(B)} = \frac{0.4}{0.6} = \frac{2}{3}$$

and of being a woman, if she is British:

$$f(W,B) = f(W|B) \cdot f(B) \rightarrow f(W|B) = \frac{f(W,B)}{f(B)} = \frac{0.2}{0.6} = \frac{1}{3}$$

so that the sum of the probabilities is 2/3 + 1/3 = 1.

The income of the British will be:

$$\text{British income} = 36 \cdot \frac{2}{3} + 30 \cdot \frac{1}{3} = 34$$

and doing the same with the Spanish:

$$\text{Spanish income} = 26 \cdot \frac{0.3}{0.3+0.1} + 20 \cdot \frac{0.1}{0.3+0.1} = 24.5$$

We can make a new table of incomes for British and Spanish. Note that in this new table (Table 5.2), the 'sex effect' has disappeared as there is no longer a classification by men and women, so sex has been *marginalized*.

Table 5.2. Annual income (thousands of euros) and proportion of individuals in two groups of Spanish and British people.

British (*B*)	Spanish (*S*)
34 (60%)	24.5 (40%)

Now we will see an example with the heritability of a trait. The additive value in the case of individual selection is:

$$a = h^2 \cdot y + e$$

We have two unknown parameters, a and h^2. The uncertainty about these parameters is represented using the joint density function $f(a, h^2|y)$. Our objective is to find the uncertainty of the additive value using the density function $f(a|y)$, we would like to *marginalize* heritability. We can apply the probability rules of section 5.1.1, adding C as an event that is always given:

$$P(A,B|C) = P(A|B,C)P(B|C)$$

which applied to heritability is:

$$f(a,h^2|y) = f(a|h^2,y)f(h^2|y)$$

We always make the inference 'given **y**', because all our inferences are conditioned to our data **y**; here, we do not imagine that we repeat the experiment infinitely many times as in the case of classical statistics. Suppose, for this example, that, given our data, the heritability can only take two values, 0.1 and 0.2. In this case, the density of the additive value would be:

$$f(a|\mathbf{y}) = f(a|\mathbf{y},\ h^2 = 0.1) \cdot P(h^2 = 0.1|\mathbf{y}) + f(a\ |\ \mathbf{y},\ h^2 = 0.2,) \cdot P(h^2 = 0.2\ |\mathbf{y})$$

that is, the density of the data assuming that the heritability is 0.1 multiplied by the probability that this is so, plus the density assuming that the heritability is 0.2 multiplied by the probability that the heritability is 0.2. This function $f(a|y)$ no longer depends on the heritability, so we say that the heritability is *marginalized*.

Actually, heritability can take all the values from 0 to 1, so we should consider all possible cases. The density $f(a|y)$ is obtained by giving to the heritability all possible values, multiplying them by the probability of those values $f(h^2 | y) \cdot dh^2$ and adding all results. As the probability rectangles (see Fig. 5.7) are infinitesimal, we add the results of the multiplications by calculating the integral:

$$f(a|y) = \int_0^1 f(a, h^2 | y) dh^2 = \int_0^1 f(a | h^2, y) f(h^2 | y) dh^2$$

> Note that we have taken into account our uncertainty when estimating the heritability, as we have given heritability all possible values, multiplied by the probability that they have these values and summed these values. We have *marginalized* heritability. To take into account the uncertainty of the genetic parameters when estimating the genetic values is something that cannot be done in the context of classical statistics, so it is necessary to apply Bayesian inference.

In the case of the mixed model:

$$f(a|y) = \iiint f(b, a, \sigma_A^2, \sigma_e^2 | y) \cdot db \cdot d\sigma_A^2 \cdot d\sigma_e^2$$

where we give all possible values to the parameters to be marginalized, multiply by their probability and sum these values; that is, we predict the additive values $\boldsymbol{a}$ marginalizing the other unknown effects and parameters. Thus, the uncertainty about the additive values is obtained, taking into account the uncertainty about the other unknowns. The uncertainty about the other effects and parameters can be obtained analogously; for example, the uncertainty around the additive variance would be:

$$f(\sigma_A^2 | y) = \iiint f(b, a, \sigma_A^2, \sigma_e^2, | y) \cdot db \cdot da \cdot d\sigma_e^2$$

To obtain the uncertainty about a specific additive value, for example the first additive value a_1, we should marginalize with respect to all other additive values as well:

$$f(a_1 | y) = \int \cdots \int f(b, a_2, a_3 \ldots a_m, \sigma_A^2, \sigma_e^2 | y) \cdot db \cdot da_2 \cdot da_3 \cdot \ldots \cdot da_m \cdot d\sigma_A^2 \cdot d\sigma_e^2$$

The need to do all these integrals, which in the case of genetic values of dairy cattle can be millions of integrals, has prevented the application of Bayesian methods. However, the development of computers and the use of numerical methods has allowed us to obtain the marginal densities approximately without solving the integrals. These methods are known as Markov Chain Monte Carlo.

5.2.3 Markov Chain Monte Carlo (MCMC) methods

Markov Chain Monte Carlo methods provide a set of random samples drawn from marginal posterior densities; thus, we do not obtain the mathematical expression of the marginal densities. The first use of random samples from a density function was proposed by someone calling himself simple 'Student', a statistician from the Guinness brewery, in his famous article in which he presented the t distribution. The use of so-called 'Markov chains' to find these random samples has its origins in the work carried out in Los Alamos at the end of World War II to solve problems related to nuclear weapons, taking advantage of the appearance of the first computer. Later, in 1990, Alan Gelfand and Adrian Smith introduced this technique to the statistical world to obtain probability distributions, and Daniel Gianola and Daniel Sorensen brought these techniques to the field of animal breeding (see McGrayne, 2011, for a short history of MCMC, and Sorensen and Gianola, 2002, for a detailed description of the method).

We know how to extract random samples from the most common probability density functions. For example, in order to extract a random number from a standardized normal distribution (mean 0 and variance 1), we extract two random numbers, x_1 and x_2, between 0 and 1 with the computer's random number generator, and perform the operation:

$$y = \sqrt{-2\log x_1} \cdot \cos(2\pi x_2)$$

The number y obtained is a random number drawn from a normal distribution. This means that if we represent the samples x extracted from the random number generator, we will obtain a horizontal straight line, because all numbers between 0 and 1 have the same probability of being extracted, but if we represent the transformed samples y, we will obtain a histogram that approximates a normal distribution such as the one in Fig. 5.7.

By extracting random samples from a posterior density function, we can make all kinds of inferences. For example, we want to estimate the response to selection for intramuscular fat in pork loin by comparing a selected population S with a control population C. Let us take a simple model:

$$y = \mu + T + e$$

where T is the treatment, with two levels: S and C. Suppose that we already know the posterior probability densities $f(\mu+S|\mathbf{y})$ and $f(\mu+C|\mathbf{y})$, and that we know how to extract random samples from these functions, which usually are normal distributions. We want to make inferences from the posterior density function $f[(\mu + S) - (\mu + C)|\mathbf{y})] = f(S - C)|\mathbf{y})$ where $\mathbf{y}$ is our data vector. To do this, we take random samples of the densities, creating two chains:

$$f(\mu + S|\mathbf{y}) : [3.1, 3.3, 4.9, 4.8, 4.1, \ldots]$$
$$f(\mu + C|\mathbf{y}) : [2.8, 3.4, 2.0, 2.6, 2.6, \ldots]$$

As they are random samples of the posterior densities, their difference sample by sample (i.e. 3.1 – 2.8 = 0.3; 3.3 – 3.4 = –0.1; 4.9 – 2.0 = 2.9; 4.8 – 2.6 = 2.2; 4.1 – 2.6 = 1.5, etc.) will also be a chain of random samples from the difference function:

$$f(S - C|\mathbf{y}) : [0.3, -0.1, 2.9, 2.2, 1.5, \ldots]$$

with which we can make a histogram and proceed to make inferences. We can make inferences not only about S – C but also about other combinations of unknowns; for example, we could be interested in knowing in what proportion the selected population exceeds the control, and make inferences about the ratio instead of about the difference: 3.1/2.8 = 1.1; 3.3/3.4 = 0.97; 4.9/2.0 = 2.45; 4.8/2.6 = 1.85; 4.1/2.6 = 1.58, etc.:

$$f\left(\frac{\mu + S}{\mu + C}|\mathbf{y}\right) : [1.10, 0.97, 2.45, 1.85, 1.58, \ldots]$$

EXAMPLE 5.1

Here, we give an example of how to make inferences from a chain. We have a chain of random samples from the posterior density function. We order it from smallest to largest, in order to facilitate inferences:

$f(S - C|\mathbf{y})$: [−0.5, −0.4, −0.2, −0.1, −0.1, 0.0, 0.0, 0.1, 0.2, 0.2, 0.2, 0.2, 0.5, 0.6, 0.6,0.7, 0.7, 1.1, 1.1, 1.3, 1.5, 1.8, 1.8, 1.8, 1.8, 2.0, 2.0, 2.1, 2.2, 2.4]

- *What is the probability that there was a response?* This is the probability that $S > C$, or the probability that $S - C > 0$, so we look at the proportion of samples in the chain that are greater than zero. There are 23 samples out of 30 that are greater than zero; which is 77%, so the probability that there has been a response is 0.77.
- *What is the probability that the response is greater than a value that we consider relevant?* Before doing the experiment, we can specify for economic or biological reasons that a response of 0.5 in intramuscular fat is the minimum value that we would consider relevant in our example. Now we can calculate the probability that the response is, indeed, relevant. To do this, we calculate the proportion of samples in the chain equal to or greater than 0.5; we see that there are 18, so the probability that the response is relevant, *the probability of relevance*, is 18/30 = 0.60.
- *What is the probability that the response is zero?* To answer this question, it is not necessary to do an experiment: the response will always be different from zero because it will never be equal to 0.000000... Rephrasing the question correctly, what is the probability that, for practical purposes, there has been no response? Or *what is the probability that the response is irrelevant?* To do this, we will count the points that are less than the relevant value, and we will also take the possibility that $C > S$ into account, so we will count the number of points between −0.5 and 0.5, which is 11. The probability that the response is irrelevant is 11/30 = 0.37. This is the *probability of similarity* between the two groups C and S; that is, the probability that both groups are the same for practical purposes.
- *What is the value of the response that we can guarantee with a probability of 70%?* As we have a total of 30 samples, 70% of them is 0.7 × 30 = 21 samples. We take the last 21 samples, and the value of the smallest of them is 0.2, so the *guaranteed value at 70% probability* is 0.2; we guarantee that at least the response has been 0.2 with a 70% probability.
- *What is the shortest confidence interval at 90% probability (90% HPD)?* This interval contains 0.9 × 30 = 27 samples. Let us take all the intervals with 27 samples; we have four intervals: [−0.5, 2.0], [−0.4, 2.1], [−0.2, 2.2] and [−0.1, 2.4]. The first has a length of 2.0 − (−0.5) = 2.5, the second of 2.1 − (−0.4) = 2.5, the third of 2.2 − (−0.2) = 2.4, and the fourth of 2.4 − (−0.1) = 2.5; therefore, we take the third interval and say that the *shortest interval with 90% probability* is [−0.2, 2.2].
- *Give an estimate of the response to the selection.* Often, the mean of the chain is given as an estimate. In our case, the mean value of the chain is 0.86, but the median (the central value of the chain) is also frequently given, which in our case, being an even number of samples, would be the mean of the two central values 0.6 and 0.7; that is, the median would be 0.65.

The median has some good properties; for example, suppose that we estimate the standard deviation of a population with a chain, and we create a new chain by squaring all samples in order to estimate the variance with the new chain. If we take the *medians* of both chains, we get estimates for the standard deviation and the variance, respectively. As the median is the central value of the chain, the estimate of the variance will be the square of the estimate of the median of the chain for the standard deviation, but this does not happen if we estimate the standard deviation and the variance with the *mean* of their respective chains; the mean of the square of the values in a chain does not match the mean of the values squared. If we estimate the standard deviation as the mean of the values in its chain and we square it, it will not coincide with the estimate of the variance obtained by taking the mean of the chain for the variance.

In standard statistical inference from samples, the median is used less than the mean because its standard error is larger; however, as we can make the chains as long as we wish, this point is irrelevant here. Both estimates, the mean and the median, are acceptable. Sometimes, the mode is given (the most probable value), but this estimate does not have good statistical properties (Blasco, 2017, p. 43), and it is rarely used when the mean or the median are available; it was popular when functions could not be marginalized or integrated, because it was enough to find the maximum of the function for obtaining the mode.

Note that the shortest confidence interval is calculated independently of the point estimate (mean, median or mode), so that any of the three estimates can be offered with the same confidence interval, which is not necessarily symmetrical around the estimated value.

Note that when we give the probability that $S > C$, or that $S - C$ is 'zero for practical purposes', *we are not doing a hypothesis test* but rather calculating a probability. This way of proceeding is more convenient than performing hypothesis tests, as it shows precisely our uncertainty about our estimate, in this case the uncertainty about the response to the selection.

The use of the word 'sample' to refer to the random samples of the posterior density function contained in the chain gives rise to confusion, as it seems that increasing the number of samples in the chain increases the precision of the estimation. Note, however, that increasing the number of samples in the chain does not improve the precision of the estimate to be made from the chain. The precision of the estimate depends on the size of the experimental sample, so a small sample will result in wide densities that will have larger confidence intervals (Fig. 5.6). The objective of the chain is to estimate the posterior density, so that a histogram made with a short chain (with few values; i.e. with few 'random samples' of the posterior density) would look like that of Fig. 5.7, while a long chain (with many 'samples') could have much narrower rectangles and better fit the shape of the curve. The precision of an estimate is given by the form of the posterior density, so that a density such as that in Fig. 5.6.A obtained with few data points will give an estimate of heritability with confidence intervals much larger than those of the density in Fig. 5.6B made with more data points, although both densities can be estimated with MCMC chains of the same length. Here, long or short chains simply represent better or worse the densities. Of course, a chain with too few 'samples' will not estimate the posterior density well, and its mean, median and confidence intervals will have a larger error; this supplementary error is called the *Monte Carlo error*. In practice, this is not serious, because if the Monte Carlo error is large, we simply have to make the chain longer to reduce it, and the only cost is that of in terms of computing.

> MCMC provides chains with random samples of the posterior marginal densities of each parameter or unknown effect, from which Bayesian inferences can be made.

Depending on the complexity of the problem, very long chains may be needed. For fixed effect estimation, a few thousand samples per chain may suffice, but when estimating multivariate genetic parameters in non-linear models, chains with hundreds of thousands or even millions of samples can be needed. The reason is that, in complex models, the consecutive samples of the chains are usually highly correlated, sampling the same area of the density function, and thus longer chains are needed to sample the whole marginal posterior distribution.

The MCMC techniques to obtain these chains (e.g. Gibbs sampling, Metropolis–Hastings) are numerical analysis techniques that fall outside the scope of a book on animal breeding and we will not detail them; we will simply indicate the available software used in animal genetic improvement. The reader can consult Sorensen and Gianola (2002) for details, and for an intuitive introduction, Blasco (2001, 2017).

It is not correct to say that a genetic parameter has been estimated or an additive value has been predicted 'by MCMC' or 'by Gibbs sampling'. The estimation or prediction has been made from the *posterior density* functions, and it is the mean or median of these functions that has been used for the statistical inference; MCMC and Gibbs sampling are mere numerical analysis methods.

EXAMPLE 5.2

The results of a Bayesian analysis with MCMC on a feed conversion index for broilers are shown in Table 5.3. Data were analysed with the mixed model of section 4.3, obtaining chains for all unknown parameters and effects. We are not interested in the fixed effects, but we are interested in additive values and genetic parameters. We obtain one chain for each additive value, and two more chains, one for the additive variance and one for the residual variance.

Table 5.3. Chains of samples of posterior marginal densities. Chains for a_1, a_2, ..., a_n, σ_A^2 and σ_e^2 have been obtained by MCMC; chains for σ_P^2, h^2 and the response, *R*, have been obtained by combining the MCMC chains.

Sample	a_1	a_2	...	a_n	σ_A^2	σ_e^2	σ_P^2	h^2	R
1	0.30	−0.05		0.20	0.01	0.05	0.06	0.17	0.41
2	0.05	−0.10		0.22	0.04	0.07	0.11	0.36	0.35
3	0.12	0.12		0.15	0.06	0.12	0.18	0.33	0.29
4	−0.01	−0.35		0.35	0.06	0.09	0.12	0.50	0.31
...	...	...		...	...	...	...	...	...

Each element in the chain for σ_A^2 is a random sample of the posterior marginal probability density of σ_A^2, and the same is true of the chain for σ_e^2. As the phenotypic variance is the sum of the additive and the residual variances, the sum of the value of σ_A^2 obtained in the first sample (0.01) and that of σ_e^2 obtained in the same sample (0.05) is also a random sample of the posterior marginal density of σ_P^2 (in this case, 0.01 + 0.05 = 0.06). We can repeat the sum for the remaining samples and create a chain for σ_P^2. Arguing in the same way, the ratio between sample 1 of σ_A^2 and sample 1 of σ_P^2 in the new chain will be a random sample of the posterior marginal density of heritability (in this case, 0.01/0.06 = 0.17), and we can therefore have a chain for heritability.

If the population has been selected, we can calculate the mean of the additive values of the last generation of selection in sample 1, which constitutes a random sample of the marginal posterior density of the response *R*, and repeating the process in each sample, we will get a chain for the *R*.

Thus, Table 5.3 shows the MCMC chains and chain combinations to estimate the phenotypic variance, heritability and response.

Using the chains, we can make inferences like those we made in the previous example: for example, we can estimate the response with the median or the mean of the chain and its precision with its shortest 95% confidence interval, we can calculate the probability of this response being greater than zero, we can give a guaranteed value of the response with a probability of 80%, and we can state whether the response has been relevant by estimating the probability exceeding a predetermined relevant value. Similarly, we can estimate the heritability with the mean or the median of the chain, calculate its 95% HPD, its guaranteed value with a given probability, the probability that the heritability is relevant, etc.

5.2.4 BLUP and REML considered as Bayesian methods

At the end of section 5.2.1, there is the joint posterior density function of all the unknown effects and parameters. If we do not estimate the variance components and assume that we know the additive and error variances, the posterior density is conditioned not only on the data but also on these variance components:

$$f\left(\boldsymbol{b},\boldsymbol{a}\,|\,\boldsymbol{y},\sigma_A^2,\sigma_e^2\right)\propto\frac{1}{\left(\sigma_e^2\right)^{\frac{n}{2}}}\exp\left[-\frac{1}{2\sigma_e^2}(\boldsymbol{y}-\boldsymbol{X}\boldsymbol{b}-\boldsymbol{Z}\boldsymbol{a})'(\boldsymbol{y}-\boldsymbol{X}\boldsymbol{b}-\boldsymbol{Z}\boldsymbol{a})\right]\cdot\frac{1}{\left(\sigma_A^2\right)^{\frac{m}{2}}}\exp\left(-\frac{1}{2\sigma_A^2}\boldsymbol{a}'\boldsymbol{A}^{-1}\boldsymbol{a}\right)$$

This is a function of $\boldsymbol{b}$ and $\boldsymbol{a}$ only, as the data and the variances are 'given' (i.e. fixed). Finding the mode (the maximum) of the function, we would have a prediction of $\boldsymbol{b}$ and $\boldsymbol{a}$ with maximum probability, and we can see that the equations for finding the maximum are the same as the mixed model equations that we saw in section 4.3 (see Appendix 5.1). BLUP is, therefore, the maximum of the joint posterior density function conditional on the additive and residual variances, when flat priors have been used for the fixed effects and a prior normal density with zero mean and $\boldsymbol{A}\sigma_A^2$ variance for the additive effects (see section 5.2.1). This joint posterior density function is conditioned to the data (as it always is in Bayesian inference), and the residuals have their usual prior density, $N(\boldsymbol{0}, \boldsymbol{I}\sigma_e^2)$.

This interpretation of BLUP is interesting because it specifies the nature of the predictor and its properties much better than in the frequentist case. From the frequentist point of view, for the BLUP to be unbiased, it is first necessary to use an ad hoc definition of bias (see section 4.2), but it is also necessary to use the *true* genetic parameters, which, as they are not known, do not allow us to ensure that the estimate is unbiased and that it is the 'best' (the one with minimum error variance) among the unbiased predictors. From a Bayesian point of view, there is no bias because we do not consider what would happen if the experiment was repeated infinitely many times; the Bayesian prediction is the one with maximum probability. Furthermore, the parameters do not need to be the true ones but rather parameters that the geneticist prefers to use when conditioning them in the posterior density function. Finally, the fixed effects do not have a different nature from the random ones; they are simply effects that have a constant prior distribution rather than a normal one.

Regarding the estimation of variance components, it can be shown (Blasco, 2001, 2017) that the maximum of the joint posterior marginal density of the variance components conditional on the data:

$$f\left(\sigma_A^2, \sigma_e^2, |y\right)$$

is the same as the restricted maximum likelihood estimate (REML) of section 4.4.2. Note that effects $\boldsymbol{a}$ and $\boldsymbol{b}$ do not appear in this function because they have been *marginalized*. Note also that the Bayesian interpretation of REML justifies the method better than the frequentist interpretation. From a frequentist point of view, REML is estimated in a restricted space, using transformed data that are always less than the original data, which implies a loss of degrees of freedom and can imply a loss of information. In fact, in the estimation of the variance of the simplest model, in which the data are distributed $N(\mu, \sigma^2)$, it can be shown that the REML estimate has a quadratic risk greater than the maximum likelihood estimate (see Blasco, 2017, for details). REML also has the same problems as maximum likelihood estimates: its good frequentist properties (being unbiased, having minimum variance) are asymptotic. From a Bayesian point of view, it is very reasonable to marginalize all effects and use the marginal function for inferences.

Note that both BLUP and REML are not optimal Bayesian estimates, as a Bayesian geneticist would normally use the posterior marginal density functions of each unknown; that is, $f(\boldsymbol{a}|\boldsymbol{y})$ in the case of BLUP and $f\left(\sigma_A^2|y\right)$ and $f\left(\sigma_e^2|y\right)$ in the case of REML.

5.3 The Multi-trait Model

We have seen the 'mixed model' for one trait in section 4.1.2; now we will see the model for several traits, in which we will predict the additive values of the animals for each trait and estimate the fixed effects and genetic parameters (heritabilities and correlations). Remember that in a Bayesian context, all unknowns are predicted, as all of them are random and have an associated probability, but we often still use the classical nomenclature of 'estimation' and 'prediction', common in animal breeding.

5.3.1 Traits having the same model

We will first address the case in which two traits such as feed conversion ratio (FCR) and weight at a certain age (W), are measured only once and simultaneously in each individual, so that the incidence matrices of the mixed model $\mathbf{X}$ and $\mathbf{Z}$ are the same for both traits. The model is:

$$\boldsymbol{y}_{FCR} = \boldsymbol{X}\boldsymbol{b}_{FCR} + \boldsymbol{Z}\boldsymbol{a}_{FCR} + \boldsymbol{e}_{FCR}$$
$$\boldsymbol{y}_{W} = \boldsymbol{X}\boldsymbol{b}_{W} + \boldsymbol{Z}\boldsymbol{a}_{W} + \boldsymbol{e}_{W}$$

which can be expressed as in section 4.1.2 as:

$$\boldsymbol{y} = \boldsymbol{X}\boldsymbol{b} + \boldsymbol{Z}\boldsymbol{a} + \boldsymbol{e}$$

where $\boldsymbol{y}' = \left[\boldsymbol{y}'_{FCR}, \boldsymbol{y}'_{W}\right]$; $\boldsymbol{b}' = \left[\boldsymbol{b}'_{FCR}, \boldsymbol{b}'_{W}\right]$; $\boldsymbol{a}' = \left[\boldsymbol{a}'_{FCR}, \boldsymbol{a}'_{W}\right]$; $\boldsymbol{e}' = \left[\boldsymbol{e}'_{FCR}, \boldsymbol{e}'_{W}\right]$.

The a priori information on the 'fixed' effects b is, as in section 5.2.1, constant, but the variance–covariance matrices of the random effects of both traits have to be expressed together because they are correlated. Sorting the data by animal:

$$a' = \left[a_1^{FCR}, a_1^{W}, a_2^{FCR}, a_2^{W}, a_3^{FCR}, a_3^{W}, \ldots\right]; \quad e' = \left[e_1^{FCR}, e_1^{W}, e_2^{FCR}, e_2^{W}, e_3^{FCR}, e_3^{W}, \ldots\right]$$

$$\operatorname{var}(e) = \begin{bmatrix} \sigma^2_{e_{FCR}} & \sigma_{e_{FCR}e_W} & 0 & 0 & \cdots & \cdots \\ \sigma_{e_{FCR}e_W} & \sigma^2_{e_W} & 0 & 0 & \cdots & \cdots \\ 0 & 0 & \sigma^2_{e_{FCR}} & \sigma_{e_{FCR}e_W} & \cdots & \cdots \\ 0 & 0 & \sigma_{e_{FCR}e_W} & \sigma^2_{e_W} & \cdots & \cdots \\ \cdots & \cdots & \cdots & \cdots & \cdots & \cdots \\ \cdots & \cdots & \cdots & \cdots & \cdots & \cdots \end{bmatrix} = I \otimes R$$

where R is the 2 × 2 variance–covariance matrix of the residuals of the traits, I is the identity matrix and $\otimes$ is what is known in mathematics as the 'Kroenecker product'. For additive effects:

$$\operatorname{var}(a) = \begin{bmatrix} \sigma^2_{A_{FCR}} & \sigma_{A_{FCR}A_W} & 2r_{12}\sigma^2_{A_{FCR}} & 2r_{12}\sigma_{A_{FCR}A_W} & \cdots & \cdots \\ \sigma_{A_{FCR}A_W} & \sigma^2_{A_W} & 2r_{12}\sigma_{A_{FCR}A_W} & 2r_{12}\sigma^2_{A_W} & \cdots & \cdots \\ 2r_{12}\sigma^2_{A_{FCR}} & 2r_{12}\sigma_{A_{FCR}A_W} & \sigma^2_{A_{FCR}} & \sigma_{A_{FCR}A_W} & \cdots & \cdots \\ 2r_{12}\sigma_{A_{FCR}A_W} & 2r_{12}\sigma^2_{A_W} & \sigma_{A_{FCR}A_W} & \sigma^2_{A_W} & \cdots & \cdots \\ \cdots & \cdots & \cdots & \cdots & \cdots & \cdots \\ \cdots & \cdots & \cdots & \cdots & \cdots & \cdots \end{bmatrix} = A \otimes G$$

where G is the 2 × 2 variance–covariance matrix of the additive values, r_{ij} are the relationship coefficients between the individuals and A is the relationship matrix.

The prediction of the additive values and the genetic parameters is carried out as in section 5.2.1. The joint density function is:

$$f(b,a,G,R \mid y) \propto f(y \mid b,a,R) \cdot f(a \mid A,G)$$
$$\propto \exp\left[-\frac{1}{2}(y - Xb - Za)'(I \otimes R)^{-1}(y - Xb - Za)\right] \cdot \exp\left(-\frac{1}{2}a'(A \otimes G)^{-1} a\right)$$

and chains of random samples are obtained by MCMC for the marginal densities:

$$f(b|y), f(a|y), f(G|y), f(R|y)$$

Thus, we obtain chains of all the unknowns with which we can build new chains to estimate heritabilities, the genetic and environmental correlations, responses to selection, etc.

EXAMPLE 5.3

As in the previous example, we will measure feed conversion ratio (FCR) and weight (W) in broilers and we will obtain MCMC chains for the additive values of the two traits, for the additive and residual variances of both traits, and for the additive and residual covariances between the two traits. Combining these chains, we can find a chain for their genetic correlation, and also for the heritabilities, responses to selection, etc. An example of chains used to estimate the genetic correlation between the two traits is given in Table 5.4.

Table 5.4. Chains of samples of posterior marginal densities. Chains of additive variances and covariance have been obtained by MCMC. Chains for standard deviations and correlation have been obtained by combining the former MCMC chains.

Sample	$\sigma_A^2(FCR)$	$\sigma_A^2(W)$	cov (A_{FCR}, A_W)	$\sigma_A(FCR)$	$\sigma_A(W)$	$r(A_{FCR}, A_W)$
1	0.01	13	−0.09	0.10	3.6	−0.25
2	0.04	15	−0.16	0.20	3.9	−0.20
3	0.06	12	−0.22	0.24	3.5	−0.26
4	0.06	10	−0.17	0.24	3.2	−0.22
...	...	...		...	...	...

As in the previous example, we can estimate the genetic correlation from the MCMC chains using the median or mean of the chain and its 95% confidence interval. We can propose that the genetic correlation is irrelevant if it is between −0.1 and 0.1 (i.e. it is zero for practical purposes), and we can calculate the *probability of similarity*, as in Fig. 5.3B. We can also give a *guaranteed value* of the correlation with a probability of, for example, 80%. We can also calculate the *probability of the correlation being relevant* by calculating the probability that the correlation is lower than the relevant value of −0.1. We can also calculate any other confidence intervals.

5.3.2 Data augmentation

In the previous model, the **X** and **Z** design matrices were the same for both traits. However, sometimes this is not the case. For example, if we analyse litter size and feed conversion rate, or milk production (M) and weight at a certain age (W), in both cases, one trait has several records in the same female and the other trait has a single one; furthermore, the 'fixed' effects (e.g. herd–year–season) are not always coincident. In this case, the model is:

$$
\begin{aligned}
\mathbf{y}_M &= \mathbf{X}_M\mathbf{b}_M + \mathbf{Z}_M\mathbf{a}_M + \mathbf{W}_M\mathbf{p}_M + \mathbf{e}_M \\
\mathbf{y}_W &= \mathbf{X}_W\mathbf{b}_W + \mathbf{Z}_W\mathbf{a}_W + \mathbf{e}_W
\end{aligned}
$$

where the first trait has permanent effects, as in the model in section 4.1.2, and the second does not, and it is not possible to describe a single model for both traits as we have done before in section 5.3.1. In order to describe a single model, each cow should have several weight records, as many as the maximum number of lactations in the population, and have them with the same fixed effect levels (e.g. herd–year–season) in which milk production is recorded, so that we could have the same model for the two traits, as in section 5.3.1. This is not possible, but we can generate 'invented' data that allows us to apply a single model but that do not take part in the estimation. This is accomplished through an ingenious procedure known as data augmentation. These new 'invented' data (which we will call 'augmented data') have to meet two conditions:

1. There should be enough data z (*augmented data*) to be able to write a single model for the two traits.
2. The new data z should not take part in the prediction. The prediction should be done only with the data of the records y.

In our example, we only need to *augment* the weight data until we have several weights for each cow. The new data vector with the augmented data z'_W would be:

$$y'^A_W = \left[y'_W, z'_W \right]$$

Thus, the new model will be:

$$y^A = Xb + Za + Wp + e$$

where:

$$y'^A = \left[y'_M, y'^A_W \right] = \left[y'_M, y'_W, z'_W \right] = [y', z']$$

With this data vector, we can now estimate the genetic parameters and predict the additive values, as we have done previously. The augmented data cannot be arbitrary; they are unknown data that should be estimated in order to meet the two conditions that we have set. In this example, we have only augmented the weight data, but there are examples where the data for both traits should be augmented, such as if certain cows have no records for weight. In this case, we would have an augmented data vector $z' = \left[z'_M, z'_W \right]$ with augmented data for both traits.

If $\boldsymbol{\theta}$ is all the parameters and effects to be predicted, then:

$$\boldsymbol{\theta}' = \left[b', a', \sigma^2_{A_W}, \sigma^2_{A_M}, \mathrm{cov}(A_W, A_M), \sigma^2_{e_W}, \sigma^2_{e_M}, \mathrm{cov}(e_W, e_M) \right]$$

and we can predict them together with the augmented data z using the posterior density function:

$$f(z, \theta | y) \propto f(y | z, \theta) \cdot f(z, \theta) = f(y | z, \theta) \cdot f(z | \theta) \cdot f(\theta)$$

We have seen in section 5.2.2 that $P(A, B|C) = P(A|B, C) \cdot P(B|C)$; therefore:

$$f(y, z | \theta) = f(y | z, \theta) \cdot f(z | \theta)$$

Substituting in the previous formula, we have

$$f(z, \theta | y) \propto f(y, z, | \theta) \cdot f(\theta) = f\left(y^A | \theta\right) \cdot f(\theta)$$

With the augmented data vector y^A, we can now use a single model and predict the additive values and genetic parameters by MCMC chains as we have done before. In the same process, we will obtain chains for the augmented data z, but we will discard them because they are not necessary; the augmented data z are only used to obtain the vector y^A that allows us to apply a single model for both traits.

Note that the augmented data **z** do not take part in the prediction of the genetic parameters and additive values, because the vector of parameters and effects $\boldsymbol{\theta}$ is predicted by the posterior density $f(\mathbf{z}, \boldsymbol{\theta}|\mathbf{y})$, which is *conditional only on the observed data* **y**, not on the augmented data **z**. Augmented data is just a mathematical device that allows transformation of two models with different effects and design matrices in a single model, in order to simplify the analysis.

Appendix 5.1

To find the maximum of the joint posterior density conditional on the variance components, we will take logarithms, as the maximum of a function and its logarithm are in the same point:

$$\log f(b,a|y,\sigma_A^2,\sigma_e^2) \propto \frac{1}{\sigma_e^2}(y-Xb-Za)'(y-Xb-Za)+\frac{1}{\sigma_A^2}a'A^{-1}a$$

$$\frac{\partial}{\partial b}\log f(b,a|y,\sigma_A^2,\sigma_e^2) \propto -\frac{2}{\sigma_e^2}X'(y-Xb-Za)=0$$

$$\frac{\partial}{\partial a}\log f(b,a|y,\sigma_A^2,\sigma_e^2) \propto -\frac{2}{\sigma_e^2}Z'(y-Xb-Za)+\frac{2}{\sigma_A^2}A^{-1}a=0$$

$$\begin{aligned} &X'X\hat{b}+X'Z\hat{a}=X'y \\ &Z'X\hat{b}+Z'Z\hat{a}+A^{-1}\frac{\sigma_e^2}{\sigma_A^2}\hat{a}=Z'y \end{aligned} \rightarrow \begin{bmatrix} X'X & X'Z \\ Z'X & Z'Z+A^{-1}\frac{\sigma_e^2}{\sigma_A^2} \end{bmatrix}\begin{bmatrix}\hat{b}\\ \hat{a}\end{bmatrix}=\begin{bmatrix}X'y\\ Z'y\end{bmatrix}$$

Appendix 5.2

One of the hypotheses of the mixed model, as we saw in section 4.1.2, is that the expectations of all additive values are zero, $E(a) = 0$. However, if there is selection for the trait, the mean of the additive values in the selected generations should be greater than zero if the selection in favour of the trait has been successful. We ignore this fact when predicting additive values and we use the same mixed model equations derived under the hypothesis that there is no selection. Unexpectedly, there are some cases in which we can ignore the fact that data are selected and predict the additive values as if there was no selection. Selection is *ignorable* if we include in the prediction or in the estimation of genetic parameters all the data used in the selection and we use the correct model (with the full relationship matrix).

EXAMPLE 5.4

Let us consider the case of a selection generation, where:

- y_0 is data from generation 0, where data are not selected;
- y_1 is data from generation 1, data of the offspring of the selected parents from generation 0;

- c is the selection threshold (as in Fig. 2.3); and
- θ is the unknown that we have to estimate or predict, which can be a single unknown or a set of unknowns (as in section 5.3.2).

Bayesian interpretation

If we want to infer using the data from generation 1, the posterior density, taking into account that the data are selected, is different from when the data is not selected:

$$f_S(\theta|y_1) \neq f(\theta|y_1)$$

where f_S is the density function under selection and f is the density when the data are not selected. These densities are not coincident because the density of selected data is generally defined in a more restricted space. To better understand the consequences, we will return to Table 5.1, in which there is a probability distribution with four cases: British men (BM), British women (BW), Spanish men (SM) and Spanish women (SW). In Table 5.5, we have their frequencies.

Table 5.5. Proportion of individuals in a group of Spanish and British, separating men and women.

	British (*B*)	Spanish (*S*)
Men (*M*)	40%	30%
Women (*W*)	20%	10%

Here the density is a discrete probability distribution, and we would represent it as:

$$f(BM, BW, SM, SW) = (0.4, 0.2, 0.3, 0.1)$$

The sum of the probabilities is 0.4 + 0.2 + 0.3 + 0.1 = 1. Selecting only the British individuals, we have a new density function in a smaller space with two cases, but its probabilities also have to add up to 1.

The probability of being British is 0.4 + 0.2 = 0.6. As we saw in section 5.2.2, we can calculate the probability of being a man, given that he is British:

$$f(M,B) = f(M|B) \cdot f(B) \;\rightarrow\; f(M|B) = \frac{f(M,B)}{f(B)} = \frac{0.4}{0.6} = \frac{2}{3}$$

And we can do the same for the probability of being a woman, given that she is British:

$$f(W,B) = f(W|B) \cdot f(B) \;\rightarrow\; f(W|B) = \frac{f(W,B)}{f(B)} = \frac{0.2}{0.6} = \frac{1}{3}$$

If from Table 5.5 we select only British people, the new discrete density function $f_S(BM, BW)$ is:

$$f_S(BM, BW) = f(M|B, W|B) = \left(\frac{2}{3}, \frac{1}{3}\right)$$

Note that the sum of the probabilities is 2/3 + 1/3 = 1.

In our example, which is a continuous case, the probability of belonging to the restricted space of those that have been selected is:

$$P = P(y_0 > c) = \int_c^{+\infty} f(y_0 | y_0 > c) dy_0 = \int_{-\infty}^{+\infty}\int_c^{+\infty} f(y_1, y_0 | y_0 > c) dy_0\, dy_1$$

Reasoning as in the discrete case:

$$f_S(y_1, y_0) = f(y_1, y_0 | y_0 > c) = \frac{f(y_1, y_0)}{P}$$

The same happens with the joint density of the data and the parameters; the only difference between f_S and f is that f_S takes into account that the data were selected so that the density after selection is defined in a restricted space:

$$f_S(y_1, y_0, \theta) = f(y_1, y_0, \theta | y_0 > c) = \frac{f(y_1, y_0, \theta)}{P}$$

Note that the density function under selection f_S integrates to one into the restricted space in which it is defined; that is, the space in which the animals have been selected:

$$\int_{-\infty}^{+\infty}\int_c^{+\infty} f_S(y_1, y_0) dy_0 dy_1 = \frac{1}{P}\int_{-\infty}^{+\infty}\int_c^{+\infty} f(y_1, y_0) dy_0 dy_1 = \frac{1}{P} \cdot P = 1$$

Now we will find the marginal posterior density of the unknowns given all the data, selected or not. To do this, we consider first the laws of probability that we saw in section 5.1.1:

$$P(A,B) = P(B|A) \cdot P(A) = P(A|B) \cdot P(B)$$

for three events:

$$P(A,B,C) = P(A|B,C) \cdot P(B,C) \rightarrow P(A|B,C) = \frac{P(A,B,C)}{P(B,C)}$$

Applying this expression to the posterior density with all the data, we have:

$$f_S(\theta | y_1, y_0) = \frac{f_S(y_1, y_0, \theta)}{f_S(y_1, y_0)}$$

and we can see that the posterior density functions with or without selection are the same:

$$f_S(\theta | y_1, y_0) = \frac{f_S(y_1, y_0, \theta)}{f_S(y_1, y_0)} = \frac{\dfrac{f(y_1, y_0, \theta)}{P}}{\dfrac{f(y_1, y_0)}{P}} = \frac{f(y_1, y_0, \theta)}{f(y_1, y_0)} = f(\theta | y_1, y_0)$$

The model used should be the correct one. In the case of the mixed model, it is necessary to enter the complete relationship matrix that relates the data from generation 0 and 1.

Generally speaking, if the data set is defined in a space $\mathbb{R}$ and the selected data is in a subset $\mathbb{R}^c \subset \mathbb{R}$, the probability of belonging to that subset is:

$$P(y \in \mathbb{R}^c) = \int f(y) \cdot Ind(y \in \mathbb{R}^c) dy$$

where *Ind* is an indicator function of values 0 or 1, depending on whether or not the data belongs to the selected data subset. The density of the selected data is, as before:

$$f_S(y) = \frac{f(y)}{P}$$

and reasoning as before:

$$f_S(y,\theta) = \frac{f(y,\theta)}{P}$$

so, following the same steps, we can conclude, as before, that $f_S(\theta|y) = f(\theta|y)$.

Frequentist interpretation

The frequentist interpretation is based on the distribution of the prediction of the unknowns when taking samples if the experiment was repeated an infinite number of times. The convenient properties for an estimator in the frequentist frame are different from the convenient properties for a Bayesian estimator. In the frequentist case, BLUP is unbiased (taking into account the peculiar definition of bias for random effects that we saw in section 4.2); in the Bayesian case, bias is not considered because the experiment is not conceptually repeated, and the properties of the estimators are derived from the examination of the posterior densities conditioned in a single sample y, not in all possible samples if the experiment was repeated.

There has been much discussion about whether or not BLUP is unbiased when there is selection. The BLUP proponent, Henderson, attempted to set the conditions under which there would be no selection bias (Henderson, 1975), but his proposition was difficult to understand and has been questioned several times since then (e.g. Thompson, 1979; Gianola *et al.*, 1988). Recently, one of the most conspicuous frequentist geneticists who have developed the mixed model, Larry Schaeffer, argued that BLUP is biased by the use of farm and season effects as fixed rather than random effects; his criticism can be found in Schaeffer (2018).

In this section we will not go into this matter, which is formally a complex matter, and we will simply see an intuitive interpretation of what is the mean of the selected additive values if we repeated the experiment an infinite number of times, without entering into adjustments for fixed effects.

Suppose that we start in generation 0 with six unrelated individuals, whose mean additive value each is zero (i.e. if we repeat the experiment infinite times, the mean additive value of each of them, taking all the repetitions, is zero). In our example, we keep offspring only from the best two individuals, 1 and 2 (Fig. 5.10). Individuals 7–12 are all the offspring of individuals 1 and 2.

If we repeat the experiment keeping the same relationship matrix, individuals 7–12 of generation 1 will continue to be offspring of individuals 1 and 2. Suppose we repeat the experiment an infinitely number of times. In the first realization of the experiment, we have chosen individuals 1 and 2 for being superior, but in the following repetitions, individuals 1 and 2 will not necessarily be superior, because, as we have said, on average individuals

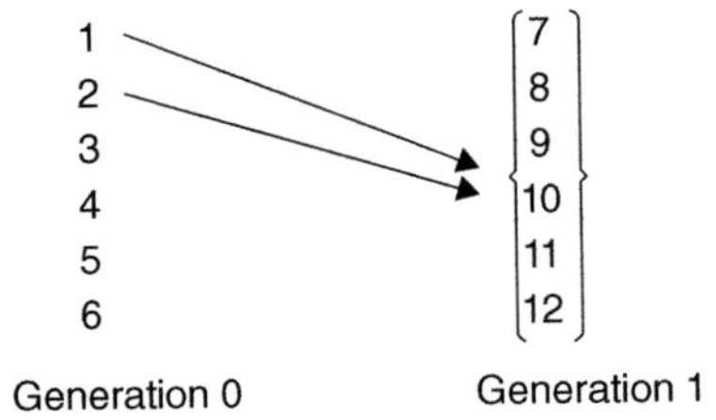

Fig. 5.10. Ignorable selection.

1 and 2 will have zero additive value after repeating the experiment an infinite number of times. Thus, in the first experiment, individuals 7–12 will have a mean greater than zero because 1 and 2 are superior, but in the other repetitions of the experiment, individuals 7–12 remain the offspring of 1 and 2, who no longer are necessarily superior individuals, so that after infinite repetitions individuals 7–12 will have a mean of zero for their additive value.

This is so because we have conditioned on the relationship matrix; it is always the same in each hypothetical repetition of the experiment. If we simply take the best individuals in each repetition, individuals 7–12 would no longer necessarily be the offspring of individuals 1 and 2. If the best animals in the next repetition are individuals 4 and 6, they could be the offspring of individuals 4 and 6, and generation 1 would be different, but as we have conditioned on the relationship matrix, this cannot happen, and individuals 7–12 will be always be the offspring of individuals 1 and 2.

Main Concepts

- *Bayes' theorem:* A theorem that allows us to infer the probability of unknown parameters of a population from a sample of that population and the a priori probability of these parameters.
- *A priori probability*: The probability of the unknown parameters before using the data sample in the inference.
- *A posteriori probability:* The probability of the unknown parameters obtained by combining the prior probability and the information from the data using Bayes' theorem.
- *Relevant value:* The minimum value of an unknown from which a decision is made. There are relevant values for parameters, effects or differences between effects even if the experiment is not performed; the relevant values depend on biological or economic characteristics related to the trait. When designing an experiment, the number of individuals required to obtain a certain precision is calculated based on the relevant value.
- *Guaranteed value:* The minimum (or maximum) value of a parameter or a difference of treatments that is guaranteed with a certain probability. We know that what we are estimating will have at least that value with that probability.
- *Probability of relevance:* The probability that an unknown (e.g. a parameter, a level of a treatment or a difference between levels) exceeds a relevant value.
- *Probability of similarity:* The probability that an unknown (e.g. a parameter, a level of a treatment or a difference between levels) is in absolute value less than a relevant value. If the probability of similarity is high, the value of the unknown is irrelevant and can be considered zero for all practical purposes.
- *Shortest confidence interval with a given probability:* The shortest of the confidence intervals that contain a given probability, usually 95%.

- ***Probability density function:*** A function in which the area contained between two values represents the probability that the unknown parameter or effect is between those two values.
- ***Marginalization:*** An operation by means of which effects or parameters of a model are eliminated considering all its possible values, multiplying by the probabilities of these values and summing up (or integrating, in the continuous case).
- ***Posterior marginal density function:*** A posterior density function of a single effect or parameter, once all the other effects and parameters of the model have been marginalized.
- ***MCMC (Markov Chain Monte Carlo):*** A numerical method that allows us to obtain random samples of the posterior marginal density functions of the unknown effects and parameters.
- ***Data augmentation:*** A technique that allows us to combine different models in a single model to estimate effects and parameters of both models. It consists of adding missing data until both models are the same, so that the resolution of the model is facilitated, but augmented data are not used in the prediction.

Problems

This problem is to determine the difference between the additive values of two bulls in terms of quantity of milk in kg. The relevant value of the difference is estimated at 100 kg. The actual chains for estimating the additive value of the bulls have hundreds of thousands of samples, but in our example, we will take two chains of only five samples. The chains for the bulls are:

Bull 1:	110	50	120	−50
Bull 2:	−110	−50	−120	50

5.1. Predict the difference between bulls with a mean and with a median.
5.2. Determine the probability that bull 1 has a higher additive value than bull 2.
5.3. Determine the probability P_R that the difference between bulls is equal to or greater than the relevant value (100 kg). Determine the probability of similarity P_S between bulls (probability that there is no difference between them, for practical purposes).
5.4. Determine the minimum value of the difference that can be guaranteed with an 80% probability.
5.5. Determine the shortest confidence interval with 80% probability (80% HPD).

Self-assessment Questions (True or False)

1. The prior probability is usually an arbitrary value.
2. The prior probability is usually a constant or a mathematical artefact (a function) that does not capture the true prior probability.
3. When the a priori probability is a mathematical artefact, it has little influence on the results, so that the results are essentially based on the data.
4. When the prior probability is a constant, the maximum likelihood results and those from the posterior marginal distributions are the same.

5. Using the data augmentation method, data are generated and used in the parameter estimation process, but this does not affect the result.
6. The Bayesian method for estimating genetic parameters is the Markov chain Monte Carlo (MCMC).
7. Marginalization takes into account the errors made when estimating the parameters that are marginalized; thus, when estimating the additive values, the estimation errors of the additive and residual variances have been taken into account.
8. BLUP can be interpreted as a Bayesian method.
9. REML gives the same result as the Bayesian estimate based on marginal posterior distributions.
10. By increasing the length of the MCMC chain (i.e. the number of samples in the chain), we increase the precision of the estimation.

References

Amrhein, V., Greenland, S. and McShane, B. (2019) Scientists rise up against statistical significance. *Nature* 567, 305–307. DOI: 10.1038/d41586-019-00857-9

Blasco, A. (2001) The Bayesian controversy in animal breeding. *Journal of Animal Science* 79, 2023–2046. DOI: 10.2527/2001.7982023x

Blasco, A. (2017) *Bayesian Data Analysis for Animal Scientists*. Springer, Cham, Switzerland.

Blasco, A., Sorensen, D. and Bidanel, J.P. (1998) A Bayesian analysis of genetic parameters and selection response for litter size components in pigs. *Genetics* 149, 301–306. DOI: 10.1093/genetics/149.1.301

Gianola, D., Im, S. and Fernando, R. L. (1988) Prediction of breeding values under Henderson's selection model: a revisitation. *Journal of Dairy Science* 71, 2790–2798. DOI: 10.3168/jds.S0022-0302(88)79873-2

Hald, A. (1998) *A History of Mathematical Statistics from 1750 to 1930*. Wiley, New York.

Henderson, C. R. (1975) Best linear unbiased estimation and prediction under a selection model. *Biometrics* 31, 423–448. DOI: 10.2307/2529430

Johnson, V.E. (2013) Revised standards for statistical evidence. *Proceedings of the National Academy of Sciences USA* 110, 19313–19317. DOI: 10.1073/pnas.1313476110

McGrayne, S.B. (2011) *The Theory That Would Not Die*. Yale University Press, New Haven, New Jersey.

Schaeffer, L.R. (2018) Necessary changes to improve animal models. *Journal of Animal Breeding and Genetics* 135, 124–131. DOI: 10.1111/jbg.12321

Sorensen, D. and Gianola, D. (2002) *Likelihood, Bayesian and MCMC Methods in Quantitative Genetics*. Springer, New York.

Sorensen, D., Fernando, R. and Gianola, D. (2001) Inferring the trajectory of genetic variance in the course of artificial selection. *Genetic Research* 77, 83–94. DOI: 10.1017/s0016672300004845

Thompson, R. (1979) Sire evaluation. *Biometrics* 35, 339–353. DOI: 10.2307/2529955

Wasserstein, R.L., Schirm, A.L. and Lazar, N.A. (2019) Moving to a World Beyond "p < 0.05". *The American Statistician* 73 (Suppl.1), 1–19. DOI: 10.1080/00031305.2019.1583913

Solutions

Solutions to the problems

5.1. The difference chain is 220 100 40 240 −100, which when ordered is −100 40 100 220 240. The mean is 100 and the median is also 100.

5.2. There are four positive numbers out of a total of five and therefore $P = 80\%$.
5.3. There are three numbers in the chain equal to or greater than 100, out of a total of five, and therefore $P_R = 60\%$. There is only one value between −100 and 100, so $P_S = 20\%$.
5.4. From the highest value, four numbers represent 80% probability, and therefore the guaranteed value is the last of these numbers, 40.
5.5. All intervals that have four numbers have 80% probability; these are [−100, 220] and [40, 240]. The first has a length of 320 and the second a length of 200, and therefore the interval [40, 240] is the shortest with an 80% probability.

Solutions to self-assessment questions

1. F, 2. T, 3. T, 4. F, 5. T, 6. F, 7. T, 8. T, 9. F, 10. F.

6 Genomic Selection

Abstract

This chapter describes the different genomic selection methods and discusses their efficacy and application in various livestock species. First, the linear model with molecular markers is presented. Then the use of genomic selection in the usual programmes based on the infinitesimal model is described, where traits are assumed to be governed by many genes of small effect each, including single-step genomic-BLUP, which integrates records of animals with or without molecular information. This is followed by a description of the genomic relationship matrix based on molecular information. Next, methods for detecting genes of main or intermediate effect are discussed, with an emphasis on what is known as the 'Bayesian alphabet'. The chapter concludes with a discussion of the practical use of the different models, describes the challenges of applying genomic selection in animal breeding programmes, and ends with a discussion of the current use of genomic selection in different livestock species.

6.1 The Model with Molecular Markers

The use of molecular markers to improve the efficiency of current selection programmes was first proposed in 1978 (Soller, 1978). At that time, few markers were available, and the expectation was to find some gene with a substantial effect linked to the marker and to increase its frequency in the population. Unfortunately, most of the production traits in livestock species are determined by a large number of genes with small effect, and the method was ineffective. When the detection of quantitative trait loci began in the 1990s and it was feasible to use more markers, it seemed that the frequencies of medium-effect alleles could be increased by marker-assisted selection and greater responses to selection could be obtained; however, there was a marked discrepancy between theoretical predictions made from simulation and the practice of marker-assisted selection, which produced relatively small improvements (Blasco, 2008).

Nowadays, the rapid development of DNA sequencing has made it possible to have a large number of markers at low cost. These markers can be associated with many of the genes controlling economically interesting traits, even if the effect of each gene is small. The simplest marker is the single-nucleotide polymorphism (SNP), which marks a location in the genome where there is variability in a single nucleotide. For example, at a specific point in the DNA chain, some animals may have the base adenine (A) or guanine (G), so in this position we have three possible genotypes, AA, AG and GG. Currently, the use of chips locating around 50,000–80,000 SNPs in a genome is common, and there are high-density chips with 850,000 SNPs or more.

The study of the structure and functions of the genome is called *genomics*. The prediction of the additive values of the animals, not only from phenotypic data but also from genomic data, has made it possible to increase the effectiveness of selection programmes from 20–30% in prolific species to more than 50% in dairy cattle, as we will see in Chapters 8 and 9. Predicting additive values for animals using genomics has two steps. First, a set of animals with phenotypic data and genomic data (SNPs) must be available to establish prediction equations that associate

DOI: 10.1079/9781836990994.0006

the values of the SNPs with the phenotypic data; later, for other animals, it is enough to specify their SNPs and use the prediction equations to predict their additive value, and it is therefore not essential for new individuals to have data. As there is recombination in each generation, these associations of SNPs with genes are lost, as we saw in sections 1.1.2 and 1.1.4; therefore, the equations have to be re-established every three or four generations, or even earlier. In practice, it is done continuously, dedicating a certain number of animals in each generation for recalculating the equations. Due to the great reduction in genotyping costs and the possibility of making imputations with low-density chips, as we will see in section 6.4.2, it is increasingly common to genotype as many animals as possible.

To establish the prediction equations with SNPs, we start from the data of a set of genotyped animals for which we know their SNPs and we have their production records – this is the so-called *reference population*. We have, for example, 4000 cows genotyped with a chip of 50,000 SNPs, for which milk production data are available. Next, we will associate these genetic markers with the production records. In the following example, we are going to consider one record per animal, ignoring fixed effects (we only include the general mean in the model). We will create a *variable* for each SNP, with an arbitrary value indicating whether the SNP *i* is 'homozygous' with two copies of a base, or 'heterozygous' (i.e. it has different bases) (Table 6.1).

Table 6.1. Genotypic values of a SNP.

AA	AG	GG
2	1	0

The values 2, 1 and 0 are arbitrary values; they could be substituted for other values (e.g. 1, 0, −1). The encoding is additive, and in this case, it relates to the number of copies (the allele *dosage*) of an arbitrary reference allele, allele A in this example. This sometimes has some consequences related to scale, but both encodings work well and both are widely used. We do not consider dominance effects, although more complex models may include this possibility.

We will use a simple and common model in which we assume that we have a sufficient number of SNPs well distributed in the genome, so that they capture the additive value of the trait through its statistical linkage with the genes that determine it, whether these genes have a small or large effect.

We will predict the genetic value of each cow in our example by multiple regression on the *variables* we have created. Suppose that the chip of 50,000 SNPs has the polymorphisms [A/G, C/G, A/T, ..., C/T] and the arbitrary reference alleles are [A, C, A, ..., C] so that SNPs [AA, CC, AA, ..., CC] have an allele dosage of 2, [AG, CG, AT, ..., CT] have a dosage of 1 and [GG, GG, TT, ..., TT] have a value of 0. The equation for cow *i* is:

$$y_i = \mu + \alpha_1 m_{1i} + \alpha_2 m_{2i} + \alpha_3 m_{3i} + \cdots + \alpha_{50,000} m_{50,000i} + e_i$$

which in matrix form is:

$$\begin{bmatrix} 10,602 \\ 11,440 \\ 9,103 \\ \cdots \\ \cdots \\ 9,657 \end{bmatrix} = \begin{bmatrix} \mu \\ \mu \\ \mu \\ \cdots \\ \cdots \\ \mu \end{bmatrix} + \alpha_1 \begin{bmatrix} 2 \\ 0 \\ 2 \\ \cdots \\ \cdots \\ 1 \end{bmatrix} + \alpha_2 \begin{bmatrix} 1 \\ 1 \\ 0 \\ \cdots \\ \cdots \\ 2 \end{bmatrix} + \alpha_3 \begin{bmatrix} 0 \\ 2 \\ 1 \\ \cdots \\ \cdots \\ 0 \end{bmatrix} + \cdots + \alpha_{50,000} \begin{bmatrix} 1 \\ 0 \\ 0 \\ \cdots \\ \cdots \\ 2 \end{bmatrix} + \begin{bmatrix} e_1 \\ e_2 \\ e_3 \\ \cdots \\ \cdots \\ e_{4,000} \end{bmatrix}$$

$$y = \mu + \alpha_1 m_1 + \alpha_2 m_2 + \alpha_3 m_3 + \cdots + \alpha_{50,000} m_{50,000} + e$$

where the regression coefficients to be estimated are $\alpha_1, \alpha_2, \alpha_3, \ldots \alpha_{50,000}$, and $\boldsymbol{m}_1, \boldsymbol{m}_2, \boldsymbol{m}_3, \ldots, \boldsymbol{m}_{50,000}$ are the variables corresponding to the markers; that is, the vectors for each animal showing the allele dosage for each SNP marker. For example, in the first row, the cow having a value of 10,602 kg of milk has the polymorphisms [AA, CG, TT, ..., CT], the second cow has a value of 11,440 kg of milk and the polymorphisms [GG, CG, AA, ..., TT], the third cow has a value of 9103 kg of milk and the polymorphisms [AA, GG, AT, ..., TT], and so on.

Regression equations can also be represented with centred variables; an equation of the type $y = \mu + bx + e$ can also be represented as:

$$y - m_y = b(x - m_x) + e$$

Centring the values of the variable is not a problem; the model could include fixed effects of year–herd–season, calving, etc., although in this simple example we will not consider them and we will present the values already pre-corrected for these effects. To centre the variables of the markers $\boldsymbol{m}_i$, we will calculate their mean. Let us take the first variable $\boldsymbol{m}_1$, which has an A/G polymorphism (Table 6.1) placed in one column and n rows. Let us call n_k the number of alleles A in row k (e.g. $n_1 = 2$ in the first row, $n_2 = 0$ in the second row, $n_3 = 2$ in the third, etc.). The mean of the variable $\boldsymbol{m}_1$ will be $\bar{m}_1 = \frac{1}{n}\sum_1^n n_k$. The total number of alleles in $\boldsymbol{m}_1$ whether A or G is two per row; that is, the total number of alleles in $\boldsymbol{m}_1$ is $2n$. The frequency of alleles A in $\boldsymbol{m}_1$ will be:

$$p_1 = \frac{\text{no. alleles A}}{\text{no. total alleles}} = \frac{\sum_1^n n_k}{2n}$$

so the mean can be expressed as $\bar{m}_1 = 2p_1$.

The model becomes:

$$\begin{bmatrix} 602 \\ -440 \\ 103 \\ \cdots \\ \cdots \\ -657 \end{bmatrix} = \alpha_1 \begin{bmatrix} 2-2p_1 \\ 0-2p_1 \\ 2-2p_1 \\ \cdots \\ \cdots \\ 1-2p_1 \end{bmatrix} + \alpha_2 \begin{bmatrix} 1-2p_2 \\ 1-2p_2 \\ 0-2p_2 \\ \cdots \\ \cdots \\ 2-2p_2 \end{bmatrix} + \alpha_3 \begin{bmatrix} 1-2p_3 \\ 2-2p_3 \\ 1-2p_3 \\ \cdots \\ \cdots \\ 0-2p_3 \end{bmatrix} + \cdots + \alpha_{50,000} \begin{bmatrix} 1-2p_{50,000} \\ 0-2p_{50,000} \\ 0-2p_{50,000} \\ \cdots \\ \cdots \\ 2-2p_{50,000} \end{bmatrix} + \begin{bmatrix} e_1 \\ e_2 \\ e_3 \\ \cdots \\ \cdots \\ e_{4,000} \end{bmatrix}$$

and in matrix form:

$$y = [M - 2p']\beta + e = W\beta + e$$

where the variable $\boldsymbol{y}$ is already centred, $\mathbf{M}$ is the matrix whose columns are the vectors $\boldsymbol{m}_1, \boldsymbol{m}_2, \boldsymbol{m}_3, \ldots, \boldsymbol{m}_{50,000}$ indicating how many alleles N_i are in each SNP, **2** is a column vector whose elements are 2 in all cases, and $\boldsymbol{p}$ is a vector with the respective frequencies of the *reference alleles* of the SNPs, $\boldsymbol{p}' = [p_1, p_2, p_3, \ldots, p_{50,000}]$.

As we have far more unknowns than data, we cannot estimate these regression coefficients by ordinary least squares. The solution for this problem is to assume that the 50,000 coefficients in $\boldsymbol{\alpha}$ are not fixed but random parameters. We have seen in section 4.1.2 that mixed models can have more unknowns than data. We can see in section 4.3.1 that for random parameters a term $\boldsymbol{I}(\sigma_e^2/\sigma_p^2)$ is added in the diagonal of the matrix $\boldsymbol{W'W}$ of the mixed model equations, allowing us to solve the equations. The random effect has in this case an a priori distribution $N(\boldsymbol{0}, \boldsymbol{I}\sigma_p^2)$. Now we need to add a priori information about the distribution of $\boldsymbol{\alpha}$, in a similar way to the a priori information in mixed models. Here, we assume that the coefficients of regression are a priori unrelated, with prior variance showing our uncertainty about the values of these coefficients.

The problem is that we do not actually have a priori information about the values of $\boldsymbol{\alpha}$ and we have to speculate about the genetic determination of the characteristic; for example, the trait may be determined by many genes of small effect each, or by some major or intermediate genes acting with many minor genes with small effects. This has led to different procedures for modelling traits, called Bayes A, B, C, D, L, R, T, Z, etc., with variants (Cπ, Dπ, D0, D1, D2, D3, RS, TA, TB, etc.). In section 6.3, we will show some examples of what Gianola *et al.* (2009) called the 'Bayesian alphabet', to see how the a priori information can be modelled. In genomic selection programmes, the infinitesimal model is usually applied, assuming that the regression coefficients contained in $\boldsymbol{\alpha}$ are a priori small and normally distributed, as we will see in section 6.2.

Once the regression coefficients have been estimated, the additive value of an individual can be predicted from their SNPs as:

$$\hat{a} = \boldsymbol{W}\hat{\alpha} = \hat{\alpha}_1 w_1 + \hat{\alpha}_2 w_2 + \hat{\alpha}_3 w_3 + \cdots + \hat{\alpha}_{50,000} w_{50,000}$$

This equation can be applied to all individuals, whether they have data or not; for example, if an untested male has the SNPs [AG, AA, AG, ..., TT] and the regression coefficients are:

$$\hat{\alpha}_1 = 0.10,\ \hat{\alpha}_2 = -0.35,\ \hat{\alpha}_3 = 2.11,\ \cdots,\ \hat{\alpha}_{50,000} = 0.22$$

the prediction of its additive value will be:

$$\hat{a} = 0.10 \cdot (1 - 2p_1) - 0.35 \cdot (2 - 2p_2) + 2.11 \cdot (1 - 2p_3) + \cdots + 0.22 \cdot (0 - 2p_{50,000}) = -160$$

where −160 kg of milk is the sum of the effects of the 50,000 markers.

Note that:

- SNPs are not genes, and therefore the values of the regression coefficients $\alpha_1, \alpha_2, \alpha_3, \ldots \alpha_{50,000}$ *are not the additive values of the genes*. By introducing new markers or deleting some of those in the equation, the regression coefficients change. The additive value of an animal should be the same when being estimated with more or fewer SNPs; thus, if we have more SNPs, the effect of each SNP should be less.
- Some SNPs may not be related to the effects of genes (in this case, the value of its coefficient would be zero), and each SNP may be correlated with the effects of several genes. *We estimate regression coefficients, not gene effects.*
- There may be genes of small effect that are not related to any SNP; in this case, a random variable $\boldsymbol{a}_0$ can be added to the model in order to collect the part of the additive value determined by these genes.

The model can be more complex and can also consider other fixed and random effects. As before, we assume that dominance and epistasis are irrelevant. Although the genetic determination of a character may be mainly dominant or epistatic, this dominance and epistasis generates additive values and additive variance useful for selection (see section 1.3.2). Nevertheless, models with dominant components have been proposed that can be useful in the analysis of crosses, which usually have a more important dominant component.

More complex models usually require greater computational facilities, and given the considerable size of the databases used in animal breeding, this is a factor to consider when applying a model. In practice, the infinitesimal model is usually applied to genetic improvement programmes, and it is assumed that a priori the coefficients contained in $\boldsymbol{\alpha}$ are small, modelling them as we will see in section 6.2. When the objective is not to perform genomic selection but to find larger or intermediate genes, several hypotheses are used for modelling the a priori information on $\boldsymbol{\alpha}$, as we will see in section 6.3.

6.2 Genomic Selection with the Infinitesimal Model

6.2.1 Random regression best linear unbiased prediction (RR-BLUP)

The term random regression underlines the fact that the parameters α_1, α_2, α_3, ... $\alpha_{50,000}$ are *random regression coefficients* with a non-flat a priori distribution, unlike the common frequentist interpretation, in which the regression coefficients are constant parameters behaving as fixed effects. The method is also called *SNP-BLUP*.

The model to be solved is the one from the previous section:

$$y = \mathbf{W}\alpha + e$$

When we define a model with random effects, we need to know the prior distribution of these effects; in our case, the random effects are the regression coefficients and the residuals. We will assume, as we usually do, that residuals are independently normally distributed with the same variance σ_e^2. For the prior distributions of the regression coefficients, we will assume that the trait is determined by many genes of very small effect each, so we will represent our a priori information assuming that each marker captures a small part of the additive value of the individual, and that all the regression coefficients are independently distributed with zero mean and the same variance:

$$f\left(\alpha_i|\sigma_\alpha^2\right) \sim N\left(0,\sigma_\alpha^2\right)\ ;\ \ \mathrm{cov}\left(\alpha_i,\alpha_j\right) = 0 \ \rightarrow\ f\left(\boldsymbol{\alpha}|\sigma_\alpha^2\right) \sim N\left(\mathbf{0},\mathbf{I}\sigma_\alpha^2\right)$$

We assume that a priori the regression coefficients $\boldsymbol{\alpha}$ are independent because if they are correlated, we do not know whether the coefficient of the first nucleotide of a SNP is correlated with the first or the second nucleotide of another SNP. Furthermore, assuming correlations would make the problem intractable, given the high number of possible covariances. The coefficients are correlated a posteriori due to the estimation process.

Note that σ_α^2 is not the additive variance of each marker but rather represents our a priori uncertainty around the additive effect captured by the marker through its association with one or more genes with a very small effect. If we increase the number of SNPs, and they are related to genes acting on the trait, σ_α^2 decreases, because we know that by augmenting the number of SNPs, the individual effect of each SNP should be less.

This variance must necessarily be small because if it were not so, some SNPs could have a large effect, and we assume that the trait is determined by many genes with small effect. In the original article where the RR-BLUP was proposed, the authors (Meuwissen *et al.*, 2001) proposed the prior value by simply dividing the additive variance of the character by the number of markers, but Fernando *et al.* (2007) argued that the variance should depend on the allele frequencies of the SNPs p_i and q_i and proposed an a priori value of σ_α^2 that for our example should be (see Appendix 6.1):

$$\sigma_\alpha^2 = \frac{\sigma_A^2}{\sum_{1}^{50,000} 2p_i q_i}$$

This is equivalent to assuming that the effects of the SNPs are equal and independent a priori, so that we are in conditions of linkage equilibrium, as if each SNP were fully associated with only a gene with a small effect. If the population is not in equilibrium (see section 1.1.4), the values of p_i and q_i of the generation in which the animals are genotyped can be used because, although incorrect, the results are very similar (Forni *et al.*, 2011). Furthermore, it is desirable that there is no linkage equilibrium between genes and SNPs so that they can be correlated; otherwise, genomic selection would be ineffective. Genes are practically never in linkage equilibrium due to multiple causes (e.g. selection, non-random mating), and there are also interactions between them (dominance, epistasis). Here, we are applying a simplified model to be able to use the information from many markers.

The model is solved as a BLUP using the equations of the mixed model of section 4.3, or using Bayesian inference with Markov Chain Monte Carlo (MCMC) as we saw in section 5.2. The solution to estimate the regression coefficients in our model is to solve the equation:

$$\left[\mathbf{W}'\mathbf{W} + I\frac{\sigma_e^2}{\sigma_\alpha^2}\right]\alpha = \mathbf{W}'y$$

If we had a model that also had fixed effects:

$$y = Xb + \mathbf{W}\alpha + e$$

then the equations of the mixed model would be:

$$\begin{bmatrix} X'X & X'\mathbf{W} \\ \mathbf{W}'X & \mathbf{W}'\mathbf{W} + I\frac{\sigma_e^2}{\sigma_\alpha^2} \end{bmatrix}\begin{bmatrix} b \\ \alpha \end{bmatrix} = \begin{bmatrix} X'y \\ \mathbf{W}'y \end{bmatrix}$$

6.2.2 Genomic-BLUP (G-BLUP)

G-BLUP involves solving a traditional mixed model like the one in section 5.2.1, using a relationship matrix **G** that takes into account the frequency in which individuals have SNPs with the same genotype. The idea is to specify the actual relationship in the concrete database that is used to estimate the additive values. The relationships of matrix ***A*** are a priori mean relationships; for example, without having data, we know that the siblings are similar and are correlated with a coefficient of 0.5. This is the expected relationship. However, by crossing two heterozygotes Aa × Aa, we could produce, for example, three siblings with genotype AA, AA and aa, respectively; in this case, the first two siblings are more similar compared with the other sibling. Taking into account all the SNPs, we can have a more precise idea about the real correlation between individuals in the specific case of our database, which would improve the estimation of additive values. The expected relationship of two siblings is 0.50, but with genomic information it could be, for example, 0.60 or 0.40. This would be the observed relationship. We obviously assume that the SNPs are well distributed in the genome and that we have SNPs enough to capture all the additive value from the many genes involved in the trait.

The model, excluding fixed effects as in the previous section, is the one we saw in section 4.3.1:

$$y = Za + e;\ a \sim \mathrm{N}\left(0, G\sigma_A^2\right);\ e \sim \mathrm{N}\left(0, I\sigma_e^2\right)$$

The model, as in the previous section, can contain fixed effects or other random effects. To predict the additive effects, we need to know the matrix **G**, and solve the equation:

$$\left[Z'Z + G^{-1}\frac{\sigma_e^2}{\sigma_A^2}\right]\hat{a} = Z'y$$

as we saw in section 4.3.1. If we have a model that also has fixed effects, then:

$$y = Xb + Za + e$$

the mixed model equations will be:

$$\begin{bmatrix} X'X & X'Z \\ Z'X & Z'Z + G^{-1}\frac{\sigma_e^2}{\sigma_A^2} \end{bmatrix}\begin{bmatrix} b \\ a \end{bmatrix} = \begin{bmatrix} X'y \\ Z'y \end{bmatrix}$$

Matrix **G** is the matrix of the genomic relationships between individuals, once we know their SNPs. To calculate it, we start assuming that the additive values are the sum of the effects of the SNPs, as we saw in section 6.1:

$$a = W\alpha = \alpha_1 w_1 + \alpha_2 w_2 + \alpha_3 w_3 + \cdots + \alpha_{50,000} w_{50,000}$$

The effects of the SNPs all have the same variance and are a priori unrelated, $\mathrm{var}(\alpha) = I\sigma_\alpha^2$, so the matrix **G** will be:

$$\mathrm{var}(a) = \mathrm{var}(W\alpha) = W\ \mathrm{var}(\beta)W' = WW'\sigma_\alpha^2 = WW'\frac{\sigma_A^2}{\sum_1^n 2p_iq_i} = \frac{WW'}{\sum_1^n 2p_iq_i}\cdot\sigma_A^2 = G\cdot\sigma_A^2$$

EXAMPLE 6.1

Let us take four animals and eight SNPs. The first polymorphism is A/G, the second one C/G, the third one A/C, etc.

	SNP1	SNP2	SNP3	SNP4	SNP5	SNP6	SNP7	SNP8
Animal 1	AA	CC	AA	TT	GT	GA	GG	TT
Animal 2	GG	CC	AA	TT	TT	GA	GG	TT
Animal 3	GG	GG	AC	TC	GG	GG	AA	CC
Animal 4	AA	CG	AC	TC	GG	GG	AA	CC

We will consider the allele with the highest frequency in each SNP as the reference allele to assign the values 2, 1 and 0 as in Table 6.1:

SNP1			SNP2			SNP3			SNP4		
AA	AG	GG	CC	CG	GG	AA	AC	CC	TT	TC	CC
2	1	0	2	1	0	2	1	0	2	1	0
SNP5			**SNP6**			**SNP7**			**SNP8**		
GG	GA	AA	GG	GA	AA	GG	GA	AA	CC	CT	TT
2	1	0	2	1	0	2	1	0	2	1	0

Allele frequencies are calculated, within each SNP, as the number of alleles with the highest frequency with respect to the total number of alleles. In this example, they are, for each SNP:

$$p_1 = \frac{4}{8} = 0.50;\ p_2 = \frac{5}{8} = 0.625;\ p_3 = \frac{6}{8} = 0.75;\ p_4 = \frac{6}{8} = 0.75;$$

$$p_5 = \frac{5}{8} = 0.625;\ p_6 = \frac{6}{8} = 0.75;\ p_7 = \frac{4}{8} = 0.50;\ p_8 = \frac{4}{8} = 0.50$$

The model is:

$$\begin{bmatrix} 10{,}602 \\ 11{,}440 \\ 9{,}103 \\ 9{,}657 \end{bmatrix} = \begin{bmatrix} \mu \\ \mu \\ \mu \\ \mu \end{bmatrix} + b_1 \begin{bmatrix} 2 \\ 0 \\ 0 \\ 2 \end{bmatrix} + b_2 \begin{bmatrix} 2 \\ 2 \\ 0 \\ 1 \end{bmatrix} + b_3 \begin{bmatrix} 2 \\ 2 \\ 1 \\ 1 \end{bmatrix} + b_4 \begin{bmatrix} 2 \\ 2 \\ 1 \\ 1 \end{bmatrix} + b_5 \begin{bmatrix} 1 \\ 0 \\ 2 \\ 2 \end{bmatrix} + b_6 \begin{bmatrix} 1 \\ 1 \\ 2 \\ 2 \end{bmatrix} + b_7 \begin{bmatrix} 2 \\ 2 \\ 0 \\ 0 \end{bmatrix} + b_8 \begin{bmatrix} 0 \\ 0 \\ 2 \\ 2 \end{bmatrix} + \begin{bmatrix} e_1 \\ e_2 \\ e_3 \\ e_4 \end{bmatrix}$$

$$\begin{bmatrix} 10{,}602 \\ 11{,}440 \\ 9{,}103 \\ 9{,}657 \end{bmatrix} = \begin{bmatrix} \mu \\ \mu \\ \mu \\ \mu \end{bmatrix} + \begin{bmatrix} 2 & 2 & 2 & 2 & 1 & 1 & 2 & 0 \\ 0 & 2 & 2 & 2 & 0 & 1 & 2 & 0 \\ 0 & 0 & 1 & 1 & 2 & 2 & 0 & 2 \\ 2 & 1 & 1 & 1 & 2 & 2 & 0 & 2 \end{bmatrix} \begin{bmatrix} \alpha_1 \\ \alpha_2 \\ \alpha_3 \\ \alpha_4 \\ \alpha_5 \\ \alpha_6 \\ \alpha_7 \\ \alpha_8 \end{bmatrix} + \begin{bmatrix} e_1 \\ e_2 \\ e_3 \\ e_4 \end{bmatrix} = \boldsymbol{\mu} + \mathbf{M}\boldsymbol{\alpha} + \boldsymbol{e}$$

$$\mathbf{2} \cdot \boldsymbol{p}' = \begin{bmatrix} 2 \\ 2 \\ 2 \\ 2 \end{bmatrix} [0.50 \quad 0.625 \quad 0.75 \quad 0.75 \quad 0.625 \quad 0.75 \quad 0.50 \quad 0.50] = \begin{bmatrix} 1 & 1.25 & 1.5 & 1.5 & 1.25 & 1.5 & 1 & 1 \\ 1 & 1.25 & 1.5 & 1.5 & 1.25 & 1.5 & 1 & 1 \\ 1 & 1.25 & 1.5 & 1.5 & 1.25 & 1.5 & 1 & 1 \\ 1 & 1.25 & 1.5 & 1.5 & 1.25 & 1.5 & 1 & 1 \end{bmatrix}$$

$$W = M - 2p' = \begin{bmatrix} 1 & 0.75 & 0.5 & 0.5 & -0.25 & -0.5 & 1 & -1 \\ -1 & 0.75 & 0.5 & 0.5 & -1.25 & -0.5 & 1 & -1 \\ -1 & -1.25 & -0.5 & -0.5 & 0.75 & 0.5 & -1 & 1 \\ 1 & -0.25 & -0.5 & -0.5 & 0.75 & 0.5 & -1 & 1 \end{bmatrix}$$

$$\sum_1^8 2p_i q_i = \sum_1^8 2p_i(1-p_i) = 2\cdot(0.50\cdot 0.50 + 0.375\cdot 0.625 + 0.25\cdot 0.75 + 0.25\cdot 0.75 + \\ + 0.375\cdot 0.625 + 0.25\cdot 0.75 + 0.50\cdot 0.50 + 0.50\cdot 0.50) = 3.56$$

$$G = WW'\frac{1}{3.56} = \begin{bmatrix} 4.38 & 2.63 & -4.88 & -2.13 \\ & 5.88 & -3.625 & -4.88 \\ & & 5.88 & 2.63 \\ & & & 4.38 \end{bmatrix} \frac{1}{3.56} = \begin{bmatrix} 1.23 & 0.74 & -1.37 & -0.60 \\ & 1.65 & -1.02 & -1.37 \\ & & 1.65 & 0.74 \\ & & & 1.23 \end{bmatrix}$$

Matrix **G** should be similar to the relationship matrix ***A***, specifying the relationship better. The numbers that appear in this example are not representative of the **G** matrices and are due to the small number of individuals and SNPs used; they serve only as an example of how to compute **G**.

Once the **G** matrix is available, the system is solved as a standard BLUP, but substituting the relationship matrix ***A*** for **G** (see section 4.3.1).

Note that we have encoded each SNP taking the allele with the highest frequency as the reference allele. This is arbitrary, and in fact the *allele with the minimum frequency* is usually taken as the reference allele, because databases are usually refined by eliminating SNPs having one of their alleles with a frequency below a certain threshold.

There are other alternatives for constructing **G**; centred genomic frequencies can be used, different weights can be given in the diagonal, or different a priori variances of the SNPs can be used; in addition, σ_α^2 can be defined in a different manner. However, the procedure we have proposed is the most common for G-BLUP. Linear combinations appear frequently between rows and between columns of **G** and consequently it does not have an inverse, which generates computing problems, usually solved by adding a small amount to the diagonal value in order to break collinearity; this procedure has little effect on the results.

> Note that the advantages of G-BLUP over BLUP come from having substituted the a priori relationship matrix ***A*** for the genomic relationship **G**, which indicates the actual relationship.

It can be shown that G-BLUP and RR-BLUP are, under very general conditions (e.g. when σ_α^2 is defined as the additive variance divided by the total number of SNPs), equivalent methods, arriving at the same solutions. This implies that the use of SNPs with RR-BLUP also bases its success on better specifying the relationships of the animals used in prediction.

6.2.3 Single-step G-BLUP

As genotyping is expensive, often not all animals are genotyped; for example, males are preferentially genotyped because they have more genetic influence on the population. However, to build the equations in the reference population or in successive generations, we need the phenotypes with which the SNPs are to be associated, but the males do not have phenotypic values for some traits of interest such as litter size or milk production. To solve this problem, males can be assigned pseudo-phenotypes based on information from relatives. In dairy cattle, one way of assigning pseudo-phenotypes is to use the mean phenotype of the daughters corrected for the environmental values of their contemporaries (known as *daughter yield deviation*), or another way of using phenotypic values estimated from the available genetic values (what is known as *deregressed values*). This way of working in multiple steps generates several problems: there are individuals with both data and pseudo-data, the creation of pseudo-data may be ineffective if the individuals have few daughters, there may be environmental covariances between pseudo-data, genotyped animals are usually pre-selected from the best (which creates biases), the residuals do not all have the same variance because the number of daughters is different for each male, etc.

To avoid these problems, Legarra *et al.* (2009) proposed the prediction of additive values in a *single step*, taking into account all the data, those of the genotyped and those of the non-genotyped animals. In order to solve this evaluation, it is necessary to have the genomic relationship matrix of all individuals, genotyped or not. To construct a global genomic relationship matrix including both genotyped and non-genotyped animals, it is necessary to estimate the genomic relationship of non-genotyped animals from those that are genotyped, which allows calculation of the joint relationship matrix, called $\boldsymbol{H}$ to differentiate it from the $\mathbf{G}$ matrix of genotyped animals and from the expected relationship matrix $\boldsymbol{A}$. Once $\boldsymbol{H}$ is obtained, the model is solved as a standard G-BLUP.

We start from the previous model, but the vector of additive values is divided into two parts, one with the additive values of non-genotyped animals and the other with those of genotyped animals.

$$y = Za + e = Z\begin{bmatrix} a_1 \\ a_2 \end{bmatrix} + e$$

where a_1 is non-genotyped animals and a_2 is genotyped animals, and where we know the genomic relationship matrix $\mathbf{G}$ of the genotyped animals.

In Appendix 6.2, we show the deduction of the joint relationship matrix $\boldsymbol{H}$ and its inverse $\boldsymbol{H}^{-1}$.

Once you have this matrix, you have to solve the system:

$$\left[Z'Z + H^{-1}\frac{\sigma_e^2}{\sigma_A^2}\right]\hat{a} = Z'y$$

or in the case of a model including fixed effects, a system such as the mixed model equations in section 4.3.1, but with $\boldsymbol{H}^{-1}$ instead of $\boldsymbol{A}^{-1}$:

$$\begin{bmatrix} X'X & X'Z \\ Z'X & Z'Z + H^{-1}\dfrac{\sigma_e^2}{\sigma_A^2} \end{bmatrix}\begin{bmatrix} b \\ a \end{bmatrix} = \begin{bmatrix} X'y \\ Z'y \end{bmatrix}$$

To calculate H^{-1}, let us first consider that all individuals, genotyped or not, have the average relationships of the expected relationship matrix $\boldsymbol{A}$, which can be partitioned as the submatrix of non-genotyped animals $\boldsymbol{A}_{11}$, that of genotyped animals $\boldsymbol{A}_{22}$, and the part of the relationship matrix that relates the genotyped to the non-genotyped animals $\boldsymbol{A}_{12}$ and $\boldsymbol{A}_{21}$. For example, if we have a genotyped sire that has two non-genotyped offspring that are half siblings, the matrix will be:

$$A = \begin{bmatrix} A_{11} & A_{12} \\ A_{21} & A_{22} \end{bmatrix} = \left[\begin{array}{cc|c} 1 & 0.25 & 0.5 \\ 0.25 & 1 & 0.5 \\ \hline 0.5 & 0.5 & 1 \end{array}\right]$$

and the inverse of the joint relationship matrix will be:

$$H^{-1} = A^{-1} + \begin{bmatrix} 0 & 0 \\ 0 & G^{-1} - A_{22}^{-1} \end{bmatrix}$$

EXAMPLE 6.2

Let us have a simple pedigree as the one in Fig. 6.1, in which animals A and C are genotyped.

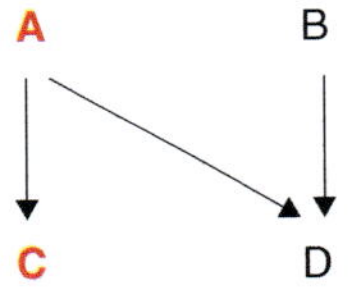

Fig. 6.1. Pedigree with genotyped (A and C) and non-genotyped (B and D) animals.

Let us assume that the genomic relationship matrix of the genotyped animals is:

$$G = \begin{bmatrix} 1 & 0.4 \\ 0.4 & 1 \end{bmatrix}$$

The full expected relationship matrix $\boldsymbol{A}$ and the expected relationship matrix of the genotyped animals $\boldsymbol{A}_{22}$ are:

$$A = \begin{bmatrix} A_{11} & A_{12} \\ A_{21} & A_{22} \end{bmatrix} = \begin{array}{c} \begin{array}{cccc} \text{B} & \text{D} & \text{A} & \text{C} \end{array} \\ \begin{bmatrix} 1 & 0.5 & 0 & 0 \\ 0.5 & 1 & 0.5 & 0.25 \\ 0 & 0.5 & 1 & 0.5 \\ 0 & 0.25 & 0.5 & 1 \end{bmatrix} \end{array} \begin{array}{c} \text{B} \\ \text{D} \\ \text{A} \\ \text{C} \end{array}; \quad A_{22} = \begin{bmatrix} 1 & 0.5 \\ 0.5 & 1 \end{bmatrix}$$

We need for H^{-1}:

$$A^{-1} = \begin{bmatrix} 1.5 & -1 & 0.5 & 0 \\ -1 & 2 & -1 & 0 \\ 0.5 & -1 & 1.83 & -0.67 \\ 0 & 0 & -0.67 & 1.33 \end{bmatrix}; \quad G^{-1} = \begin{bmatrix} 1.19 & -0.48 \\ -0.48 & 1.19 \end{bmatrix}; \quad A_{22}^{-1} = \begin{bmatrix} 1.33 & -0.67 \\ -0.67 & 1.33 \end{bmatrix}$$

The inverse of the joint relationship matrix is:

$$H^{-1} = \begin{bmatrix} 1.5 & -1 & 0.5 & 0 \\ -1 & 2 & -1 & 0 \\ 0.5 & -1 & 1.83 & -0.67 \\ 0 & 0 & -0.67 & 1.33 \end{bmatrix} + \begin{bmatrix} 0 & 0 & 0 & 0 \\ 0 & 0 & 0 & 0 \\ 0 & 0 & -0.14 & 0.19 \\ 0 & 0 & 0.19 & -0.14 \end{bmatrix} = \begin{bmatrix} 1.5 & -1 & 0.5 & 0 \\ -1 & 2 & -1 & 0 \\ 0.5 & -1 & 1.69 & -0.48 \\ 0 & 0 & -0.48 & 1.19 \end{bmatrix}$$

Single-step G-BLUP has become the usual method in animal breeding when there are genotyped animals. It replaces BLUP, which has been the preferred method in genetic improvement when genomic selection was not performed. There are two public software packages for single-step G-BLUP, Ignacy Mizstal's BLUPf90 family of programmes that we saw at the end of section 4.3.2.D (http://nce.ads.uga.edu/software; accessed 9 July 2025), and the DMU package (https://dmu.ghpc.au.dk; accessed 9 July 2025).

Single-step G-BLUP has the fundamental advantage of including all the information available for evaluation in a single step. Furthermore, being a form of BLUP, the models that we saw in section 4.3.2.C can easily be adapted.

The single-step G-BLUP assumes, like BLUP, that $E(\boldsymbol{a}) = \boldsymbol{0}$ whether the animals are genotyped or not, but in dairy cattle, for example, genotyped animals are pre-selected and are better on average, and thus this hypothesis is not fulfilled. There is considerable literature proposing ways to address these and other local problems, which is beyond the scope of this book. The interested reader can consult Legarra *et al.* (2024), an excellent manual dedicated to genomic selection.

6.3 Gene Detection

6.3.1 Genome-wide Association Studies (GWAS)

Gene localization is a complex problem that has been approached in several ways. For example, candidate genes have been proposed based on existing knowledge about the metabolic pathways affecting a trait, or on knowledge about the pathology of a disease, or because these genes are known from homologous areas in humans or laboratory mice. Another way of detecting genes has been to relate molecular markers to phenotypes. When the use of SNPs was feasible, multiple markers were available economically throughout the genome, which allowed the performance of GWAS. The simplest way of carrying out these studies has been to perform individual regressions of each SNP on the characteristic of interest, observing whether any of the regression coefficients was particularly relevant. This means doing, in our example, 50,000 regressions, but by mere chance some coefficients will be significantly different from zero if we apply a frequentist test with the usual value of $P < 0.05$, and thus the signification level should be raised. How to do it is a matter of debate; if the P values are assumed to be independent, the threshold is very high but unrealistic, as the SNPs are frequently in linkage disequilibrium. Based on considerations about the density of the chips and the linkage disequilibrium between them, it is common to use a P value of 10^{-8} in humans and 10^{-4} in animals, as linkage disequilibrium is greater in animals than in humans for a number of reasons (e.g. closer relationships, selection). The *Bonferroni correction* is frequently used, which divides the significance threshold of a test by the number of tests performed (e.g. if the threshold is 0.05 and there are 20 tests, the proposed threshold is 0.05/20 = 0.0025). As this test is very rigorous (a lot of data is needed for a SNP to be significant), a less demanding methodology is common, the *false discovery rate* (FDR), which estimates the proportion of 'false discoveries' (i.e. declaring that a SNP has an effect on a trait when it does not, which is called a false positive) among all tests in which SNPs were significant. FDR is estimated with frequentist or Bayesian techniques (Sorensen, 2025). All these procedures and these P value thresholds for significance should be understood as merely indicative, and the subsequent examination may include SNPs that have not passed the significance barrier or discard some that have passed it; GWAS should be understood as an exploratory analysis.

These association studies have been criticized (Blasco and Toro, 2014; Bernardo, 2016), mainly because the most important traits from an economic point of view are generally determined by many genes each of small effect. Furthermore, when there are larger genes, the most favourable allelic version is usually fixed. In human medicine, most of the important diseases are also multi-factorial, but in humans a rare disease affected by a gene is always important, so GWAS have received support despite their limitations; in animals, individuals with harmful rare genes are simply eliminated. The use of GWAS in animal breeding is rather limited; even if a segregating gene with a significant effect is detected with SNPs in linkage disequilibrium with it, this would not guarantee that in another population the same SNPs would have the same correlation; indeed, if there is no correlation, these markers will be useless in this population. The same would happen after a few generations; the linkage disequilibrium would be lost and the predictive capacity of the SNPs would radically decrease, unless their linkage disequilibrium with the gene of interest was very strong. On the rare occasions that larger genes have been found segregating, which we will discuss in Chapters 8 and 9, the proportion of the additive variance of the trait explained by the gene has generally been low, at most around 10%.

So far, we have argued that most traits are determined by many genes each of small effect. However, it could happen that some traits also have genes with a large effect and genes with a medium or moderate effect. The models that we are going to examine in this section have been used for the detection and localization of genes with relevant effects, associating SNPs with productive traits. They assume that traits are genetically determined by many genes with small effects and some genes with medium or large effects. These models can also be used in genomic selection, but G-BLUP and single-step G-BLUP are usually preferred because productive traits generally respond to the infinitesimal model; moreover, they have a much higher computing speed and there is usefully free software for them.

6.3.2 Bayesian regression for gene detection

In multiple regression, when the variables are correlated, the coefficients are not proportional to the importance of the variables. For example, if we predict the weight of pig meat with the variables live weight and carcass weight, the first variable can have a coefficient of 0.05 and the second one of 0.50, which does not mean that live weight has no relationship to the amount of meat on the animal, but it is already predicted for the most part by the variable carcass weight. The regression coefficients of the SNPs are independent a priori, but the SNPs, the variables $\boldsymbol{m}_i$ of the equation in section 6.1, are not. Close

SNPs are usually correlated, but as we saw in section 1.1.4, they can be correlated even if they are on different chromosomes, so in a multiple regression, the coefficients that multiply the variables $\boldsymbol{m}_i$ are not proportional to the importance of the corresponding SNP. It is usual to find groups of SNPs that are correlated with each other, in which a SNP may have a high coefficient due to its association with a larger gene, but close SNPs may have very low coefficients even though they are equally associated with the same larger gene. Therefore, the coefficient of a SNP in a multiple regression is not the same as the coefficient that can be found when this SNP is adjusted in isolation.

One way of approaching this problem has been to group correlated SNPs once the regression has been done, so that these groups (called *windows*) are independent of each other (as far as possible), and they estimate the part of the additive variance explained by the SNPs contained in the window. Ideally, the windows should group SNPs in linkage disequilibrium, which are usually contiguous SNPs, and the windows should be unrelated to each other, explaining independent parts of the additive variance. In practice, you have to play with various window sizes, even with sliding SNP windows containing SNP 1 to n, SNP 2 to $n + 1$, SNP 3 to $n + 2$, etc., which are in this case dependent because they contain common SNPs.

Next, we will see some of the models allowing SNPs to take medium to large values.

A. Bayes A

If the trait is determined by genes with large, medium and small effects, this implies that we are not sure that all SNPs have a small effect, so we increase our uncertainty about some of them; that is, we increase the prior variance of the regression coefficient of the SNP. One way to do this is to assume that the effects are distributed with a Student's t distribution rather than a normal one as we have assumed so far. Representing the model with Student's t distribution allows some effects to be large, others medium and others small because Student's t distribution has 'fatter' tails than normal (see Fig. 6.2), depending on its degrees of freedom.

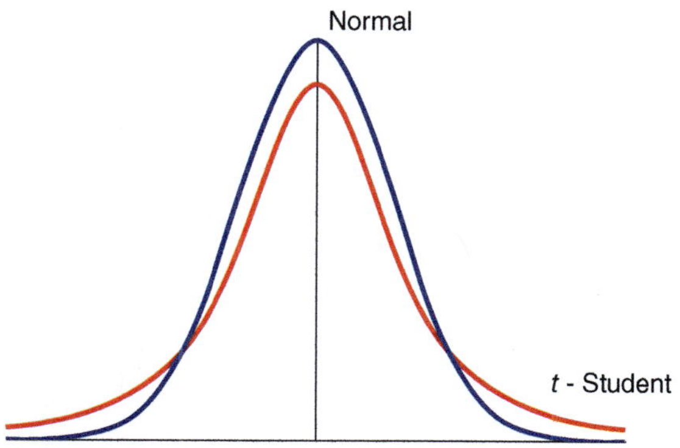

Fig. 6.2. Normal density (blue) and Student's t density (red). Student's t density allows us to find more genes with higher effect.

B. Bayes B

We generally assume that traits are genetically determined by far fewer causal genes than the SNPs we have on a chip, and we also assume that causal genes are not evenly distributed throughout the genome. As the SNPs on our chips are distributed throughout the entire genome, we assume that most SNPs will have little or no effect on a trait; that is, they will not be associated with any of the genes that determine the trait, or this association (its correlation with the gene) will be very small. We can model it like Bayes A; that is, assuming that the effects are distributed with a Student's t distribution, but allowing a π percentage of the SNPs to have a zero effect.

The value of π is given a priori and should be based on our previous information about the percentage of SNPs having a zero value. This creates a problem, because different values of π can lead to different results, so a robustness analysis is sometimes performed by testing multiple values of π.

We assume the coefficients are unrelated a priori, as in the previous models.

Bayes A and Bayes B were originally proposed by Meuwissen *et al.* (2001) having a more complex justification, but it was shown by Gianola *et al.* (2009) that this justification was not necessary.

C. Bayes C and Bayes Cπ

Bayes C (Habier *et al.*, 2011) is the same model as RR-BLUP, in which all SNPs are normally distributed with the same prior variance, but with the possibility of assigning a null value to a π percentage of SNPs, as in Bayes B.

In the variant Cπ, it is the data that will decide the values of π, so we consider that π is a random variable that has a prior density. Frequently, a flat prior is given for π unless we can obtain prior information from previous experiments allowing us to propose a different a priori density. Often, the data do not allow a reliable estimate of π, which forces us to use Bayes C.

D. Bayes L

In Bayes L (de los Campos *et al.*, 2009), also called Bayesian LASSO according to the name of the original statistical procedure, we consider that the prior density of the effects of each SNP (i.e. of the regression coefficients) is the distribution known as the Laplace distribution:

$$f(\alpha_i|\lambda) = \frac{\lambda}{2}\exp(-\lambda|\alpha_i|)$$

where $|\alpha_i|$ is the absolute value of the regression coefficient. We also assume that the regression coefficients are unrelated.

Bayes L allows a certain number of genes with greater effect, as in Bayes A. In Fig. 6.3 it can be seen that the distribution of Bayes L has, like that of Bayes A (Fig. 6.2), 'fat' tails that allow detection of larger genes than when we use RR-BLUP with the normal distribution.

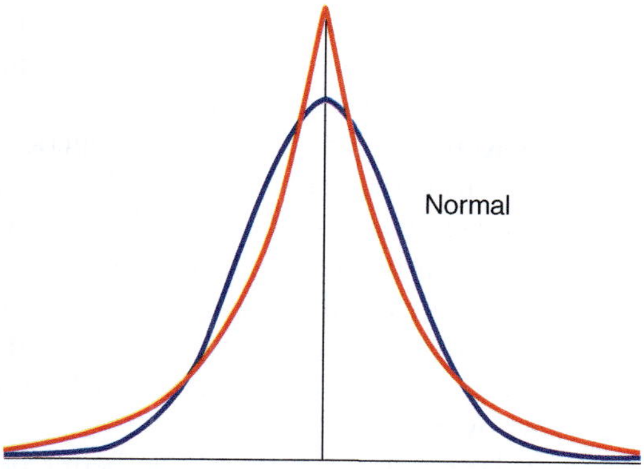

Fig. 6.3. Normal density and Laplace density for Bayesian LASSO.

The parameter λ determines how sharp the distribution is, allowing more or fewer genes with large effect. As in Bayes C, it can be fixed by the researcher or can be assigned a prior density, as in Bayes Cπ.

E. Bayes R

In Bayes R, we consider that the a priori density of the effects of each SNP is a mixture of four normal variables:

$$
\begin{aligned}
&f(\alpha_i \mid \sigma_i^2) \sim \mathrm{N}(0, \sigma_i^2) \quad ; \quad \mathrm{cov}(\alpha_i, \alpha_j) = 0 \\
&\sigma_i^2 = 0 \quad \text{with probability } \pi_1 \\
&\sigma_i^2 = 0.0001\, \sigma_A^2 \quad \text{with probability } \pi_2 \\
&\sigma_i^2 = 0.001\, \sigma_A^2 \quad \text{with probability } \pi_3 \\
&\sigma_i^2 = 0.01\, \sigma_A^2 \quad \text{with probability } \pi_4
\end{aligned}
$$

The π_i values can be estimated from the data, as in the case of Bayes Cπ, using a flat prior or an informative one based on prior knowledge. Its authors (Erbe *et al.*, 2012) maintain that this method exhibits good behaviour for detecting larger genes when there are any.

6.4 Genomic Selection in Practice

6.4.1 The Bayesian alphabet in practice

Genomics has been used to predict the genetic merit of selection candidates and also to search for genes with large or medium effects. In section 6.3.2, we described the results of several models with different prior information on the effects of SNPs. The success of one or the other model depends to a great extent on the true genetic determination of the trait:

- If the trait is determined by many genes with small effects, RR-BLUP or G-BLUP are suitable models for genomic prediction of additive values, but even when a trait is determined by many genes, they are not expected to be uniformly located throughout the genome, so Bayes C should perform better than RR-BLUP or G-BLUP. Bayes A, Bayes B, Bayes L and Bayes R allow us to detect large genetic effects, so when there are indeed genes with large effect, they should work well in genomic prediction of additive values. Nevertheless, Fernando and Garrick (2013) noted that, in simulation studies, Bayes A and Bayes B had a higher prediction precision than RR-BLUP, but in real applications RR-BLUP works as well as Bayes A, Bayes B or Bayes C, and sometimes even better. This is probably because in practice, even if there are genes with medium to large effects, they are not closely associated with only a few markers, but their effect is captured by many markers.
- When there are genes with large effects, they can be better detected with Bayes B, using a high value of π, and therefore giving a zero value to a large proportion of SNP effects, with the danger of finding results that are excessively adapted to our base of concrete data. For example, in a divergent selection of a trait that actually follows the infinitesimal model, we might decide that all SNPs have a zero value with the exception of one SNP that has different alleles in each line; in this case, we will assign the value of the response to the selection not to many genes of small effect but to this SNP, obtaining a false discovery.
- As Bayes C is sensitive to the value of π used, Bayes Cπ avoids having to do robustness analysis on the value of π; however, when the data set is small, π is not determined with precision and there are estimation problems with the MCMC procedures, so Bayes B or C are preferable to Bayes Cπ.
- Bayes L allows us to find genes with larger effects than the other models (see Fig. 6.3), and is computationally more effective, particularly when there are many SNPs with zero or near-zero effect.
- Proponents of the Bayes R method affirm that, due to its flexibility, working with several distributions simultaneously allows us to find genes with large effects and also smaller contributions associated with other SNPs.
- There are a large number of studies comparing these Bayesian models, with both real and simulated data. The choice of one or another is determined by various circumstances, such as the objective of the work (prediction of the genetic value or detection of genes), the computing facilities when handling large data sets, the appropriate method when the sample is small. Factors such as the actual genetic determination of the data, the actual association of genes with markers and the number of markers considered determine the best model, and in many cases, there is no indication of what the real situation is, so it is sometimes useful to use several models and compare the results.

6.4.2 Difficulties in implementing genomic selection in breeding programmes

As we will see in Chapter 9, the impact of genomic selection on dairy cattle breeding programmes has been spectacular, increasing the overall response to selection by more than 50%. This has been obtained essentially by reduction of the generational interval, because bulls can be genetically evaluated before being tested with the production of their daughters. The benefit of using genomic selection in other species has been lower and its implementation more complicated, but the overall response to selection has increased by 20–30% due to genomic selection. The difficulties of implementing genomic selection in current breeding programmes, particularly in the case of prolific species, have been highlighted by Blasco and Toro (2014), Jonas and de Koning (2015) and Blasco and Pena (2018), and we will examine them now:

- *Reference population.* to create the equations to be used in the prediction of additive values, we need a large 'reference population' of several thousand animals. This is not a problem in dairy cattle, but in other species, it can be a serious problem. In prolific species, for example, selection is done in small farm nuclei, sometimes having only a few hundred females. Some alternatives can be considered, such as the use of siblings in multiplication farms or the use of animals of several generations, or even the use of crossbred animals, but the efficiency of the equations declines rapidly; therefore, alternative strategies must be carefully examined. A second problem is the need to generate new equations every three or four generations, because the associations between SNPs and causal genes are lost over time. This is due to recombination and also because many of the associations are due to SNPs located on chromosomes other than the causal gene (we already saw in section 1.1.4 that genes can be related even if they are not physically linked in the same chromosome). Ibáñez-Escriche and Blasco (2011) showed that the precision of the equations is rapidly lost generation after generation, which means that new reference populations are needed to re-estimate the effects of SNPs. In practice, instead of having large reference populations every three or four generations, phenotypes are collected every generation to update the equations. This is not a problem for routinely recorded traits (e.g. litter size), but can be a problem for traits that are more expensive or difficult to measure.
- *Cost of genotyping.* Another major problem with genomic selection is the cost of genotyping. This cost has been drastically reduced in recent years but is still important for species in which the individual value of the animal is small and the generation interval is short (e.g. pigs, poultry and rabbits), which implies frequent genotyping with a high cost compared with the value of the animal. One way to tackle this problem is to use low-density chips with only a few hundred or thousand SNPs, assigning the missing SNPs from the high-density chips. This technique, called *imputation*, is based on the fact that recombination is low in a single generation. Imputation has generally worked effectively, predicting additive values with acceptable precision. Imputation must be repeated every three to four generations, because recombination leads to imputation errors. As in the case of the reference populations, high-density chips are used every generation to repeat the imputation.
- *Use of genomic selection for specific traits.* Genomic selection was considered as a possible procedure to improve traits whose measurement cost is high, or which are difficult to record. For example, expensive-to-measure meat-quality traits were considered natural candidates for genomic selection, in the hope that after collecting data in a reference population, many animals could be evaluated using their genomic data without needing to collect phenotypic data or data from relatives. However, as the equations have to be reformulated every three or four generations (or even every generation), continuous data collection is needed to recalculate the prediction equations, so genomic selection became less attractive, at least for species with a short generation interval such as pigs, birds and rabbits.

Nevertheless, genomic selection can contribute to the efficiency of current selection programmes, if implementation strategies are carefully studied and the cost of genotyping is low enough. Current animal genetic improvement programmes have introduced genomic selection as a main tool in the programme (in the case of dairy cattle) or as a useful complement (in the case of pigs), and the question nowadays is not whether or not to use genomic selection but rather how to use it.

6.4.3 The use of genomic selection in animal breeding programmes

Genomic selection is implemented in almost all livestock species but with uneven development and application due to the characteristics of each species. The selection method that has been imposed on all species is the single-step G-BLUP (section 6.2.3), because it combines available phenotypic information with genomic information.

The first commercial chip with thousands of SNPs was available in 2007 in dairy cattle, and more chips have progressively appeared in other species. Table 6.2 shows the characteristics of the main chips currently marketed for different livestock species. Below, we will see the practical application of genomic selection in animal breeding programmes.

A. Dairy cattle

Genomic selection has had a high impact in dairy cattle breeding programmes, as we will see in section 9.1, because dairy cattle are particularly suitable for genomic selection. This is because dairy cattle have a long generation interval (6 years) due to the need for progeny testing, the traits of interest (milk production and quality) cannot be measured in the bull, selection pressure must be applied essentially on bulls because the average number of calvings per cow is only 2.7, and the spread of genetic progress is cheap and easy through artificial insemination. Furthermore, a bull with a high additive value is a very valuable animal that produces great benefits; note that in dairy cattle, particularly in

Table 6.2. Main commercial chips for genomic selection. From Blasco and Pena (2018), used with the permission of Springer Nature.

Animal	Chip name	No. of SNPs	Interval[a]	MAF	Supplier
Chicken	Axiom Chicken array	580,961			Affymetrix
Cattle	BovineSNP50	54,001	50.6	0.26	Illumina
	BovineHD	777,000	3.43	0.25	Illumina
	BovineLDv2	7,931	383	0.31	Illumina
	Bovine3K	2,900			Illumina
	Axiom BOS1 array	648,855			Affymetrix
Horse	Axiom Equine Array	650,000			Affymetrix
Pig	PorcineSNP60	64,232	43.4	0.28	Illumina
	GGP Porcine HD	>65,000	43.0		GeneSeek
	Axiom Porcine 650K	658.692	3.34	0.32	Affymetrix
Sheep	OvineSNP50	54,241	50.9	0.28	Illumina
	OvineHD	603,350	5	0.30	Illumina
Rabbit	Axiom OricunSNP	200,000	15–20	0.20	Thermofisher

MAF: allele with minimum frequency averaged across tested populations.
[a]Average interval between SNPs (kb).
Table used with the permission of Springer Nature

Holstein cattle, a small number of males supply semen to the all cows in the world. This implies that the cost of genotyping is not an impediment to developing genomic selection compared with other species. It is also not a problem to create large reference populations and maintain a continuous record system to update the prediction equations, as all farms constitute a global nucleus linked by artificial insemination.

B. Beef cattle

The success of genomic selection in dairy cattle has led us to consider introducing genomics in beef cattle programmes. However, beef cattle are organized into many breed associations, which are much smaller in size than dairy cattle breeds. Moreover, beef cattle are not always well connected by artificial insemination. Because of this, most beef cattle associations cannot have a large reference population and continuous data recording as in dairy cattle. It has been proposed to use reference populations with several breeds for predictions, but the problem is that the effectiveness of the genomic value prediction is greater when the reference population is genetically close to the animals to be predicted; otherwise, the prediction is poor. The reason is that markers do not necessarily have the same linkage disequilibrium in some populations as in others, as we saw in section 1.1.2, and a marker may be associated with the favourable allele of a gene in one population and the unfavourable one in another population, so the use of multi-breed reference populations is now under discussion.

While it is true that commercial companies have used genomics as a marketing tool (Rolf *et al.* 2014), genomics could improve the precision of traits that are not directly measured, for example when the target is slaughter weight but only weaning weight is measured, or for carcass traits. Even in these cases, a careful study must be carried out taking into account the large reference populations required and the ongoing cost of genotyping in relation to the expected benefits. In section 9.2.2.C, we discuss in more detail the possible application of genomic selection in beef cattle.

C. Sheep and goats

Sheep and goats raised for milk production have the same scheme as dairy cattle, albeit on a much lower scale, limiting the application of genomic selection. Sheep meat shares most of its problems with beef cattle in terms of the efficient use of genomic selection. In both sheep and goats, the low price of the animals limits the application of genomics due to the relatively high cost of genotyping.

Rupp *et al.* (2016) reviewed the application of genomics in sheep and goats, concluding that the gains in precision when applying genomic selection were quite modest, around 10–20%, even for milk production traits. Taking into account genotyping costs, Shumbusho *et al.* (2016) estimated that the economic advantages of using genomic selection in sheep meat were only 15% in the best scenario, and similar results have been found in Australian merino sheep. Despite this, genomic selection is starting in some productive milk breeds, as we will see in section 9.3.1.

D. Pigs

There is no progeny test in pigs, so the generation interval is short (about 1 year). Selection objectives are traits expressed in both males and females, with the exception of litter size. As we will see in section 8.1.1, the dissemination of genetic progress is carried out through a pyramidal structure of farms (nucleus to multipliers to commercial farms) where genetic improvement is carried out only in the nucleus. There is no global nucleus, but rather several companies compete in a free market, with small nuclei of approximately 25–50 males and 300–2000 females per line, limiting the size of the reference populations. Litter size is an obvious candidate trait for genomic selection because females are selected before expressing the trait, but the heritability of litter size is low, so large reference populations are needed and a strategy to obtain them is not obvious. Information from the multipliers can be used, or even information from the commercial crossbred females, as we described earlier. The price of the selected animals is much lower than that of dairy cattle, which makes the cost of genotyping a critical factor. Furthermore, as pigs are normally produced in a three-way crossover scheme, genotyping costs are three times higher than the cost of species using only one breed in their breeding programme. Imputation is a common strategy to reduce costs, as the precision of the estimates of additive values obtained with imputation is also high in pigs.

Due to all these restrictions, the application of genomics has had less spectacular results in pigs than in dairy cattle; however, the increase in gains when using genomic selection has been evaluated from 10% to 50% depending on the method of implementing genomic selection (see Ibáñez-Escriche *et al.*, 2014, for a detailed description of some strategies).

E. Poultry

Poultry has similar restrictions to those of pigs in their genetic improvement programmes to implement genomic selection. Companies with small nuclei (compared with dairy cattle) also compete, the generation interval is also equally short, the economically relevant characters are expressed in both sexes in broilers with the exception of reproductive traits, and they are expressed only in females in laying hens. Genomics was implemented in 2013 in both production systems. A careful imputation procedure has given very good results, although, due to the continued decrease in the cost of genotyping, medium-density SNP chips are being used, eliminating the need for imputation.

The efficiency of genomic selection has been similar to that of pigs. In a selection experiment in laying hens, the response to selection using genomic selection was evaluated in comparison with a line in which selection was made on the same traits without genomic selection. The results were variable depending on the characteristic used in the selection index; traits such as the number of eggs showed little advantage, but for some traits such as egg weight, the use of genomic selection was much more efficient (Wolc *et al.*, 2015). In broilers, in a line selected mainly for growth rate, precision increased 20% for body weight and 17% for breast ultrasound measurements when genomic selection was used (Chen *et al.*, 2011).

In general, the greatest advantage of using genomic selection is, as in other species, in traits that are not available at the time of selection.

F. Rabbits

Genomics has yet to be implemented in rabbits, mainly due to the cost of genotyping. A high-density chip (200,000 SNPs) appeared in 2015 and low-density chips have yet to be produced. Rabbit selection schemes are three-way crosses with the same structure as in pigs, but the nuclei are even smaller (around 20 males and 150 females per line). The lines to produce the crossed doe are selected primarily for litter size and those producing the terminal sire for growth rate. The generation interval is very short (6–9 months) and the price of the animals is low, which represents the main restriction for the application of genomic selection. In rabbits, careful research is currently under way to find the best strategy for implementing genomic selection (Mancin *et al.*, 2021).

Appendix 6.1

As the additive variance of a locus is $2pq\alpha^2$ (see section 1.3.2), the additive variance of the sum of 50,000 independent loci, all with the same effect, is:

$$\sigma_A^2 = \sum_1^{50,000} 2p_i q_i \alpha^2$$

If σ_α^2 is the variance of the effect α at one locus, and taking into account that the mean of the α effects is zero, $\sigma_\alpha^2 = E(\alpha - 0)^2 = E(\alpha^2)$. We do not know the effect of each SNP, so we can a priori assign to α^2 the value of its mean $E(\alpha^2)$ (i.e. $\alpha^2 = E(\alpha^2) = \sigma_\alpha^2$) and therefore:

$$\sigma_A^2 = \sum_1^{50,000} 2p_i q_i \alpha^2 = \sum_1^{50,000} 2p_i q_i \sigma_\alpha^2 \quad \rightarrow \quad \sigma_\alpha^2 = \frac{\sigma_A^2}{\sum_1^{50,000} 2p_i q_i}$$

Appendix 6.2

To calculate $\boldsymbol{H}$, we will make the prediction of the additive values of the non-genotyped animals $\boldsymbol{a}_1$ from the values of the genotyped animals $\boldsymbol{a}_2$. This is like applying the selection index that we saw in section 3.1, but instead of predicting $\boldsymbol{a}_1$ from the observed values $\boldsymbol{y}$, we will make the prediction from the additive values $\boldsymbol{a}_2$:

$$\boldsymbol{a}_1 = \text{cov}(\boldsymbol{a}_1, \boldsymbol{a}_2) \cdot [\text{var}(\boldsymbol{a}_2)]^{-1} \cdot \boldsymbol{a}_2 + \boldsymbol{\varepsilon}$$

We can break down the relationship matrix $\boldsymbol{A}$ into the submatrix of the non-genotyped animals $\boldsymbol{A}_{11}$, that of the genotyped animals $\boldsymbol{A}_{22}$ and the part of the relationship matrix that relates the genotyped to the non-genotyped animals $\boldsymbol{A}_{12}$ and $\boldsymbol{A}_{21}$:

$$\boldsymbol{A} = \begin{bmatrix} \boldsymbol{A}_{11} & \boldsymbol{A}_{12} \\ \boldsymbol{A}_{21} & \boldsymbol{A}_{22} \end{bmatrix}$$

Then, the prediction of $\boldsymbol{a}_1$ is:

$$\boldsymbol{a}_1 = \boldsymbol{A}_{12} \cdot \boldsymbol{A}_{22}^{-1} \cdot \boldsymbol{a}_2 + \boldsymbol{\varepsilon}$$

The residual and additive variances are calculated, taking into account that $\boldsymbol{A}'_{12} = \boldsymbol{A}_{21}$ $\text{var}(\boldsymbol{a}_1) = \boldsymbol{A}_{11}\sigma_A^2$, and $\text{var}(\boldsymbol{a}_2) = \boldsymbol{A}_{22}\sigma_A^2$:

$$\varepsilon = a_1 - A_{12} \times A_{22}^{-1} \cdot a_2$$

$$\text{var}(\varepsilon) = \text{var}(a_1) - A_{12} \cdot A_{22}^{-1} \cdot \text{var}(a_2) \cdot A_{22}^{-1} \cdot A_{12}' = \left(A_{11} - A_{12}A_{22}^{-1}A_{21}\right)\sigma_A^2$$

$$\begin{aligned}\text{var}(a_1) &= \text{var}\left(A_{12} \cdot A_{22}^{-1} \cdot a_2 + \varepsilon\right) = A_{12} \cdot A_{22}^{-1}\,\text{var}(a_2) A_{22}^{-1} \cdot A_{12}' + \text{var}(\varepsilon) = \\ &= A_{12}A_{22}^{-1}GA_{22}^{-1}A_{21}\,\sigma_A^2 + \left(A_{11} - A_{12}A_{22}^{-1}A_{21}\right)\sigma_A^2 = \\ &= A_{11}\,\sigma_A^2 + A_{12}A_{22}^{-1}GA_{22}^{-1}A_{21}\,\sigma_A^2 - A_{12}A_{22}^{-1}A_{21}\,\sigma_A^2\end{aligned}$$

If we multiply the last term by the identity matrix $I = A_{22}A_{22}^{-1}$, it does not change, and we can take $A_{12}A_{22}^{-1}$ as a common factor as well as its symmetric $A_{22}^{-1}A_{21}$; then, var(a_1) becomes:

$$\begin{aligned}\text{var}(a_1) &= A_{11}\sigma_A^2 + A_{12}A_{22}^{-1}GA_{22}^{-1}A_{21}\,\sigma_A^2 - A_{12}A_{22}^{-1}A_{22}A_{22}^{-1}A_{21}\,\sigma_A^2 = \\ &= A_{11}\sigma_A^2 + A_{12}A_{22}^{-1}\left(G - A_{22}\right)A_{22}^{-1}A_{21}\,\sigma_A^2\end{aligned}$$

We know that var(a_2) = G, and thus we can calculate the covariances between a_1 and a_2:

$$\text{cov}(a_1, a_2) = \text{cov}\left(A_{12} \cdot A_{22}^{-1} \cdot a_2, a_2\right) = A_{12} \cdot A_{22}^{-1}\,\text{var}(a_2) = A_{12} \cdot A_{22}^{-1} \cdot G$$

$$\text{cov}(a_2, a_1) = G \cdot A_{22}^{-1} \cdot A_{21}$$

Finally, we arrive at:

$$H = \begin{bmatrix} \text{var}(a_1) & \text{cov}(a_1, a_2) \\ \text{cov}(a_1, a_2) & \text{var}(a_2) \end{bmatrix} = \begin{bmatrix} A_{11}\sigma_A^2 + A_{12}A_{22}^{-1}(G - A_{22})A_{22}^{-1}A_{21}\sigma_A^2 & A_{12} \cdot A_{22}^{-1} \cdot G \\ G \cdot A_{22}^{-1} \cdot A_{21} & G \end{bmatrix}$$

which, after some algebraic transformations, becomes:

$$H = A + \begin{bmatrix} A_{12}A_{22}^{-1} & 0 \\ 0 & I \end{bmatrix}\begin{bmatrix} I \\ I \end{bmatrix}[G - A_{22}][I\ \ I]\begin{bmatrix} A_{22}^{-1}A_{21} & 0 \\ 0 & I \end{bmatrix}$$

However, the inverse of H has a very simple expression:

$$H^{-1} = A^{-1} + \begin{bmatrix} 0 & 0 \\ 0 & G^{-1} - A_{22}^{-1} \end{bmatrix}$$

In order to prove this, the reader can multiply the last two formulas and obtain the identity matrix. A derivation of H^{-1}can be found in Aguilar *et al.* (2010).

Main Concepts

- *SNP (single-nucleotide polymorphism):* The simplest molecular marker, which occurs at a specific position in the genome where there is variability in a single nucleotide.

For example, at a specific point in the DNA chain, some animals may have the base guanine (G) or adenine (A), so we have three possible genotypes at this position, GG, GA and AA.

- ***Coding of SNPs:*** This is the code used to identify the heterozygote and each of the homozygotes of a SNP. There are several possible codes; a common one is (0, 1, 2) to indicate, respectively, the most frequent homozygote, the heterozygote, and the least frequent homozygote. The encoding (1, 0 −1) is also often used. This coding facilitates the implementation of additive models.
- ***Reference population:*** The population from which the phenotypes and SNPs of genotyped animals are used to establish the prediction equations. These equations should be re-established every three or four generations because recombination affects predictions, although at present it is more common to re-establish a fraction of the reference population in each generation.
- ***Genomic selection:*** The selection that uses genomic markers (SNPs) in the prediction of the additive value of individuals, in addition to phenotypic data. It can dramatically increase the response to selection (e.g. >50% in dairy cattle) or simply dramatically increase the response before using genomic selection (around 20–30% in prolific species).
- ***Genomic matrix:*** A kinship matrix based on the actual similarities between genotyped individuals according to the common SNP alleles they share.
- ***RR-BLUP (random regression best linear unbiased prediction):*** A prediction method that assigns to all the markers (SNPs) the same a priori uncertainty, defined with a normal distribution.
- ***G-BLUP (genomic-BLUP):*** A BLUP that uses the genomic matrix instead of the kinship matrix. It gives the same results, under certain general conditions, as RR-BLUP.
- ***Bayes A:*** A prediction method that assigns to all markers (SNPs) the a priori uncertainty defined with a t distribution.
- ***Bayes B:*** The same method as Bayes A but assigning zero value to a high proportion π of the SNPs (e.g. $\pi = 95\%$).
- ***Bayes C:*** The same method as RR-BLUP but assigning zero value to a high proportion π of the SNPs, as in Bayes B.
- ***Bayes Cπ:*** The same method as Bayes C but assigning a prior density to the value of π instead of fixing it.
- ***Bayes L:*** A prediction method that assigns to all markers (SNPs) the same prior uncertainty, defined with a Laplace distribution.
- ***Bayes R:*** A prediction method that assigns a priori to all the markers (SNPs) a mixture of four normal distributions.

Problems

6.1. Calculation of the genomic matrix $\boldsymbol{G}$. In the example of section 6.2.2, we assign the values 2, 1 and 0 to the allele with the highest frequency in each SNP. However, it is more common to consider the lowest frequency allele for the 2, 1, 0 assignment, for practical reasons in file management (to debug low-frequency SNPs). Repeat the analysis assigning 2, 1, 0 to the lowest frequency allele.

6.2. Starting from the four genotyped and phenotyped animals, where the first column is the phenotype and each of the following six columns the values of the SNPs, predict the additive values of the four animals that do not have data.

	Phenotype	SNP1	SNP2	SNP3	SNP4	SNP5	SNP6
Animal 1	100.2	1	1	1	1	0	1
Animal 2	102.3	1	2	1	1	2	2
Animal 3	108.7	1	0	2	2	2	0
Animal 4	99.0	0	1	1	2	2	0
Animal 5		0	1	2	1	1	2
Animal 6		0	2	1	0	0	0
Animal 7		1	1	2	2	1	1
Animal 8		1	1	1	0	1	1

The heritability is $h^2 = 0.3$ and the phenotypic variance is 1. Calculate the prior variance σ_α^2 of each marker as the additive variance divided by the number of markers (see section 6.2.1).

6.3. If we have the individual effect of each SNP, could we select animals having all the favourable versions without the need to recalculate the prediction equations every few generations?

6.4. If we detect large- and medium-effect genes by GWAS, can we keep the favourable versions of all of these genes?

6.5. Genomic selection adds more information to the classic BLUP, for example by better specifying the relationship matrix. Why is genomic selection not used in all species?

Self-assessment Questions (True or False)

1. Genomic selection with chips of 50,000 SNPs is much more efficient than with chips of 3000 SNPs.

2. Genomic selection with 1,200,000 SNP chips is much more efficient than with 50,000 SNP chips.

3. Genomic selection, if the whole genome sequence is readily available, would be much more efficient than current selection with chips of 50,000 SNPs.

4. The effects of SNPs are the effects of the genes involved in the trait (they are 'causal variants').

5. A SNP always has the same effect, whether the analysis is done with chips of 50,000 SNPs or with chips of 850,000 SNPs.

6. The variance of the SNP describes our a priori uncertainty about the effect of the SNP.

7. In Bayes B, we give a priori value zero to SNPs with a given probability, so we cannot ensure a priori that some specific SNPs have zero value.

8. The efficiency of genomic selection is expected to be similar in dairy cattle, pigs and poultry.

9. Genomic selection is a new paradigm in selection, and selection methods have radically changed since SNPs have been used.

10. Genomic selection has contributed to an overall increase of the response to selection in poultry and pigs by around 20–30%.

References

Aguilar, I., Misztal, I., Johnson, D.L., Legarra, A., Tsuruta, S. and Lawlor, T.J. (2010) Hot topic: a unified approach to utilize phenotypic, full pedigree, and genomic information for genetic evaluation of Holstein final score. *Journal of Dairy Science* 93, 743–752. DOI: 10.3168/jds.2009-2730

Bernardo, R. (2016) Bandwagons I, too, have known. *Theoretical and Applied Genetics* 129, 2323–2332. DOI: 10.1007/s00122-016-2772-5

Blasco, A. (2008) The role of genetic engineering in livestock production. *Livestock Science* 113, 191–201. DOI: 10.1016/j.livsci.2007.03.012

Blasco, A. and Toro, M.A. (2014) A short critical history of the application of genomics to animal breeding. *Livestock Science* 166, 4–9. DOI: 10.1016/j.livsci.2014.03.015

Blasco, A. and Pena, R. (2018) Current status of genomic maps, genomic selection/GBV in livestock. In: Niemann H. and Wrenzycki C. (eds). *Animal Biotechnology 2: Emerging Breeding Technologies*. Springer, Cham, Switzerland, pp. 61–80.

Chen, C.Y., Misztal, I., Aguilar, I., Tsuruta, S., Meuwissen, T.H.E. *et al.* (2011) Genome-wide marker-assisted selection combining all pedigree phenotypic information with genotypic data in one step: an example using broiler chickens. *Journal of Animal Science* 89, 23–28. DOI: 10.2527/jas.2010-3071

de los Campos, G., Naya, H., Gianola, D., Crossa, J., Legarra, A. *et al.* (2009) Predicting quantitative traits with regression models for dense molecular markers and pedigree. *Genetics* 182, 375–385. DOI: 10.1534/genetics.109.101501.

Erbe, M., Hayes, B.J., Matukumalli, L.K., Goswami, S., Bowman, P.J. *et al.* (2012) Improving accuracy of genomic predictions within and between dairy cattle breeds with imputed high-density single nucleotide polymorphism panels. *Journal of Dairy Science* 95, 4114–4129. DOI: 10.3168/jds.2011-5019

Fernando, R. and Garrick, D. (2013) Bayesian methods applied to GWAS. In: Gondro, C., van der Werf, J. and Hayes, B. (eds) *Genome-wide Association Studies and Genomic Prediction*. Springer, Cham, Switzerland, pp. 237–274.

Fernando, R.L., Habier, D., Stricker, C., Dekkers, J.C.M. and Totir, L.R. (2007) Genomic selection. *Acta Agriculturae Scandinavica A* 57, 192–195. DOI: 10.1080/09064700801959395

Forni, S., Aguilar, I., and Misztal, I. (2011) Different genomic relationship matrices for single-step analysis using phenotypic, pedigree and genomic information. *Genetics Selection Evolution* 43: 1. DOI: 10.1186/1297-9686-43-1

Gianola, D., de los Campos, G., Hill, W.G., Manfredi, E. and Fernando, R.L. (2009) Additive genetic variability and the Bayesian alphabet. *Genetics* 183, 347–363. DOI: 10.1534/genetics.109.103952

Habier, D., Fernando, R.L., Kizilkaya, K. and Garrick D.J. (2011) Extension of the Bayesian alphabet for genomic selection. *BMC Bioinformatics* 12: 186. DOI:10.1186/1471-2105-12-186

Ibañez-Escriche, N. and Blasco, A. (2011) Modifying growth curve parameters by multitrait genomic selection. *Journal of Animal Science* 89, 661–668. DOI: 10.2527/jas.2010-2984

Ibáñez-Escriche, N., Forni, S., Noguera, J.L. and Varona, L. (2014) Genomic information in pig breeding: science meets industry needs livestock science. *Livestock Science* 166, 94–100. DOI: 10.1016/j.livsci.2014.05.020

Jonas, E. and de Koning, D.J. (2015) Genomic selection needs to be carefully assessed to meet specific requirements in livestock breeding programs. *Animal Frontiers* 6: 49. DOI: 10.3389/fgene.2015.00049

Legarra, A., Aguilar, I. and Misztal, I. (2009) A relationship matrix including full pedigree and genomic information. *Journal of Dairy Science* 92, 4656–4663. DOI: 10.3168/jds.2009-2061

Legarra, A., Lourenco, D.A.L. and Vitezica, Z. (2024) *Bases for Genomic Prediction*. Available at: https://learnanimalbreeding.com/files/Legarra_Book.pdf (accessed 10 July 2025).

Mancin, E., Sosa-Madrid, B.S., Blasco, A. and Ibáñez-Escriche, N. (2021) Genotype imputation to improve the cost-efficiency of genomic selection in rabbits. *Animals* 11: 803. DOI: 10.3390/ani11030803

Meuwissen, T.H., Hayes, B.J. and Goddard, M.E. (2001) Prediction of total genetic value using genome wide dense marker maps. *Genetics* 157, 1819–1829. DOI: 10.1093/genetics/157.4.1819
Rolf, M.M., Decker, J.E., McKay, S.D., Tizioto, P.C., Branham, K.A. *et al.* (2014) Genomics in the United States beef industry. *Livestock Science* 166, 84–93. DOI: 10.1016/j.livsci.2014.06.005
Rupp, R., Mucha, S., Larroque, H., McEwan, J. and Conington, J. (2016) Genomic application in sheep and goat breeding. *Animal Frontiers* 6, 39–44. DOI: 10.2527/af.2016-0006
Shumbusho, F., Raoul, J., Astruc, J.M., Palhiere, I., Lemarié, S. *et al.* (2016) Economic evaluation of genomic selection in small ruminants: a sheep meat breeding program. *Animal* 6, 1033–1041. DOI: 10.1017/S1751731115002049
Soller, M. (1978) The use of loci associated with quantitative traits in dairy cattle improvement. *Animal Production* 27, 133–139. DOI: 10.1017/S0003356100035960
Sorensen, D. (2025) *Statistical Learning in Genetics*, 2nd edn. Springer, Cham, Switzerland.
Wolc, A., Zhao, H.H., Arango, J., Settar, P., Fulton, J.E. *et al.* (2015) Response and inbreeding from a genomic selection experiment in layer chickens. *Genetics Selection Evolution* 47: 59. DOI: 10.1186/s12711-015-0133-5

Solutions

Solutions to the problems

6.1. The same results are obtained.

6.2. Predictions come from RR-BLUP (section 6.2.1). The model is:

$$y = \mu + \alpha_1 SNP_1 + \alpha_2 SNP_1 + \alpha_3 SNP_1 + \alpha_4 SNP_1 + \alpha_5 SNP_1 + \alpha_6 SNP_1 + e$$

As the data are not centred, we have the mean μ as a fixed effect, so we will apply (section 6.2.1):

$$\begin{bmatrix} X'X & X'W \\ W'X & W'W + I\dfrac{\sigma_e^2}{\sigma_\alpha^2} \end{bmatrix} \begin{bmatrix} b \\ \alpha \end{bmatrix} = \begin{bmatrix} X'y \\ W'y \end{bmatrix}$$

$$X = \begin{bmatrix} 1 \\ 1 \\ 1 \\ 1 \end{bmatrix},\ y = \begin{bmatrix} 100.2 \\ 102.3 \\ 108.7 \\ 99.0 \end{bmatrix},\ W = \begin{bmatrix} 1 & 1 & 1 & 1 & 0 & 1 \\ 1 & 2 & 1 & 1 & 2 & 2 \\ 1 & 0 & 2 & 2 & 2 & 0 \\ 0 & 1 & 1 & 2 & 2 & 0 \end{bmatrix},\ \begin{matrix} \sigma_A^2 = h^2\sigma_P^2 = 0.3 \\ \sigma_e^2 = \sigma_P^2 - \sigma_A^2 = 0.7 \\ \sigma_\alpha^2 = \dfrac{\sigma_A^2}{6} = 0.05 \end{matrix}$$

The results are:

$\alpha_1 = 0.252,\ \alpha_2 = -0.359,\ \alpha_3 = 0.372,\ \alpha_4 = 0.119,\ \alpha_5 = 0.264,\ \alpha_6 = -0.107$

Animal 5: $0.252 \cdot 0 - 0.359 \cdot 1 + 0.372 \cdot 2 + 0.119 \cdot 1 + 0.264 \cdot 1 - 0.107 \cdot 2 = 0.554$. Using the same procedure, animal 6: −0.346, animal 7: 1.033, animal 8: 0.423

6.3. SNPs are not genes but markers, which implies that there may be a crossover between the gene and the marker, and sometimes the unfavourable version is the one that is linked to the marker, as shown in Fig. 1.2 of section 1.1.2, which is the reason why the relationships between traits and markers have to be recalculated from time to time. Moreover, two very close SNPs will capture the effect of the same genes, or almost the same ones, so their effects cannot be added as if they were markers for the effects of independent genes.

6.4. GWAS have contributed little to animal genetic improvement, essentially because there are no genes with a greater effect, or because if there are, they can often be captured without the need for markers (see section 10.1.2.C). Genomic selection makes it possible for the effects of genes with a major or intermediate effect to be captured at the same time as that of genes with a minor effect, as we saw in section 6.3.2, but in practice and for most traits, the methods that consider that the effects of genes are small and that many genes affect a character (section 6.2) usually give the best results.

6.5. Because it is expensive to implement. In some species, such as dairy cattle, the cost of a chip is very small compared with the cost of a breeding stock, but in others, such as rabbits, the cost of a chip is very high compared with the cost of a breeding female (see section 8.3.2), so genomic selection is, at the moment, unaffordable. Through imputation, the costs of a chip are significantly reduced; although currently the cost is still high in the case of rabbits, in the near future it may be possible to implement this.

Solutions to self-assessment questions

1. T, 2. F, 3. F, 4. F, 5. F, 6. T, 7. T, 8. F, 9. F, 10. T.

7 Breeds and Biodiversity

Abstract
This chapter deals with the concept of breed and its use in animal breeding, and with biodiversity and the problem of conservation of endangered breeds. It begins by highlighting the difficulties in defining what a breed is, the limited usefulness of this concept in animal production, and why breed is not a technical term and has no biological meaning. The rationale for biodiversity and breeds conservation is then examined, discussing some clichés that bear little resemblance to reality. The classification of endangered breeds according to the Food and Agriculture Organization of the United Nations is presented. The issue of sustainability and the use of local breeds in animal production is discussed. The chapter ends with a presentation of conservation procedures, tools and programmes.

7.1 The Concept of Breed in Animal Genetic Improvement

Unlike the genetic improvement of plants, the use of a popular concept in livestock, the concept of breed, has had an important role in the beginnings of animal breeding, and even today is widely used in some areas such as conservation of genetic resources. In some animal productions such as companion animals, the breed is still an important concept, and the breed used is also decisive for the definition of some products with a protected designation of origin. However, the concept of breed is much less useful for industrialized animal genetic improvement, even when certain breeds are used preferentially, such as in milk or white-egg production. In fact, the demands on compliance with certain morphological standards and the obligation of the animals of a breed to be inscribed in herd books caused problems and serious delays in the development of industrial animal genetic improvement in the 1950s and 1960s. Currently, commercial companies use the better-defined terms of *lines* or *populations* in their breeding programmes, and the use of mixtures of animals of different breeds to create *synthetic breeds* is frequent in animal genetic improvement.

7.1.1 Breeds

There is no consensus on the definition of what a breed is. Rodero and Herrera (2000) compared the common elements of 16 definitions of breed and the only requirement common to all of them was genetic homogeneity, which applies essentially to external traits. This is common when defining what a breed is, and is implicit in the definitions collected by several authors in different articles. The simplest definition of breed comes from two of the most important animal breeding geneticists:

> A population of farm animals is a breed, for practical purposes,
> a) when it has some identifying features, or
> b) when it has a formal association of breeders, or
> c) when certain government officials say it is.
>
> Lerner and Donald (1966)

DOI: 10.1079/9781836990994.0007

This definition emphasizes that the external characteristics defining a breed should be proposed by someone, normally an association that wants to govern the records in a herd book authorized by the government and get subsidies for the preservation of the breed and management of the herd book. The organizer and former director of the Food and Agriculture Organization of the United Nations (FAO)'s international genetic conservation programme, Keith Hammond, gave an even broader definition:

> A breed is a breed if enough people say it is.
>
> Cited as personal communication in Woolliams and Toro (2007)

Without going to the Hammond extreme, it seems that a breed is simply a set of animals with external characteristics defined by some organism, company or association; these are animals that are kept in closed reproduction, and to which certain productive performances are associated as an orientation. A breed requires some people to decide what the external characteristics of the breed are. These animals should generally be kept in closed reproduction, with the herd book only allowing animals of the breed. In the most extreme case, only descendants of animals that are previously inscribed in the herd book are admitted as animals of this breed.

The main problem with this definition is that it depends too much on external characteristics, which can be very useful for dog breeds but not necessarily for animals whose main objective is to produce meat or milk efficiently. Some average performances are often attributed to the breed, but average performances change with time, place, feeding, handling, facilities, selection programmes, etc., and thus these attributions are not particularly useful.

This dependence on external characteristics produces some paradoxes; for example, sometimes an attempt is made to recover an extinct breed starting from animals supposedly close to this breed and selecting the external characteristics that the extinct breed had (e.g. external morphology in pigs or a certain colour of plumage in poultry). The problem is that the productive characteristics for which the extinct breed was known, such as an extraordinary quality of meat, are not included in the recovery process, so in the end we have a set of animals 'disguised' as the extinct breed, only having an external appearance of this breed.

'Breed' is not a technical term nor does it have any biological meaning, something that has been stressed for many years; for example, in 1915, a professor at the University of Iowa published in the prestigious scientific magazine *Journal of Heredity* an article entitled 'What is a breed?', in which he said:

> The word breed has no biological meaning… A breed is whatever farmers want to call it, there are not natural boundaries, and no arbitrary ones that are universally accepted.
>
> O. Lloyd-Jones (1915)

> *'Breed' is not a technical term nor does it have any biological meaning.* In genetic improvement, the technical term *line* is used to indicate a group of animals in closed reproduction selected by one or more traits.

Let us consider some important issues related to breeds:

- *Breeds are a human product.* Breeds are human creations destined to be tools to produce milk, meat, eggs, work or supply any other human need. In nature, there are no breeds; rather, there are subspecies of a certain species. The breeds were created by farmers and most of them are very recent, dating from the 19th century onwards.

Breeds are human creations whose objective is the production of food, work, etc. They are *tools* created by humans to meet their needs. They should not be confused with species found in nature.

Breed conservation must take into account its usefulness. If a breed is a tool for making meat, milk or eggs, conservation should focus on whether this tool works well or not, or if there are expectations for using this tool in the future. In other areas, we do not maintain tools that are not going to be useful; old computers or steam engines are not kept, except in museums of technology, which would be the equivalent of zoological parks of breeds. We will deal with the problem of conservation later.

- *New breeds can be created without difficulty*. It is not difficult to create a breed. Today we continue creating new breeds. Aside from pets, many pig, rabbit and poultry companies use synthetic breeds with no regard for external characteristics other than functional traits, and synthetic breeds are also used in dairy sheep and beef cattle. This is an important point to remember, because the extinction of a breed is totally different from the extinction of a species. The extinction of a species by humans is a catastrophe, but we cannot compare the extinction of the breed with the irrecoverable loss of a species created by natural evolution as part of a particular ecosystem. The breeds can be recreated or improved if it is convenient for their use, but the species are definitively lost.
- *How many breeds do we have?* This is a difficult question, given the imprecision in the definition of what a breed is. Creating a new breed is easy, and defining breeds by external characteristics leads to the creation of many breeds that are intermediate types, with almost the same genetic background for productive traits, and with the frequent purpose of obtaining subsidies or administrative advantages. According to the FAO programme (Domestic Animal Diversity Information System (DAD-IS), https://fao.org/dad-is; accessed 10 July 2025), we have more than 15,000 breeds of domestic animals, but this figure is very approximate because it depends on what each country declares a breed to be. In addition, some breeds that receive a particular name in some countries are listed as different breeds (e.g. the Brown Swiss cow is called Bruna dels Pirineus in Catalonia and has its own herd book). The problem is how many of them are in danger of extinction and whether they should be conserved. The reasons for investing in endangered breeds are discussed below.
- *What is the value of a breed?* The main value of a breed lies in its usefulness, usually in its productive capacity in certain environments. For example, in intensive white-egg production, the Leghorn breed is practically the only one used because its production exceeds any other; in dairy cattle, the predominant breed is the Holstein-Friesian for the same reason. In extensive and unfavourable productive circumstances, other breeds (not necessarily local breeds) may be more productive, but determining this requires information, something that is often lacking. For example, in Ethiopia, of the 35 beef cattle breeds listed in the FAO DAD-IS programme, productive information is lacking in 90% of cases. A specific breed is not necessarily the most suitable in a given environment, and alternatives could be proposed, such as mixtures of different breeds, crosses, creation of synthetic breeds, etc.

Before maintaining a breed or importing a breed, it is worthwhile doing a small experiment to compare the productive performances between alternatives.

7.1.2 Pure breeds and pure lines in animals

The term '*pure line*' refers to a group of animals (not necessarily of a breed) that are kept in closed reproduction; it has a different meaning from the expression pure line applied to plants. In plants, a pure line is made up of individuals homozygous at all loci (Fig. 7.1). These lines are achieved by mating relatives and, where possible, by self-fertilization. Mating relatives increases inbreeding, reduces the number of heterozygotes (see section 1.2.1), and increases homozygotes leading to allele fixation. The reason for creating these lines is that the crossing of two of these lines sometimes produces heterozygotes with reproductive advantages, such as maize with ears having more grains and better resistance.

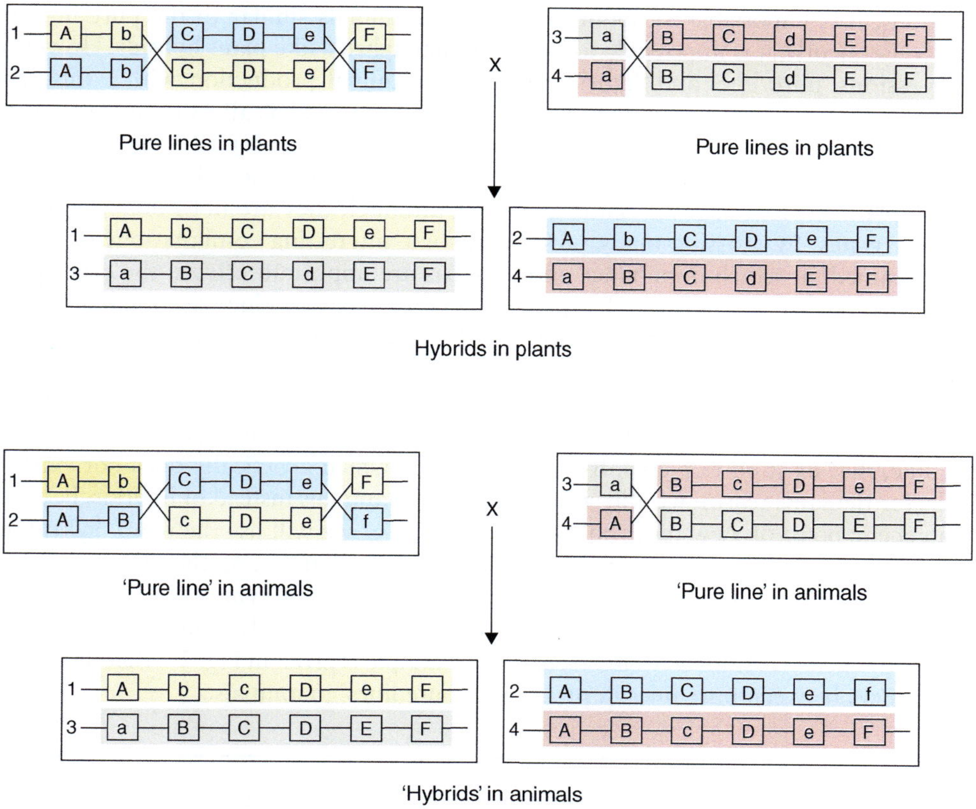

Fig. 7.1. Pure lines and hybrids in plants and 'pure lines' and 'hybrids' in animals. Note that in plants, the pure lines are totally homozygous and all hybrids are genetically identical, which is not the case with 'pure lines' and 'hybrids' in animals.

In general, the crossing of two populations with high inbreeding, not related to each other, increases the proportion of heterozygotes (see section 1.2.4). Some traits, especially reproductive traits, are affected by inbreeding, and have worse yields when inbreeding increases. As inbreeding disappears in the cross and the proportion of heterozygotes increases, individuals with better yields for these traits are produced. This is the phenomenon that we call heterosis, and which is popularly known as *hybrid vigour*, because the product of the cross, which in plants is called a *hybrid*, is generally more robust than the average of the parental lines (e.g. it is more resistant to diseases). In addition, as all hybrids are genetically the same, the product that is sold is more uniform, because it only varies due to environmental causes. There is an additional advantage for plant-breeding companies, as farmers must always buy the seeds of the pure lines from the same company to make a particular hybrid; if two hybrids are crossed, the offspring are no longer genetically uniform. This has led to the production of many highly inbred populations in plants, to later cross them and commercialize the best cross. In animals, the same system was tried in the 1930s and 1940s, but the results were disappointing due to the sharp decrease in fertility and viability when inbreeding increased. An ear has many grains, but a domestic animal, even in the case of poultry, does not have enough offspring to withstand the drop in fertility that occurs when crossing close relatives.

Today, inbred populations are no longer used in animal breeding. By not using inbred populations, it is impossible to obtain pure lines of animals similar to the pure lines of plants; consequently, it is impossible to create animal hybrids like plant hybrids. However, the expressions 'pure line' and 'hybrid' are used in animal production with a different meaning to their use in plant production. A pure line in plants has all its genes fixed; that is, it has the same homologous chromosomes, it is homozygous for all its genes and all individuals have the same genotype, so the product of the cross (the hybrid) is always the same. All plant hybrids have the same genotype, although they are not homozygous for all of their genes. By crossing many pure lines, some interesting combinations of alleles can be obtained, in which dominance and epistasis are important effects; the other pure lines are then discarded and those that produce a particularly productive combination are sold. This cannot be done in animals not only because of the impossibility of obtaining pure lines but also because of the cost of crossing many lines with each other. A 'pure line' in animals is simply a closed breeding population, and a 'hybrid' animal is the crossing of two such 'pure lines'. Here we will use quotation marks to distinguish what is genetically a pure line, that existing in plants, from a 'pure line', which genetically does not exist in animals; for the same reason we will speak of crosses in animals instead of hybrids, because from a genetic point of view these only exist in plants, or we will use the word 'hybrid' in quotation marks. Another reason that has contributed to abandoning the programmes of inbred lines in animals is a problem that also occurs in plants: a hybrid is a fixed product that cannot be improved, while the non-inbred lines are constantly being improved, and consequently their crossings as well. A schematic example is given in Fig. 7.1 that illustrates the differences between pure lines in plants and 'pure lines' in animals.

> *There are no pure lines in animals* like those in plants, with their genes fixed. Consequently, *there are no hybrids in animals* either. The use of expressions like 'pure line' and 'hybrid' referring to animals is simply a popular way of speaking, not a scientific expression.

Over time, in a 'pure line' of animals, some alleles will be fixed by genetic drift (see section 1.1.3). It is to be expected that the alleles that are fixed in a line by chance will not be exactly the same as in other lines, and therefore the crossing of two 'pure lines' will have more heterozygotes than the 'pure lines' and their performance in the traits affected by inbreeding will be greater. Even if the alleles are not fixed, genetic drift will make the frequencies of these alleles in the two populations different, which also produces more heterozygotes in the cross (see section 1.2.4). The longer each line has been in close reproduction, the more heterozygous the 'hybrid' will be and the higher its performance will be; two different breeds have been in closed reproduction for a long time, so this is the reason for using breeds as 'pure lines' in animals. Only when a breed has performances much higher than others is it not crossed with other breeds, as the 'hybrid' would be inferior to this breed. In this case, crosses between 'pure lines' of that breed are used to produce the 'hybrids'. As there are no true 'pure lines' in animals, we can make a new 'pure line' by crossing several breeds or several 'pure lines' and then keeping the new 'pure line' in closed reproduction.

The term 'pure breed' refers, in animals, to members of a breed that show the external characteristics of the breed, and it is also sometimes required to come from animals registered in herd books. The animals registered in these books should also have their parents registered in the book, although there are open books that allow registration of animals meeting the racial characteristics even if their parents are not registered animals. The traits that make up the external appearance depend on few genes, so they are easier to fix by increasing consanguinity, without reaching extremes in which the survival of the breed is compromised. This is useful when pursuing the achievement of animals with morphological characteristics that are transmitted exactly from generation to generation, such as fur animals, pets, etc. In the case of industrial animal production, the concept of 'pure breed' is of little use, as within each 'pure breed' there are 'pure lines' with different productive characteristics. For example, there are Large White pigs that are lean, others with a high fat content, some with a low conversion rate of feed into lean meat and others with a high rate, some with a high litter size and others not, etc.

7.2 Conservation of Endangered Breeds

The phenomenon of globalization – intensive animal production at an industrial level and facilities in the transport and storage of gametes (semen and frozen embryos) – has made 'pure lines' and 'hybrids' with high productivity available to many farmers around the world. These lines have progressively replaced local breeds as animal production has been intensified, and, in the case of extensive productions, other breeds from other countries have even replaced local breeds because they have been shown to perform better in the local conditions. This has led to the disappearance of a number of local breeds that were not sufficiently productive, even in their own environment. This should not be a problem, as breeds are human creations for satisfying our needs for food, clothing, etc., and the creation of new breeds does not represent a special difficulty. However, attention has been drawn to the fact that disappearing breeds may have genes that are useful in the future, or genes for specific production niches, and therefore they deserve to be preserved. In the developed world, with urban populations far from any rural environment, local breeds have been considered part of the local culture and it has been proposed that they should be conserved as a part of the local cultural heritage. For all these reasons, conservation programmes have been proposed at local, national and global levels. We will now examine the rationale for conservation programmes and the procedures commonly used to conserve breeds.

7.2.1 Breeds in danger of extinction

The FAO has a global programme for the conservation of genetic resources, DAD-IS. Its database provides information on known breeds, as well as contacts with national coordinators dealing with the management of animal genetic resources. The FAO has classified breeds according to their status in terms of danger of extinction; the simplified categories are:

- *Extinct.* A breed is classified as extinct when there are no breeding males and any cryopreserved genetic material that may be available is insufficient for reconstituting the breed.
- *Cryopreserved.* These are breeds that do not have live animals, but there is enough cryopreserved material (gametes, embryos and even tissues capable of providing material for somatic cloning) to allow the reconstitution of the breed.
- *In critical condition.* These are breeds where any of the following conditions are met:
    - The total number of reproductive females is less than or equal to 100 (300 in the case of species with low reproductive capacity, such as ruminants (cattle, sheep, goats, etc.); we will put the numbers necessary for these species in parentheses).
    - The total population size is less than or equal to 80 (240) and the population is increasing.
    - The total size of the population is less than or equal to 120 (360) and the population trend is stable or decreasing.
    - The total number of breeding males is less than or equal to five.
- *In danger of extinction.* These are breeds where any of the following conditions are met:
    - The total number of breeding females is greater than 100 (300) and less than or equal to 1000 (3000).
    - The total population size is greater than 80 (240) and less than 800 (2,400) and the population is increasing.
    - The total size of the population is greater than 120 (360) and less than or equal to 1200 (3600) and the population trend is stable or decreasing.
    - The total number of breeding males is less than or equal to 20 and greater than five.
- *Vulnerable.* A breed is classified as vulnerable if:
    - The total number of breeding females is between 1000 and 2000 (3000 and 6000).
    - The total population size is greater than 800 (2400) and less than or equal to 1600 (4800) and the population is increasing.
    - The total population size is greater than 1200 (3600) and less than or equal to 2400 (7200) and the population trend is stable or decreasing.
    - The total number of breeding males is between 20 and 35.
- *Not at risk.* A breed is classified as not at risk if the status of the population is known and the breed does not fall into the above categories.
- *Unknown.* Status unknown, or status known but the breed is presumed to be in critical, endangered or vulnerable condition.

A low value for the least favourable parameter is enough to result in the breed being assigned to the highest risk status category. If the evolution of the population is not known, it is assumed that it is stable. It is generally assumed that when the population is growing, the percentage of females that interbreed with males of the same breed is higher than 80%.

7.2.2 Reasons for keeping breeds in danger of extinction

There is an extensive bibliography (e.g. Ruane, 1999, or Gandini and Oldenbroek, 2007) that sets out the reasons for conserving endangered breeds. There is a general idea that breeds must be conserved in order to conserve the planet's genetic biodiversity, or to maintain genes that may be useful in the future, or that must be conserved for cultural reasons.

> There would be no problem keeping all possible breeds if we had a large amount of funding that did not compete with other priorities. However, keeping a breed that is not profitable is expensive and takes funds from other relevant activities such as feeding the population at a reasonable cost, something particularly important in developing countries.

We must admit that we cannot afford to maintain all the breeds that live on our planet, and even if we could afford it, other more urgent priorities could displace that of the maintenance of this or that breed. However, extinction of many of the endangered breeds is not a great loss. We will now critically examine the arguments for maintaining these breeds, including arguments beyond technical or scientific considerations but which are related to the ways of seeing our role in the world.

The main reasons for keeping breeds in danger of extinction can be grouped into three types of arguments: genetic, productive and ideological reasons, and cultural reasons:

- *By losing breeds, we lose genetic diversity.* This argument presupposes several points:
  - Genetic variability is a good thing that must be preserved.
  - Genetic variation between breeds is much greater than within breeds.
  - Selection leads to extinction of genetic variability.
  - There are genes that are not useful now but may be useful in the future, and which are maintained in unselected breeds.
- *Breeds that are not useful in intensive industrial systems can be used in local productions in poor areas.* This argument implies that in certain poor areas, it is easier, cheaper or more efficient to use local products for food, housing, etc. than in the case of industrial systems. This argument also contains several elements:
  - Local breeds tend to be better adapted to the local environment, mainly when local products are used for food, housing, etc.
  - Local breeds often produce better-quality products.
  - Intensive industrial systems have several environmental drawbacks that must be avoided (e.g. greenhouse emissions, excess of slurry).
  - Intensive industrial systems produce very poor animal welfare.
  - Sustainable systems (excluding sustainable intensive industrial systems, which also exist) allow the independence of farmers from large multinational companies.

- *Breeds are part of the cultural heritage.* This declaration has some implicit arguments:
  - Some groups of people (nations, tribes or groups of people who have some sense of identity) have or had in the past some special eating habits, lived in special houses, wore distinctive clothes, etc., and all of this, including some specific animal breeds from these communities, make up what we consider to be a heritage that must be preserved.
  - By maintaining a herd of these specific animals, we help to maintain this heritage.

  Below, we will examine these arguments in detail.

7.2.3 Maintaining genetic diversity

To deal with the problem of genetic diversity, we must first define what is understood by genetic variability and establish why it is so important to maintain it.

- *Genetic variability:* In section 2.1.1, we saw how phenotypic variance could be broken down into genetic and environmental variance, and in section 1.3.2, we defined genetic variance and its additive and dominant components. Finally, in section 2.3.3.A, we saw that the response to selection is directly proportional to the additive standard deviation, and hence the maintenance of genetic variability is important. However, how to do it and for which traits to do it is a matter that should be examined in detail.
  - *Should breeds or genes be maintained?* An alternative to keeping many breeds with separate programmes could be to create several synthetic breeds with the most important genes we want to preserve, which would reduce the number of conservation programmes and make them more viable by having more unrelated founder individuals.
  - *Should all genetic variability be maintained?* The conservation of genetic variability is usually limited to productive traits, on the one hand because the effective number of a population to maintain all genetic variability is too high and on the other hand because in general productive traits are the only traits of interest in livestock breeds. Even in this case, it makes no sense to maintain the variability of certain traits; for example, for survival, the ideal would be to have 100% survival without any variability. However, it is true that some traits that are not useful today could be useful in the future; we will consider this later in this section.
  - *Can genetic variability be increased?* It can, and is not difficult to do. We could prepare divergent selection programmes to increase and decrease the value of a productive trait on two separate lines, or we could select and mate the extreme individuals (the best and the worst). Except for experiments for specific purposes, this is something that is not done because no one is interested in having low-survival or low-productivity breeds.
  - *Is it useful to maintain the genetic variability of productive traits?* This is needed for selection. If all individuals are genetically identical, as *pure lines* in plants, selection is not possible.
- *Is genetic variability between breeds much higher than within breeds?* As we said earlier, we should first state what type of genetic variability we are talking about. Genetic variability for milk production is not the same as variability for resistance to ticks. Measuring the number of single-nucleotide polymorphisms (SNPs) per kilobase in chicken, the International Chicken Polymorphism Map Consortium (2004) detected

'surprisingly little differences in diversity in comparisons between red jungle fowl and domestic lines, between different domestic lines, and within domestic lines'. For productive traits, it is generally accepted that around 50% of genetic variability occurs between breeds and 50% within breeds (Hall, 2004), but in the classic beef breed comparison experiment carried out at the Animal Breeding Research Organisation in Edinburgh, UK, the variation between breeds in feed conversion ratio was 25% of the total variation (Thiessen *et al.*, 1985); therefore, we cannot say as a general rule that genetic variation between breeds is much greater than within breeds, but we can still emphasize that there is useful variation between breeds that deserves to be conserved.

- *Are selection programmes depleting genetic variability?* The classical theory of quantitative genetics predicts the extinction of genetic variability by selection. The fixation of genes or the proportion of genes that reach high frequencies would reduce genetic variability, which would not be compensated by mutation, as favourable mutant alleles are rare. However, this image of extinction of genetic variability is not in agreement with the current data. Hill (2008) underlined the fact that there is no loss of genetic variability in any commercial breeding programme (see section 10.2). Furthermore, long-term selection experiments in quail and mice do not show any signs of depletion of genetic variance after more than 100 generations of selection. Moreover, the heritability of milk production in dairy cows not only does not decrease over time but actually increases! And this is not only simply to better control of environmental variance or new methods of correcting environmental effects, as the response to selection has been maintained continuously over the years, even increasing in recent years due to the use of genomic selection.

> Animal breeding programmes in intensively exploited species do not appear to be depleting genetic variability. The productive traits continue progressing without showing signs of decline in genetic variability.

The reasons for this seemingly unlimited response to selection are selection pressure on genes produced by mutation and epistatic interactions that generate additive variance useful for selection, as we saw in section 1.3.2. We will discuss the response to long-term selection in more detail in section 10.2.

- *Should we keep genes for the future?* The basis of the discussion for maintaining genetic variability between breeds is that some breeds have genes that other breeds do not have even in low frequency, and that may be useful in the future. Here we must distinguish two types of arguments.
  - *A gene reserve is needed when a rapid change of selection targets is needed.* A gene pool is useful when rapid change of selection targets is demanded. For example, rapid action was required to produce dark-egg hens as productive as white-egg hens, or to reduce fat content in the carcass or increase intramuscular fat in pigs. It is also a safeguard against potential emerging new difficulties, such as the appearance of new diseases.
  - *Preserve all possible genes for an uncertain future.* It can be argued that genes of endangered breeds are not useful today, but that they may play an important role

in the future. The problem with this argument is that it cannot be refuted; if the conditions we refer to in the future are not defined, it is always possible that some current gene will be useful in an uncertain future, and therefore, as it cannot be refuted, this is not a scientific argument. There is nothing wrong with keeping everything with an unlimited number of financial resources, but normally we would organize spending in some order of preference. Keeping breeds is not different from other human activities. We do not save everything just in case we might need it in the future, unless we have a good reason to save it or a clear idea about its future use. There is a pathology called Diogenes syndrome, which involves compulsive hoarding, the indiscriminate collection of a large number of objects. People who suffer from this disorder keep everything they buy, useful or not, for the future, and die with their houses cluttered with all kinds of worthless objects that they have kept throughout their lives. If a breed is a tool, old tools that are no longer useful can be kept in zoos or other types of 'museums', but there is no reason to keep them in productive activity, unless they have some alternative use, such as avoiding desertification of an area that for some reason we do not want to be deserted, cultural reasons that someone wants to finance, etc. We will discuss these reasons below. If there are no such reasons, the breed can become extinct without actually losing much.

7.2.4 Sustainable breeds and systems

A. Intensive and non-intensive sustainable systems

There are many definitions of sustainable systems. In general terms, we understand sustainable agricultural systems to be systems that are capable of maintaining their productivity indefinitely without damaging the environment. This definition does not preclude having intensive sustainable systems with highly productive animals integrated in an industrial food chain, but in most cases a sustainable system is referred to when there is a traditional small-scale agricultural exploitation, based on feeding local breeds with local sources of food, in which the waste is recycled, and in which a fairly high amount of labour is required. Local breeds play a key role in this second type of sustainable system, particularly when environmental conditions are harsh or food resources are not particularly good. This type of sustainable system is generally much less efficient for producing products of animal origin than intensive systems, and in practice it is more contaminating and produce more greenhouse emissions per kilogram of product, as we will see in section 10.3.1. However, there are some reasons to establish them. We will now examine some of the reasons for and against non-intensive sustainable systems:

- *There are environments where no other system will function properly*. A common example is beef cattle in swampy tropical areas. This and other examples can be used to defend non-intensive sustainable systems, but this premise applies essentially to cattle, sheep and goats and not necessarily to pigs, rabbits and poultry, which have traditionally been kept in much less harsh conditions. Crossbreds (commercial 'hybrids') of prolific species from breeding companies tend to work well in developing countries, as pigs, poultry and rabbits tend to have far less environment-dependent farming conditions.

- *By using these systems, desertification is avoided.* There are poor areas where, due to the climate and local conditions, life is particularly hard. Under these conditions, land abandonment and emigration to urban areas is common. The use of breeds adapted to the environment has been promoted, which, together with subsidies for production, try to prevent the desertification of these areas. The arguments to promote sustainable systems in areas that may end up deserted must be made with caution, because they are strongly influenced by ideological premises. Some people may think that urban misery is much worse than rural misery, and that some sustainable systems can help poor people to have a better life. Other people may think that, even in this case, this type of sustainable system can be considered a temporary solution, because these poor people have the right to have a better life for themselves and their descendants; for example, they should have easy access to a hospital, have the opportunity for their children to go to university, be able to enjoy an opera, or have a culture that allows them to read Kant (who, by the way, wrote not only for East Prussians but for the whole of humanity). In this case, non-intensive sustainable systems would only be a temporary solution for people who deserve a better life. However, some people may also think that assuming that university and knowledge of Mozart and Kant is a better life is an ethnocentric consideration, and that people from poor areas and their descendants have a different life project that makes them happier. We should even avoid asking them what they prefer, as classical Western values can be a mirage and can destroy the roots of their civilization; for example, some missionaries with good intentions introduced foreign habits harmful to tribes in Africa and America who were happier before the advent of civilization. These discussions are not technical but ideological, so they are outside the scope of this book.
- *Do non-intensive sustainable systems produce better animal welfare?* This may or may not happen, and we should critically examine it on a case-by-case basis. For example, Australian merino sheep in extensive systems go through periods of starvation and weight loss in a harsh winter, and are attacked by parasites and insects in hot summers; another example is extensive Nelore cattle farming in Brazil in the dry season, which has difficulties in terms of an appropriate intake; their lives are far from what we would consider good well-being, even though they are extensively exploited. As a specialist in poultry welfare, John Webster has stressed that free-range hens in non-intensive systems are not necessarily 'happier' than hens in enriched cages in intensive systems (Webster, 2005, p. 121). Seeking better animal welfare is not a particular task of industrial systems; it affects non-intensive systems as well.

> An intensive industrial system can lead to better animal welfare than certain extensive systems, depending on the conditions of each system.

Improvements in welfare have to be based on scientific experiments that determine when and how animal welfare is affected, and on knowledge about animal biology. A detailed analysis of these issues is beyond the scope of this book, and can be found in texts on animal welfare (e.g. Broom, 2021, and Blasco, 2013, for ethical aspects).

- *Some of the non-intensive sustainable systems provide farmers with independence from large multinational companies.* This may be true, but it is not necessarily a good thing. Feeding people must be a priority for poor countries, and the cheapest way to achieve this may be to buy genes from multinational companies. Genetics is very cheap; the genetics of 1 kg of pork, chicken or rabbit costs less than 1% of the total cost of the meat, as we will see in section 8.1.1.B, and the same can be said of the cost of the genetics of 1 kg of milk, which we will see in section 9.1.2.A. On the other hand, what are the benefits of independence from multinational companies? Some companies offer the cheapest animal protein in the world (eggs, chicken meat and, to some extent, pork), and dairy cattle genetics are now managed in a single global nucleus. Nobody proposes that poor countries grow in autarky, producing their own vehicles, computers, mobiles, etc., so why should this autarky be sought for livestock? Obviously, as we have just said in the first point that there are situations that cannot be covered by multinational companies because, particularly in ruminants, only animals that are very well adapted to the environment can be profitable in those situations; but this is more the exception than the rule. When in doubt as to whether or not foreign genetics can work better than local genetics, the most appropriate thing is to do a comparative test between the local and the foreign breeds.
- *Are intensive industrial systems more polluting and do they affect global warming more?* There can be intensive systems where waste is not properly treated and energy is wasted, but this need not be the case. In fact, per unit of product (e.g. per kilogram of meat, per litre of milk), an intensive system is less polluting and leaves a lower carbon footprint than an extensive system, as shown by a study at Cranfield University (Williams *et al.*, 2006). If an industrial broiler can reach commercial weight in 6 weeks and a local breed takes 4 months to reach the same weight, during the time between 6 weeks and 4 months, the local-breed chicken has been increasing the carbon footprint and has been polluting.

> An intensive industrial system is usually less polluting per unit of product than an extensive system.

In poultry, organic production produces 45% more global warming, 75% more pollution by nitrates and phosphates in the water and ammonia in the air, and 53% more acidification, when measured per kilogram of meat produced (Williams *et al.*, 2006). We will deal with this subject in more detail in section 10.3.1.

B. The use of local breeds

A non-intensive sustainable system is usually associated with the use of local breeds, arguing that local breeds are better adapted to the local environment and that, due to their racial characteristics and non-intensive production, they produce better-quality products. This raises certain questions that must be examined in detail;

- *Are local breeds better adapted to local environments?* Today, we know that this is not necessarily true. We have spectacular examples of particularly well-adapted foreign breeds, such as the case of extensive use of Nelore cattle from India, which are well adapted to the conditions in Brazil. However, adaptation is a bigger problem in some species than in others; poultry, pigs and rabbits have been raised in better environments than sheep or goats, and therefore commercial intensive livestock breeds have fewer adaptation problems than breeds of other species.

  Local food sources may be of poorer quality than the usual food of highly productive breeds, and it has been said that local breeds, essentially ruminants, can then make better use of local food. In monogastric animals, the differences in feeding are not so significant, and local food sources usually work well; in swine, Gibson et al. (2006) showed that commercial swine crosses using local food sources in a developing country on well-managed small farms obtained good productive yields. In ruminants, the advantages of local breeds for local food sources are highly speculative, as the information we have on these local breeds is usually scarce if not absent. Local breeds of cattle, sheep and goats may be better adapted in some environments, although it would be wise to check whether this is true. Before introducing a foreign breed or promoting the use of a local one, a brief and not necessarily expensive experiment can pinpoint the benefits and drawbacks of using one or the other.

> Some foreign breeds may be better adapted than local breeds to the local environment. Their importation depends on their productivity in the local environment being higher than the productivity of the local breeds. It is convenient to do an experiment to compare the local breed with the foreign one before making a decision.

  Unfortunately, decisions are often taken without a minimum of information, and the consequences are often much more costly than what an experiment of this type would have cost to carry out.

- *Do local breeds produce better-quality products?* This question is too general to give a simple answer. It is obvious that an Iberian pig (local breed) produces a much better ham than an intensive commercial pig, regardless of the fact that its diet and handling are different and that the curing process of the ham is especially long and careful. The production of high-quality products is among the main reasons for keeping breeds that are less efficient at producing meat or animal products. However, it is convenient to check whether this quality is detectable by the consumer, as being produced by a local breed does not guarantee a better product.

It is important to note that comparisons between breeds in meat quality should be made not only at commercial weight for practical reasons but also at the same stage of maturity in order to avoid attributing higher meat quality to breeds that arrive at the slaughterhouse in a more mature state. If a local breed of poultry reaches commercial weight near its adult state, its meat will have different characteristics from those of the intensive broilers slaughtered at the same weight but much younger. Clearly, it is interesting to underline the advantages of the local breed if its meat quality is better at commercial weight, but it is convenient to know whether this is due mainly to the fact that it is slaughtered at a higher degree of maturity, or to other properties of the breed. In sections 8.1.2.A4, and 9.2.1.B2, we will discuss this issue in more detail.

7.2.5 Which breeds should be kept?

There is a continuous generation of new breeds and productive lines; however, new breeds are often being registered without information other than morphological information, a fact that is not unrelated to the frequent subsidies to aid conservation. In the first edition of the FAO DAD-IS programme in 1993, it was presumed that there were around 4000 breeds in the world, but today the number is around 15,000, and continues to grow. Many of these breeds are not in danger of extinction, and do not need special aid because they generate an economic benefit that makes their exploitation interesting. However, there are several thousand breeds in different endangered states and, particularly in developing countries, there are not funds enough to maintain them; these countries should first feed their population and use their available funds efficiently. We are faced with what Simianer (2002) called the 'Noah's Ark Dilemma': how many breeds should we take on to the Ark? Decisions on conservation should take into account some of the following considerations:

- *Why are there breeds in danger of extinction?* A definition of an endangered breed could be 'An endangered breed is a set of morphologically peculiar domestic animals that are not profitable even in their own environment.' As we have said before, breeds are tools created by humans; some of these tools made sense in some periods of history to some people but may no longer make sense today. This has also happened with other agricultural tools that are currently kept in museums of agriculture or ethnology. However, despite some breeds not being profitable now, we have good reason to think that some of their genes may be profitable in the future. In addition, some breeds represent the extremes of genetic variation for certain productive traits (e.g. lowest growth rate, higher back fat, higher milk fat content), and it may be interesting to keep them to have the possibility of producing rapid changes in certain traits, even if we do not have a clear idea about how this will be used in the future; Ruane (1999) and Bennewitz *et al.* (2007) have given several examples of these singularities.

> The key point in conserving breeds in danger of extinction is that *to conserve a breed, a specific reason is needed.*

  Most of the breeds in danger of extinction are in developing countries. If we wish to preserve these breeds from developed countries, we must provide the aid and programmes necessary for this, because this will not be a priority for investment of available funds in these countries and may lead to extinction of these breeds.

- *Are breeds part of the cultural heritage?* Are there cultural reasons for conserving endangered breeds? Breeds are sometimes considered part of cultural heritage, and an objective assessment of their cultural value has been proposed (Gandini and Villa,

2003). Considerations about the cultural value of breeds would probably be better made by anthropologists than by animal breeders, and funding to maintain rare breeds should probably come from the Department of Culture, rather than the Department of Agriculture. There is extensive discussion about what can be considered as part of culture. A definition restricted to animal breeding is that of Durham (1991), which uses the meme concept proposed by Richard Dawkins. The argument is the following: some groups of people (e.g. nations, tribes or simply locals who have some sense of identity) have elementary units of values, beliefs and ideas, which are transmitted to the next generation; they are the so-called *memes*. Memes are under pressure from evolutionary processes in a way similar to genes, but here cultural evolution is a much faster process. Cultural products that today are less appreciated than in the past for their usefulness (e.g. Gothic churches) are appreciated due to the action of a meme that contains a value of artistic pleasure or a value of reinforcing the common identity. Scotland's Highland cattle are not particularly profitable, but Scots like to see them in their landscape, or they just like to know that these cows are still grazing in the hills. Gandini and Villa (2003) emphasized the ability of breeds to document the past world in which they lived; however, this is only one of the products of the cultural memes that can be considered, although the scale for evaluating this capacity is largely arbitrary. In any case, *in situ* conservation for cultural reasons is a classical concern of a developed country. In developing countries, cheaper solutions could be found (e.g. cryopreservation).

- *How much do we lose when a breed becomes extinct?* Not much, in many cases. Breeds are easy to create, and many of the existing breeds are intermediate states between other breeds or simply a mixture of genes that can be found in many other breeds. For example, the Manchado de Jabugo pig breed is, according to the FAO DAD-IS programme, a 'hybrid of Iberico Negro and Iberico Retinto, with pigs imported from the United Kingdom'. It was created by farmers in the 1920s (therefore it has no 'historical cultural value') and is in danger of extinction; so why should we keep the product of an experiment carried out by farmers in the 1920s? The Spanish Giant rabbit breed was created by a farmer in 1914 by mixing Flemish Giant rabbits with local common rabbits; it was a popular breed from the 1940s to the 1960s and almost disappeared when the industrial production of rabbits came along. The breed Giant of Flanders has rabbits larger than the Spanish Giant, and the synthetic breeds used today in industrial rabbit farming as terminal sires are also giant breeds, but with better productive characteristics; so what are the reasons for keeping the Spanish Giant breed, apart from curiosity?
- *Are there objective criteria for determining which breeds to keep?* Several scientists have listed reasons for keeping breeds in danger of extinction (e.g. Ruane, 1999, Gandini and Oldenbroek, 2007; FAO, 2013). In addition, there have been some attempts to objectively evaluate the benefits of breed conservation. Simianer (2002) proposed utility functions based on the information available on the valuable characteristics of the breeds. Roosen *et al.* (2005) listed some of the 'diversity values' that can be considered, and proposed the development of simulation models and other methods to better specify the values to be used for making decisions about genetic resources. However, a main problem is that for the most threatened breeds we do not have any information, or the information is so scarce that it is of little use; for example, in Ethiopia for 29 of 34 breeds of cattle, we only know their names and the number of horns (two). Current information is so vague that it is difficult to make proper

decisions in a global context. Maintaining breeds with extreme characteristics, with clearly established cultural values, or with a clearly defined possible use for the future may be some of the decision rules needed for conservation; if we have to divert funds used to feed people for breed conservation, we will need very good reasons.

7.3 Conservation Programmes

It is difficult to talk about conservation programmes in general, because they depend on the endangered status of the breed, and on whether or not is possible to obtain an economic benefit from the breed. It also depends on the financial means available to the conservation programme, the interest of current and future farmers in conserving the breed, and many other factors. Decisions about conservation programmes often have to be made on a case-by-case basis. We will now examine important features of conservation programmes.

7.3.1 Procedures for breed conservation

There are three main procedures for breed conservation:

- *Cryopreservation:* This consists of freezing or vitrifying tissues, gametes or embryos. Cryopreservation of tissues hypothetically allows the animal to be cloned in the future, a technique that is currently difficult and ineffective (see section 10.1.1). Cryopreservation of gametes would allow a breed to be reconstituted by backcrosses (see section 10.1.2.C and Fig. 10.2). Gamete cryopreservation is generally effective, but the breed's recovery process is slow, particularly in species with a high generational interval such as cattle. Cryopreservation of embryos would be the preferred method, but this is not very successful in some species (pigs) and is simply not feasible in others (poultry). Keeping a sufficient number of embryos is the quickest procedure for recovering a population.
- *Ex situ conservation:* This consists of maintaining a small population in one or more experimental farms in which a sufficient effective number of animals can be kept, while collecting productive performances to better characterize the breed and determine better what is being conserved. *Ex situ* conservation may allow investigation into whether it is possible to derive attractive commercial products that would allow the breed to become profitable. In the event that the breed is to be reintroduced, *ex situ* conservation serves as a basis for the diffusion of gametes through artificial insemination or embryos through embryo transfer.
- *In situ conservation:* This consists of keeping the population in the areas where it can be exploited. This is the best solution, but breeds are usually in danger precisely because there are more efficient breeds in the same area. In these cases, subsidies are used, but the efficient use of funds must be considered; it may be that *ex situ* conservation coupled with research to make the breed profitable is more effective than direct subsidies to maintain the breed. When *in situ* conservation is the solution adopted, a selection programme can be established to make the breed more competitive. This last solution is complex, not only because of the difficulties of implementing a selection programme but also because these programmes increase consanguinity and decrease

the effective number of the population, which must be taken into account in the case of breeds in danger because they have a small population size.

7.3.2 Tools for breed conservation

Assuming that the breed can be conserved *in situ*, actions can be established to facilitate its conservation. Depending on the endangered status of a breed, one can simply propose to conserve it, to increase its genetic diversity, or to establish a selection programme that can contribute to make the breed more productive in its environment:

- *Control of inbreeding*. Conservation is carried out preferably in breeds of small numbers with a high danger of extinction. In such cases, the limiting factor is inbreeding, which can not only reduce variability but may also lead to serious reproductive problems. In section 1.2.2, we discussed inbreeding control. A frequent problem in breeds in danger of extinction is the lack of data; often, the relationship between the animals of the small breeding stock is not known. In these cases, molecular relationships can be used (see section 6.2.2), although this is more expensive. In general, it is recommended that inbreeding should not increase beyond 1% per generation.
- *Increasing genetic diversity*. Total genetic diversity (that present in all the DNA) cannot be maintained because it would require a very large number of individuals, and furthermore, it probably makes no sense to maintain it. As a breed is a productive instrument, the question is which parts of the genetic diversity do we want to maintain. Normally, it will be that associated with productive traits, but it may be interesting to conserve diversity in other traits, although there must *always be some reason for it*. Ideally, with the help of genomics, it should be possible to identify genes causing variability in productive traits; unfortunately, traits usually depend on many genes of small effect. Furthermore, genomics is expensive to implement, and in many developing countries, it will not be possible to use it for conservation programmes.
- *Genetic distances between breeds*. Genetic variability can be established between breeds and within breeds. With the advent of molecular genetic methods, *genetic distance* between breeds became popular as a measure of distance between gene frequencies; that is, as the average of the squared differences in gene frequencies. The problem with these measurements is that they depend on which markers are used to determine them, although nowadays, with the generalization of SNPs, this is less of a problem. A more serious problem is that the breeds with a reduced effective number, as is usually the case for breeds in danger of extinction, lose variability due to genetic drift and move away from other breeds to which they are closely related by the simple fact that, by chance, the alleles that are fixed in one are not those that are fixed in the other. Besides, it is not unusual for geographically close breeds to have migrated from one to another throughout their history, which makes them genetically close even though historically they are not.

In general, it is recommended that efforts to conserve variability should be split into between-breed variability and within-breed variability. When molecular genetic instruments are not available, sometimes a high number of morphological traits is taken with the intention of using morphological variability as a criterion for the distance between breeds, but this is not practical unless the morphological variability is linked to productive traits (which is not usually the case; see sections 8.1.2.A5, 9.1.2.B2 and 9.2.1.B2), in addition to the fact that the heritability of some of these traits may be null or their genetic correlations with other traits may not be of interest. There is a copious bibliography on the maintenance of genetic diversity; (e.g. Oldenbroek, 2007, 2017).

7.3.3 Programmes for breed conservation

The ideal form of conservation is *in situ*, producing enough profits to make investments for conservation unnecessary. In other words, the best way to preserve a breed is to eat the breed or take advantage of its products. This is usually difficult, otherwise the breed would not be in danger of extinction. Breeds in danger of extinction can be promoted due to current characteristics if they are valuable; if this is not possible, selection programmes can make them become more competitive in the future. Let us examine both possibilities:

- *Economic use of the breed.* There are several ways to take advantage of a breed with special characteristics:
    - *Promote special products that other races do not provide.* These can be products of special quality, linked to designations of origin, or simply products that identify the use of the breed with some tradition or product that is not usually found on the market, such as a type of cheese or a special preparation of the meat (e.g. smoked, cured).
    - *Use the cultural argument.* If a breed can be associated with some tradition or social group, or with some characteristic activity, it can be conserved for cultural reasons; for example, breeds of donkeys are conserved in certain areas by using them for tourist walks.
    - *Promote their adaptation in difficult areas.* With the argument of avoiding desertification, certain breeds can occupy difficult areas, taking advantage of local resources. These breeds can be local or can be imported breeds well adapted to the environment.
- *Selection programmes.* Ideally, an unproductive breed can be selected in order to be productive. In practice, this is not easy for a number of reasons: breeding programmes are expensive to carry out and their results are only seen in the long term, other breeds continue to improve economically and it is difficult to reach a comparable profitability, markets have established channels in which the breed in danger is often not included, etc. However, sometimes it may be interesting to establish a breeding programme that is better adapted to the requirements of certain local markets. The most spectacular example has been the increase in productivity lines for brown eggs, which were previously significantly less productive than white-egg lines. Other examples include the increase in production in small-format dairy cattle breeds such as the Jersey breed, with much lower maintenance costs than the Holstein breed, and the reduction of feed conversion ratio and fat content of the Duroc pig breed, which is appreciated for its intramuscular fat, etc. Without trying to go so far, increasing productivity through selection programmes for breeds to be conserved can be a solution or an aid for their conservation. In the case of ruminants, these programmes can be established by promoting the collection of data on the farms; the technician collecting the information can also advise farmers on good practices, and can help to facilitate contacts among farmers for common projects. Given the peculiarities of the breeds in danger of extinction and their environment, market, etc., it is necessary to examine the possibilities and feasible programmes on a case-by-case basis.

Main concepts

- *Breed:* A set of animals of a domestic species with certain common external morphological characteristics defined by a group of people or by the administration, generally conserved in closed reproduction, avoiding migration of genes from other populations. Average performances are assigned as a guideline, which may change with time, place, food, handling, etc. Breed is not a technical term, nor does it have a biological meaning.
- *Pure line:* In plants, a pure line is a set of plants with identical genotypes that is homozygous for all genes. In animals, there are no pure lines.
- *'Pure line':* In animals, a 'pure line' is the way of describing a set of domestic animals of a species kept in closed reproduction. The maintenance of stable morphological characteristics is not required here. The name 'pure line' is made by analogy with pure lines in plants, but in animal production, it is not possible to obtain animals with homozygosity for all genes and an identical composition of all genotypes as in plants.
- *Hybrid:* In plants, a hybrid is a cross between two pure lines, so that all hybrids have the same genetic composition, although they are not homozygous for all their genes as in pure lines. As there are no pure lines in animals, consequently there are no hybrids in animals in the genetic sense as in plants.
- *'Hybrid':* In animals, it is common for commercial companies to call 'hybrids' the result of crossing of two of their 'pure lines'. In this case, the descendants are not genetically identical as in the case of plants.
- *Genetic diversity:* Genetic variability, both between races and within races. The term is vague, so it is convenient to refer to the genetic variability of specific traits.
- *Genetic distance between breeds:* The average of the squared differences between breeds in gene frequencies of certain molecular markers. It does not take into account genetic variability within breeds or the fact that breeds can differ in gene frequencies due to genetic drift, not necessarily due to having a different genetic history. It depends on the markers used, although the generalized use of SNPs facilitates comparisons because the number of SNPs available is much higher than the number of other types of markers.
- *Sustainable system:* In agriculture, sustainable systems are systems capable of maintaining their productivity indefinitely without damaging the environment. There are both intensive and extensive sustainable systems. Damage to the environment is measured per unit of product, which is why many extensive systems produce, per kilogram of product, greater environmental damage (carbon footprint, pollution) than intensive ones.

Problems

7.1. The Guirra breed of sheep (red cinnamon colour) coexists with the Segureña breed (white). When proposing a conservation programme, farmers suggest dividing the breed into two subpopulations, dark red and light red. Should the two populations be in separate closed reproduction with two conservation programmes?
7.2. The Latxa breed of sheep has two varieties, one with a black face and the other with a white face. Both are used for the production of the same cheese and both share the same environment in the Basque Country. Is it appropriate to keep them in separate closed reproduction?

7.3. A natural park was supervised on horseback before the introduction of off-road vehicles, so the horses were abandoned and live freely in the park. Creation of a 'Park horse' breed has been proposed. Should public funds be given for this creation?
7.4. The Chulilla hen was considered extinct, but it has been possible to recover it from mixed populations, by selecting its external characteristics. Has the breed really recovered?
7.5. Does cryopreservation of breeds whose productivity is ignored make sense?
7.6. Breeding companies of pigs often recommend that the maternal line, usually a crossed line, should be made by crossing only two lines of the company, because their 'hybrid' is a particularly favourable combination of genes from their 'pure lines'. Is this true?
7.7. Can a 'pure line' be made from the crossing of several 'hybrids'?

Self-assessment Questions (True or False)

1. All existing breeds must be conserved because there are facilities for it.
2. A breed is a set of animals with specific external characteristics and productivity.
3. Herd books for a breed are not required for genetic improvement.
4. Sustainable systems cannot be intensive production systems.
5. Cryopreservation is the most effective method for the preservation of a breed.
6. Genetic distances always reflect how far apart two races are in their evolution over time.
7. If a selection programme is applied to an endangered breed, strict control of inbreeding must be observed, because selection tends to increase inbreeding.
8. It is possible to generate a 'pure line' from the crossing of two 'hybrids' in animals.
9. There are races whose extinction is not a problem economically, ecologically culturally, or for the use of difficult areas.
10. It is important to conserve races with extreme values for certain traits, for possible use in gene migration to commercial populations if selection objectives change.

References

Bennewitz, J., Eding, H., Ruane, J. and Simianer, H. (2007) Selection of breeds for conservation. In: Oldenbroek, K. (ed.) *Utilisation and Conservation of Farm Animal Genetic Resources*. Wageningen Academic, Wageningen, Netherlands, pp. 131–146.

Blasco, A. (2013) *Animal Ethics: A Scientific Perspective*. Akal, Madrid, Spain (Kindle edition).

Broom, D.M. (2021) *Broom and Fraser's Domestic Animal Behaviour and Welfare*, 6th edn. CAB International, Wallingford, UK.

Durham, W.H. (1991) *Coevolution: Genes, Culture and Human Diversity*. Stanford University Press, Stanford, California

FAO (2013) In vivo *Conservation of Animal Genetic Resources.* FAO Animal Production and Health Guidelines. No. 14. FAO, Rome.

Gandini, G. and Oldenbroek, K. (2007) Strategies for moving from conservation to utilisation. In: Oldenbroek, K. (ed.) *Utilisation and Conservation of Farm Animal Genetic Resources*. Wageningen Academic, Wageningen, Netherlands, pp. 29–54.

Gandini, G. and Villa, E. (2003) Analysis of the cultural value of local livestock breeds: a methodology. *Journal of Animal Breeding and Genetics* 120: 1–11. DOI: 10.1046/j.1439-0388.2003.00365.x

Gibson, J.P., Steane, D.E., Plastow, G.S. and Warkup, C. (2006) Meeting the needs of rich and poor: the needs opportunities and trends in pig genetic improvement. In: *8th World Congress on Genetics Applied to Livestock Production, 13–18 August, Belo Horizonte, MG, Brazil*, Communication 06.12.

Hall, S.J.G. (2004) *Livestock Biodiversity.* Blackwell, Oxford, UK.

Hill, W.G. (2008) Estimation, effectiveness and opportunities of long-term genetic improvement in animals and maize. *Lohmann Information* 43, 3–20.

International Chicken Polymorphism Map Consortium (2004) A genetic variation map for chicken with 2.8 million single-nucleotide polymorphisms. *Nature* 432, 717–722. DOI: 10.1038/nature03156

Lerner, I.M. and Donald, H.P. (1966) *Modern Developments in Animal Breeding*. Academic Press, Cambridge, Massachusetts.

Lloyd-Jones, O. (1915) What is a breed? *Journal of Heredity* 6, 531–537. DOI: 10.1093/oxfordjournals.jhered.a109038

Nagel, A. and Wood, G. (2010) *Anachronic Renaissance*. Zone Books, New York.

Oldenbroek, K. (ed.) (2007) *Utilisation and Conservation of Farm Animal Genetic Resources*. Wageningen Academic, Wageningen, Netherlands.

Oldenbroek K. (ed.) (2017) *Genomic Management of Animal Genetic Diversity*. Wageningen Academic, Wageningen, The Netherlands.

Rodero, E. and Herrera, M. (2000) El concepto de raza. Un enfoque epistemológico. *Archivos de Zootecnia* 49, 5–16.

Roosen, J., Fadlaoui, A. and Bertaglia, M. (2005) Economic evaluation for conservation of farm animal genetic resources. *Journal of Animal Breeding and Genetics* 122, 217–228. DOI: 10.1111/j.1439-0388.2005.00530.x

Ruane, J. (1999) Selecting breeds for conservation. In: Oldenbroek, K. (ed.) *Genebanks and the Conservation of Farm Animal Genetic Resources*. DLO Institute for Animal Science and Health, Lelystad, Netherlands, pp. 59–70.

Simianer, H. (2002) Noah's dilemma: which breeds to take aboard the ark? In: *7th World Congress on Genetics Applied to Livestock Production, 19–23 August, Montpellier, France*, Communication 26-02.

Thiessen, R.B., Taylor, C.S. and Murray, J. (1985) Multibreed comparison of British cattle. Variation in relative growth rate relative food intake and food conversion efficiency. *Animal Production* 41, 193–199. DOI: https://doi.org/10.1017/S0003356100027859

Webster, J. (2005) *Animal Welfare: Limping towards Eden.* Blackwell, Oxford, UK.

Williams A.G., Audsley E. and Sandars D.L. (2006) *Determining the Environmental Burdens and Resource Use in the Production of Agricultural and Horticultural Commodities.* Main Report, Defra Research Project IS0205. Cranfield University and Defra, Bedford, UK.

Woolliams, J. and Toro, M.A. (2007) What is genetic diversity? In: Oldenbroek, K. (ed.) *Utilisation and Conservation of Farm Animal Genetic Resources*. Wageningen Academic, Wageningen, Netherlands, pp. 55–74.

Solutions

Solutions to problems

7.1. The light red colour will almost certainly have come from crossing with Segureña (white). Both herds are managed jointly and it is not unlikely that there has been a migration of genes from Segureña giving rise to the light colour. It makes no sense here to establish two programmes.

7.2. If the differences between the races are limited to the colour of the face, it does not make sense to have different selection programmes, except for cultural considerations. If at a given moment their survival is compromised, it would not make sense maintaining separate conservation programmes.
7.3. No, because the creation of this breed has no apparent reason, as it does not have particularly useful characteristics.
7.4. A breed with the external form of the Chulilla hen has been recovered. This hen can be kept for cultural reasons, as a reminder of what the Chulilla hen was, but if the interest in the Chulilla hen was the quality of her meat or eggs, the breed has not really recovered, as the external form is not related to the traits of interest. Using an example from another conservation field, in the Forbidden City in Beijing, the woods have been replaced as they had deteriorated, so little remains of the original but the shape survives. Similarly, the Athenians kept the ship in which Theseus returned after his fight with the Minotaur, and exchanged the damaged boards for new boards, so that in a few dozen years the ship only retained the shape of the original, but for them the age of the boards was accidental and the shape was the essential property (Nagel and Wood, 2017).
7.5. This question depends on whether the cost of cryopreservation is so low that it is worth saving anything. However, the cost is not that low (liquid nitrogen must be always provided); moreover, it is unlikely that thawing in the future will reveal anything that is not known at present, unless cultural motives are argued. If there are no data, there should be some specific reason to keep them.
7.6. This is a confusion with plant pure lines and hybrids. As there are no pure lines in animals, there are no hybrids, and the company's recommendation cannot be based on these arguments.
7.7. If several hybrids are crossed in plants, the resulting pool of segregating genes do not form a pure line. In animals, the situation is different because there are no pure lines; the 'pure lines' of animals are non-homozygous populations, which when crossed can constitute 'pure lines' simply by keeping them in closed reproduction. It is common in pigs and rabbits to generate synthetic lines from the crossing of several 'pure lines' or commercial 'hybrids'.

Solutions to Self-assessment questions

1. F, 2. F, 3. T, 4. F, 5. F, 6. F, 7. T, 8. T, 9. T, 10. T.

8 Genetic Improvement in Monogastrics

Abstract
In this chapter, the genetic improvement of prolific animals is presented. In animal production, the main prolific animals are monogastric, and the main source of animal protein consumed in the world comes from pigs and poultry and from the consumption of eggs. In some countries, industrial rabbit production is also relevant. Selection schemes and dissemination of genetic progress are similar in all prolific species. The selection objectives are also similar for all species, focusing on feed efficiency and prolificacy, and nowadays carcass and meat quality are also important. Particular emphasis is placed on the different ways of estimating feed efficiency. This chapter examines these traits, their genetic parameters and the consequences of selection. Other traits of interest such as longevity and disease resistance and resilience are also examined. The different crossing schemes are discussed and the practical application of breeding by commercial companies is presented.

Animal genetic improvement can be practised on a small scale by farmers or by associations of small farmers, but this can only be done with local breeds, and its efficiency is low. Today, the animal genetic improvement of the main livestock species is generally in the hands of multinational companies or large cooperatives, although there are still medium-sized companies that carry out genetic improvements on a smaller scale.

In general terms, there are two types of improvement schemes, one based on the recording of data in commercial production farms, and the other on concentrating genetic improvement in certain advanced farms called selection nuclei and disseminating the progress through multiplication farms, which provide improved crossbred animals to commercial farms. The first scheme applies mainly to ruminants, especially dairy cattle, and the second to pigs, poultry and rabbits, and also to the main species in aquaculture. There are a number of minor variants that we will see throughout the chapter.

8.1 Pig Breeding

8.1.1 Organization of genetic improvement and dissemination of genetic progress

A. Nuclei and multipliers

When all genetic improvement is concentrated in one farm, it is called a *nucleus*, and this is extended to commercial farms through multiplication stages. This is the most common scheme for prolific species (Fig. 8.1).

- *Three-way cross.* Commercial farms need males and females. To produce females for commercial farms, usually two lines in closed reproduction (called maternal lines) are selected. The males of one of the lines and the females of the other line are sent to

DOI: 10.1079/9781836990994.0008

farms called *multipliers*, in which they are crossed and produce the *crossed females* that are sent to the farmers (Fig. 8.2), discarding the crossed males, which go to the slaughterhouse. These maternal lines are generally selected by prolificacy but can also be selected for other traits, as we will see later. Each multiplier usually receives females of line A and males of line B, but not males and females of the same line, to facilitate management and to prevent multipliers reconstituting the lines. To produce males from commercial farms (called *terminal sires*), a third line is selected; in this case, the line is not selected for prolificacy because litter size is mainly a trait of the female over which males have little influence.

There is usually one selection nucleus in each company, and multipliers are established in several countries. The multipliers are often independent companies with a contract with the genetic improvement company; they buy the animals from the lines

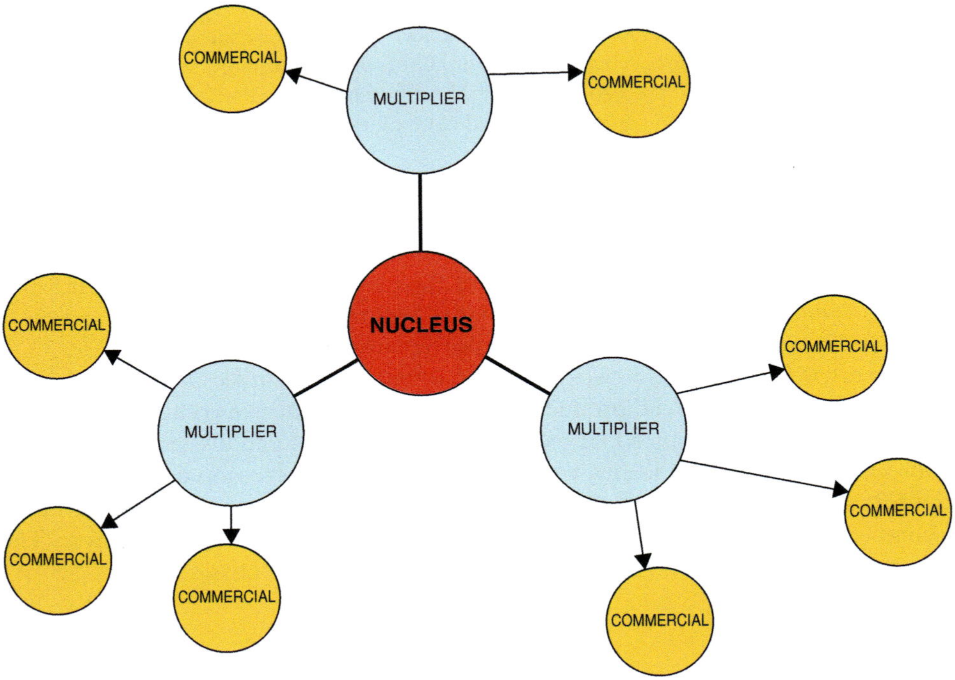

Fig. 8.1. Dissemination of genetic progress in prolific species from the nucleus to multiplier and commercial farms.

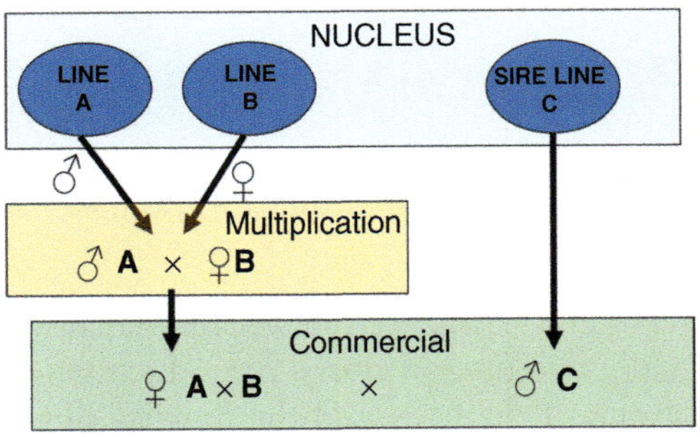

Fig. 8.2. Three-way crossing scheme used for poultry, pigs and rabbits.

selected in the nucleus, use their own facilities for crossing the lines, and are in charge of supplying the cross female (the 'hybrid' in popular language) to the commercial farms. This system has allowed the rapid development of genetic improvement businesses, as a substantial part of the investment and commercialization is carried out by the multiplying companies.

- *The system has been criticized because genetic improvement is carried out only in the nucleus*, which is a farm that takes special care of the environment and has high health standards. The best animals in these conditions may not the best in commercial farms; i.e. there may be genotype–environment interactions as discussed in section 2.1.1. In practice, this does not usually happen, because, unlike other species such as cattle, sheep and goats, the rearing conditions in commercial farms for prolific species are relatively closer to those of the nucleus. In pigs, the correlations between breeding values in the nucleus and in commercial farms for growth traits are around 0.7, which allows the genetic improvement of the nucleus to be transmitted efficiently to commercial farms (see Wientjes and Calus, 2017, for a review).
- *A small nucleus can produce genetically improved animals for a large number of commercial farms*. For example, Fig. 8.3 shows how a small pig nucleus of 300 females can provide the genes for farms with more than 30,000 sows in production. A nucleus of pigs, rabbits, chicken or fish cannot have fewer than 150 females and 25 males per line for two reasons: (i) inbreeding (see section 1.2.2), and (ii) selection pressure, as having a small number of animals to select makes selection less efficient.

The 300 dams of the nucleus in Fig. 8.3 produce a minimum of 20 piglets per year, of which ten are females; one will be kept to give rise to the next generation, and four are likely to be culled due to diseases, leg problems, poor performance, etc. Therefore, each female of the nucleus sends around five females annually to the multipliers. In our example, the nucleus sends 5 × 300 = 1500 females to the multipliers every year. As the productive life of the females is usually around 2 years, they need to have a total of 2 × 1500 = 3000 places available.

Each female in the multipliers also produces a minimum of ten females per year, of which on average four are culled and six are transferred to commercial farms. In our example, this assumes that commercial farms receive 6 × 3000 = 18,000 females each year, and commercial farms must have 2 × 18,000 = 36,000 places, because productive life is around 2 years.

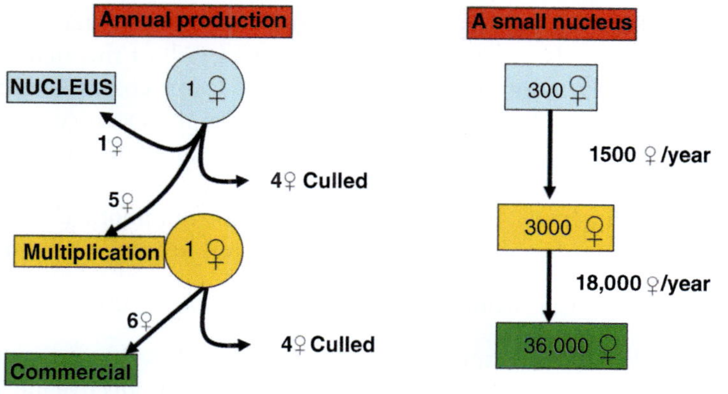

Fig. 8.3. Dissemination scheme for genetic progress in pigs.

All these figures are approximate, and maternal pure lines in pigs nowadays can produce more than 12 piglets per parity, and 'hybrid' sows 13 or more; moreover, these figures do not take into account the fact that production is not uniform, that the sale of animals may be temporarily stopped for pathological reasons, and other practical incidents. Nevertheless, these calculations serve to give a fairly realistic idea of the size of a prolific species improvement programme.

As a thumb rule, the number of females of the nucleus shoud be multiplied by ten to calculate the number of places needed for multiplication, and the number of females in multiplication should be multiplied by ten to calculate the number of places needed in commercial farms to absorb the females produced by the multiplier. For example, in Fig. 8.3, we would say that a nucleus of 300 females requires approximately 3000 places in multiplication and 30,000 in commercial farms. Thus, a group of pig farmers who currently have 5000 sows could set up a multiplier, for which they would need a farm nucleus of approximately 500 females; the nucleus would provide new sows to replace their current sows when they die or are culled. With artificial insemination, the number of males required has been greatly reduced, but it is advisable to have a larger group than needed to reduce risks, because a single good male can have poor offspring by chance (see section 2.3.1).

- *It is common for multipliers or breeders' associations to also have an insemination centre* to provide semen from terminal sires to production farms. A set of males, not necessarily from the same commercial company, can be used to mix semen for insemination on commercial farms, reducing the genetic risk that a particular male will have poorer offspring than expected.

B. The cost of genetic improvement

The system of nucleus and multipliers dramatically lowers the cost of sires and dams on the commercial farm. The cost of genetic improvement is high, not only due to the records taken of the animals but also due to business costs. For example, a geneticist and a specialist in numerical calculation and programming techniques are needed, and there are expenses derived from commercialization of the product such as advertising, vendors, etc. Moreover, breeding stock is a product that cannot be kept in storage, and it is also susceptible to hazards such as diseases that may prevent sale of the animals. If for any reason (e.g. lack of orders, diseases) animals are not sold and must be slaughtered, the extra costs are passed on to the animals to be sold as breeding stock. All of this makes the animals from nucleus significantly more expensive. One way to lower costs is multiplication. In prolific species, multiplying the improved animals divides the extra cost due to genetic improvement among the offspring. Figure 8.4 shows an example of the reduction in the cost of the breeding stock due to the multiplication process.

In pigs, the price of a 'pure' female (popularly called 'grandmother') is around €600. If the production cost of a female on a commercial farm is around €100, then:

$$\text{Genetic extra cost of a gilt from the nucleus} = 600 - 100 = €500$$

Each 'pure' female in the multiplier (each 'grandmother') has an average of four to five parities, and can produce around 40–50 piglets throughout her productive life, half of which will be female. Up to 40% may be eliminated due to disease problems, conformation, etc.

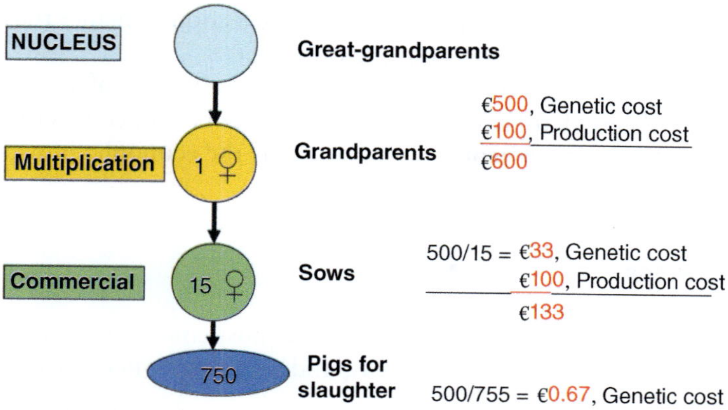

Fig. 8.4. Cost of genetic improvement in pigs.

(we will talk about this later), which leads to a 'pure' female of the multiplier producing at least 15 'hybrids' throughout her productive life, which are transferred to commercial farms. Each commercial 'hybrid' has around five parities throughout its productive life, in which it can produce around 50–60 pigs for the slaughterhouse; therefore, the number of descents produced per 'grandmother' going to the slaughterhouse is at least 15 × 50 = 750, so:

$$\text{Genetic cost per pig sold to the slaughterhouse} = 500/750 = €0.67$$

For a slaughter weight of 110 kg:

$$\text{Cost of genetic improvement per kg of live weight} = 0.67/110 = €0.006$$

These figures are calculated assuming that the multiplier sells the crossbred females without any genetic extra cost apart from the genetic cost of the nucleus, but in practice there is some extra cost for producing the 'hybrids'. The multiplier's 'pure line' females do not produce as much as the 'hybrids' and are more sensitive to disease and stress. If the company is not the owner of the multipliers, which is common, the 'hybrids' are bought by the multiplier, and the cost is higher; the multiplier, as an intermediary company, charges a commission on the animals it sells. Moreover, the multiplier should also sell animals with a high health-status, and should send 'hybrids' to the slaughterhouse in case of disease, as the nucleus does, which also increases the price of the 'hybrid'. A multiplier can sell their 'hybrids' for a price around €180, of which the cost of genetic improvement would be around €80. Even in this case, as each hybrid produces 50–60 pigs for slaughter, the cost of genetic improvement would be very low:

$$\text{Genetic cost of buying the 'hybrid' to a company} = 180 - 100 = €80$$

$$\text{Genetic cost per pig sold to the slaughterhouse} = €80/50 = €1.60$$

$$\text{Genetic cost per kg of live weight} = €1.60/110\,\text{kg} = €0.015/\text{kg}$$

This difference in cost has encouraged many farmers to join together to set up a multiplier, and has also encouraged large companies to multiply their animals themselves.

Multiplying is a simple task, as in the multiplier no selection is made: the males of one line are simply crossed with the females of the other line.

These figures are only approximate, but they are indicative of the cost of genetic improvement, and indicate that farmers should preferably be guided by criteria of quality and not price when buying their reproductive stock, as the economic impact on the final product is not important.

C. Crossing schemes

The three-way crossbreeding scheme (Fig. 8.2) is the most common scheme, but there are a few variations, such as two-way or four-way, and also other alternatives:

- *Two-way crossing.* This has been widely used in pigs, particularly in the USA, and does not use a terminal male but rather a male from one of the nucleus lines, which does not allow females to specialize in litter size traits and males in growth traits (Fig. 8.5, left). The main advantage is the use of the males produced in the nucleus, which can be used as terminal sires, so it is not necessary to maintain a third selected line.
- *Four-way crossing.* The terminal sires come from the crossing of two lines selected for growth traits (Fig. 8.5, right). The crossed males show a superior libido and some advantages derived from heterosis, such as more resistance to adverse environmental conditions.
- *Increasing the number of multiplication stages of a three-way crossing.* One example is to have more selection nuclei in which no selection is made but instead the company lines are simply multiplied (Fig. 8.6, left). This is frequent when animals have to be supplied to places far from the main nucleus, such as on another continent. This scheme causes a certain genetic delay, as the improvement of the nucleus takes longer to reach commercial farms.
- *Crossing the 'hybrid' with another breed.* The purpose of this is to supply crossbred females to the multiplier (Fig. 8.6, right). This is often implemented in countries where the environment is less favourable, and where multipliers prefer to avoid 'pure lines', which are more sensitive to diseases and stress. Remember that these schemes do not have higher heterosis as a result of using more breeds, as heterosis is not cumulative and will not increase when crossing many lines, but rather depends exclusively on the distance of the genetic frequencies from the parents, as we saw in section 1.2.4. This scheme also causes a certain genetic delay.

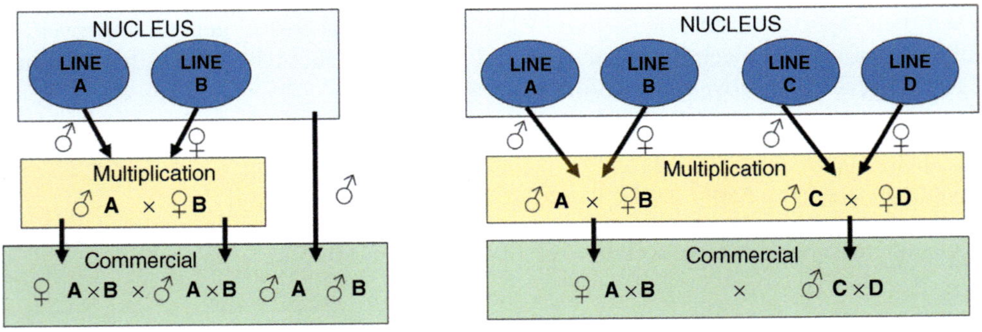

Fig. 8.5. Two-way (left) and four-way (right) crossing schemes.

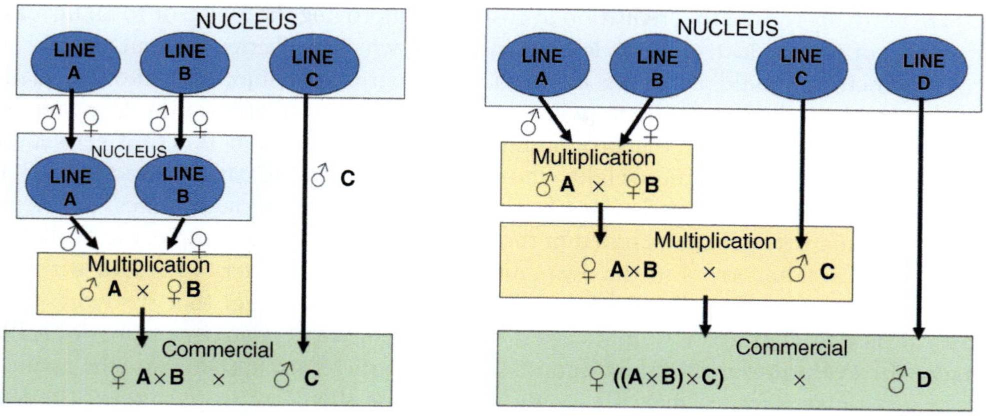

Fig. 8.6. Alternative crossing schemes.

- *Self-replacement.* Despite the fact that genetic improvement does not represent a significant production cost, sometimes a cash flow shortage encourages farmers to self-replace animals from their commercial ones, instead of buying females in the multiplier and also buying semen or males. This system is cheap, but it leads to stagnation of the genetic progress on the farm. While in the nucleus, the animals become increasingly more efficient, and this improvement is transmitted to the new crossbred females, while the animals that the farmer self-replaces have not increased their efficiency. Nevertheless, if temporarily self-replacement is the only solution, animals that would be taken to the slaughterhouse should not be used, as the lines that are used as the terminal sire in commercial farms have litter sizes that are smaller than the lines used to create the 'hybrid' female, and the result would be to have three-way-crossed females with a low litter size. It is preferable to buy males of maternal lines A or B into the nucleus or A × B crossed males into the multipliers, and mate them with the crossed females from the farm. Only one parity is needed for self-replacement of some sows, and the other parities can be made with the usual terminal sires. Terminal sires should not be recovered from animals that would be taken to the slaughterhouse, because they would lose efficacy in food conversion when coming from a three-way crossing with maternal lines. Terminal sires should always be bought into the nucleus, or semen should be bought from insemination centres having these terminal sires, as males do not represent an economic problem because their number is much lower than that of females.

8.1.2 Selection objectives

In the chain that begins on a farm and ends at the consumer's table, diverse interests take place, and they are not always coincident. As we saw in section 3.2.1, the farmer that sells weaned piglets for fattening is interested in litter size but not in growth traits; the farmer growing pigs may be interested in the feed conversion ratio (FCR), but not in litter size or carcass yield; and the slaughterhouse is interested in the carcass yield, but this does not interest the butcher, who will be interested in the meat yield of the carcasses, which is not of the interest of the consumer. The objectives of an improvement company are set by the market.

They are normally related to a reduction in costs (e.g. improving the FCR), or to an increase in the value of the product sold (e.g. low-fat carcasses, which are better value in pigs).

A distinction should be made between the programme objectives and the traits included in the selection indices (what we called selection criteria in Chapter 3). Regarding the objectives of the programme, ideally a production function should be constructed and the economic weights should be derived as indicated in section 3.2.1, including all the relevant traits in the selection index. However, some technical problems appear when many traits are included in the selection indices; for example, the probability of making serious errors in the estimation of genetic parameters increases with the number of traits (see section 3.3). In addition, we must not forget that we work with statistical models that are a simplification of reality, so it is convenient to periodically measure the evolution of the traits under selection. If there are many traits, the estimation of genetic progress is difficult, as the more traits there are in the index, the less progress will be made in each of the traits, especially if there are negative correlations between them. This is why it is recommended to set only a few and clear objectives for the indices, as we will see below.

A. Growth, carcass and meat-quality traits

The profits in animal production depend, as we saw in section 3.2.1, on costs and returns. Among the costs, we have the fixed costs, which are always the same regardless of the quantity of product (e.g. maintenance of the breeding stock, farm facilities, human labour), and the variable costs, which vary with the quantity of product (e.g. number of animals sent to the slaughterhouse). In poultry, pigs and rabbits, feed consumption of animals sent to slaughter represents, by far, the main cost. This cost represents more than 50% of the total cost of production in rabbits, more than 60% in pigs and more than 75% in poultry. *Feed efficiency* is therefore the first trait to consider in selection to reduce costs.

A1. FEED INTAKE One way to select for feed efficiency is to select against feed intake; the animals consuming less to reach the same commercial weight are the most efficient. This has some drawbacks:

- *It is expensive to measure*, because it must be measured in each individual.
- *Individual feed intake can have genotype–environment interactions* because it is measured in individual cages or in individual electronic feeders, whereas animals are collectively fed on commercial farms, where the competition for food is higher.
- *It is not a trait recorded in a commercial farm*, so the farmer may prefer animals that grow more, a trait observed by the farmer, even if they are not as efficient.
- *Carcass and meat composition changes when reducing feed intake*, generally reducing fat content. This is a factor of quality for the carcass, as lean meat is preferred, but intramuscular fat content is a meat-quality factor sometimes affecting the final price of the meat.
- *Feed intake should not fall below certain levels that could compromise the productivity of the animal.* For example, animals with poor appetite can have difficulties in accumulating the fat necessary to maintain lactation, or, in the case of pigs, to maintain a firm fat that facilitates the processing of the retail cuts of the carcass.

A2. FEED CONVERSION RATIO *FCR is the kilograms of feed intake per kilogram of live weight gained over a determined period*. It is an index with no dimensions, the lower the better. This or its inverse (kilograms of live weight per kilogram of feed intake, also called *feed efficiency*) have been the most common way of measuring feed efficiency, although today both measurements are criticized for the following reasons:

- *FCR is a ratio of two traits*, so that the best individuals can have a low FCR by reducing consumption, increasing growth or some intermediate situation. This should not be a problem in practice, but interpretation of results of selection is less clear than for individual consumption or growth rate. In practice, selection tends to have a greater impact on the more variable component of the ratio, in this case feed consumption, which is why selection for FCR tends to decrease appetite rather than increase growth rate.
- *It is equivalent to a selection index with non-optimal weights*. As a selection index is the economically optimal method to select traits, as we saw in section 3.2, other methods are suboptimal. However, we also saw in section 3.3 that selection indices are sensitive to errors in the calculation of economic weights and estimation of genetic parameters, so an index may not necessarily be the best option.
- *It does not have a normal distribution* because it is the quotient of two normal variables, which makes predictions of the response to selection subject to bias.
- *Feed consumption has to be recorded*, so it has the problems associated with measuring feed intake that we have discussed before.

A3. RESIDUAL FEED INTAKE To avoid the problems associated with the quotient, it was proposed to measure feed intake at constant weight gain. For this, a regression equation of feed intake on individual weight gain over a determined period is made, and the residual is interpreted as feed intake at constant weight gain. The original proposal (Koch *et al.*, 1963) also adjusted by mid-weight during the test to take into account the feed intake for the maintenance of the animal. Often, instead of mid-weight, the weight to the power of 0.75 (called *metabolic weight*; Kleiber, 1961) is used. The residual is then interpreted as *the difference between actual feed intake and feed intake predicted from production and maintenance requirements*, and is used to measure the excess or deficiency of feed consumed by a growing animal with respect to the other animals in the test. In pigs, another regression variable is often added to correct the fat content, and this is interpreted as measuring feed intake at constant lean weight and constant maintenance requirements. Residual feed intake (RFI) has become popular because it measures a trait with an understandable biological and economic interpretation. However, it also has certain drawbacks:

- *RFI is not a residual but rather the estimate of a residual*. There may be estimation errors; for example, the regression between consumption and weight may not be linear, or if linear, the fit may be poor (e.g. having a large coefficient of determination). These estimation errors are not taken into account when using RFI in statistical models, such as in a single- or multi-trait best linear unbiased prediction (BLUP). Furthermore, the residuals are unrelated, but the estimates of the residuals are not, which is not taken into account either. It is true that there is a Bayesian solution to this problem (e.g. Blasco, 2017, section 8.1, or Shirali *et al.*, 2018), but it is not often used because it is complex to implement.

- *The weight of the animal to the power of 0.75 does not measure the maintenance requirements during growth.* This coefficient was calculated only for maintenance requirements of adult animals, and is different (and variable) for growing animals (Brody, 1945), which means that the maintenance energy is not well calculated, or its calculation is not optimal (other authors have estimated a coefficient of 0.60 for pigs; Noblet *et al.*, 1999). Furthermore, caution must be exercised when interpreting the regression coefficients, because, as we said in section 6.1, if the variables are related (weight, maintenance, fat content), the regression coefficients are not proportional to the effect of the covariate.
- *RFI is equivalent to an index of selection on weight and consumption with non-optimal weights* (Kennedy *et al.*, 1993), as in the case of the FCR.
- *Consumption has to be measured*, so it has the disadvantages that we mentioned earlier related to the measurement of feed consumption.
- *Genotype–environment interactions can appear.* Even if it does not measure maintenance energy correctly, the weight to the power of 0.75 can capture part of this maintenance energy. Commercial farms have similar environments in prolific species, but they are not so uniform as to have the same maintenance energy, which can be quite different from that of the selection nucleus, where animals are well kept with environmental conditions as constant as possible. This produces genotype–environment interactions, and the best animals in the nucleus may not be the best in commercial farms. In pigs Knap and Wang (2012) calculated the correlation between breeding values in the nucleus and in commercial farms and found that they were relatively high (~0.7) for growth rate and FCR, but for RFI they were practically null.

A4. GROWTH RATE *Growth rate is the daily weight gain* between two specific moments in the animal's life, for example between weaning and the time of slaughter. It is usually measured as the difference between the weights in those two moments divided by the number of days that have elapsed (measured as kg/day or g/day). A similar trait, highly correlated with growth rate, is simply *weight at a given age*, usually close to slaughter. Another closely related trait is the *time elapsed between birth and commercial weight.*

- *Growth rate is easy and cheap to measure*, which makes it a common trait in selection indices. The sooner the animals reach the commercial weight, the more space they leave in the facility, so the less labour and other fixed costs are used in their care. However, the main interest of this trait is its indirect relationship with feed efficiency. By sacrificing animals earlier, their maintenance cost is lower, because if one animal is sacrificed a week earlier, it does not consume feed during that week (Fig. 8.7).
- *Selection for growth rate usually leads to increasing fat content along the growth period.* This should not happen, because fat is a late tissue (Fig. 8.8) and the animal is slaughtered at an earlier physiological age (i.e. at an earlier stage of maturity) (Fig. 8.7). In addition, 70% of the meat is water and a gram of meat contains four times less energy than a gram of fat tissue, so when selecting for growth rate, animals with more meat and less fat should be favoured, because they weigh more for the same amount of feed intake. However, selection for growth rate strongly increases appetite, and it has been seen experimentally that after a certain level of ingestion, there is a tendency to deposit more fat (Fig. 8.9).
- *Growth rate is genetically negatively correlated to FCR, so selection for growth rate increases feed efficiency.* The increase in fat content of the animal throughout its

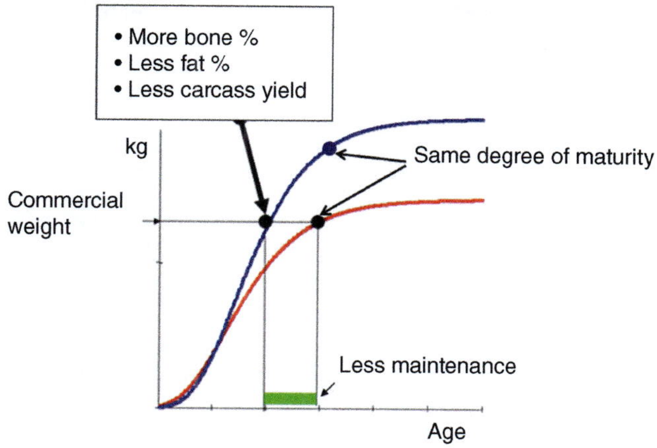

Fig. 8.7. Growth and commercial weight. Curves of two animals, one having high growth and the other low growth. At the same commercial weight, the animals that grow the most have fewer maintenance needs because they are slaughtered earlier, whereas the animals with low growth still eat feed until they reach the commercial slaughter weight. The animals that grow the most have a more juvenile carcass composition.

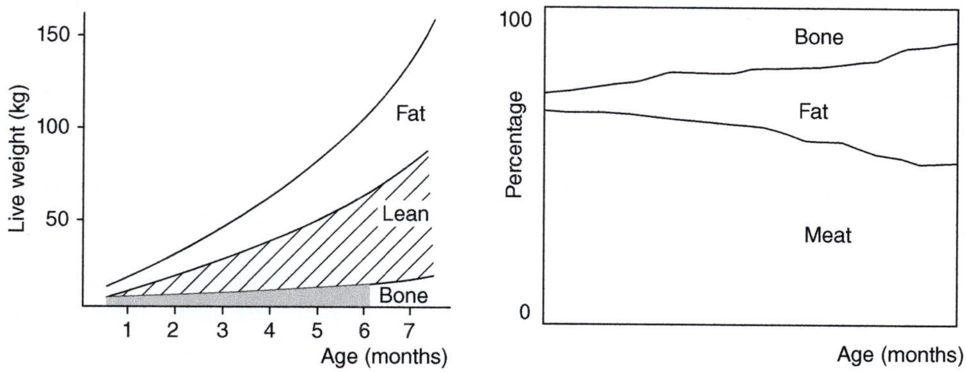

Fig. 8.8. Carcass growth and composition. Bone is an early tissue and fat is a late tissue. Left: pig (Kyriazakis and Whittemore, 2006). Figure used with permission of Wiley. Right: beef cattle. Adapted from Fig 2.4, *New Concepts in Cattle Growth*, by Roy T. Berg and Rex M. Butterfield, Sydney University Press, 1976.

growth could impair this efficiency, but the effect of reducing maintenance by sacrificing the animal earlier largely compensates, so the balance remains favourable (see Table 8.2).

- *At each point of the growth curve, at the same age, the animals growing more are larger and have better thermoregulation.* This means that they lose less energy per kilogram of live weight because they have a lower surface:volume ratio. For a sphere, the volume increases with the cube of the radius, and the surface with the square of the radius, so if two spheres of different sizes contain hot water, the smaller one will cool down sooner.
- *Selection for growth rate increases adult weight.* This leads to a number of drawbacks (Fig. 8.7). First, maintenance of the adult livestock is more expensive because their

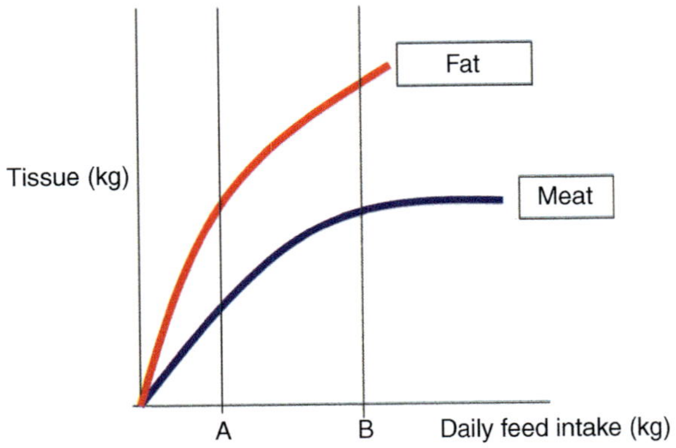

Fig. 8.9. Appetite and carcass composition. If the appetite is as shown at point A, fat and muscle are deposited; if it is as shown at point B, almost all of the deposit is fat. Adapted from Whittemore (1986).

maintenance needs are higher than the needs of growing animals. In prolific species, this is less important, as the costs of the adults is divided among the products that go to slaughter. If a sow produces 50 piglets during its productive life, its maintenance costs are divided by 50, so they become relatively irrelevant. A second inconvenience is the management of larger adult animals for reproduction, although the use of artificial insemination mitigates this problem.

- *Selection for growth rate modifies the composition of the carcass at the time of slaughter.* This is because the slaughtered animals are younger and less mature (Fig. 8.7), so they have a higher percentage of bone, a worse carcass yield (the intestinal content is an early growth development) and could have more fat content, as discussed earlier. These drawbacks are largely outweighed by the increase in food efficiency, which is by far the largest cost of production.
- *The composition of the retail cuts is also modified by slaughtering less mature animals.* Different parts of the carcass develop throughout the life of the animal, changing proportions over time. For pigs, the most expensive parts of the carcass, the ham and loin, develop proportionally more with time (e.g. Landgraf *et al.*, 2006).
- *Differences in carcass and meat quality may appear between breeds or treatments comparing fast- with slow-growing animals at slaughter weight.* This is because the fast-growing animals are slaughtered at an earlier stage of maturity (also called degree of maturity); that is, when their weight with respect to the adult weight is lower in one breed or treatment than in the other one. This point is important when comparing breeds or treatments, because differences between groups in terms of carcass and meat quality may be attributed to special qualities of the breed or the treatment, when they are actually due to their different degrees of maturity. For example, organic animals have a lower growth rate compared with commercial ones, so at the same slaughter weight they are more mature (Fig. 8.7), which makes their meat quality different (e.g. more intramuscular fat, more flavour). It is important not to attribute these differences to the different handling and feeding, or to differences in breeds; these differences may largely disappear when comparing both types of pigs at the same degree of maturity.
- *Meat quality could be altered by selection.* This occurs essentially because the slaughtered animals are younger at the same commercial weight. The meat of the younger animals is more tender and juicier, but has less flavour; these differences disappear

when comparing them at the same age, which is an approximation to the comparison at the same degree of maturity (Fig. 8.7). Even in pigs slaughtered at the same weight, differences in meat quality due to selection do not seem to be relevant. Schwab *et al.* (2006) compared control animals with animals selected for high growth and low-fat content for 25 years; the selected animals showed some quality parameters measured in the laboratory that were different from those not selected, but the meat-quality sensory panels did not find relevant differences in juiciness, tenderness or chewiness, and only slight differences in flavour.

A5. CONFORMATION Conformation, like type (which we will see in the next chapter), refers to certain external traits that can have consequences for production. Here, it is necessary to distinguish between functional traits (e.g. related to feet and legs) and external traits related to the apparent musculature. In Fig. 8.10, two classic problems of feet and legs in pigs that affect the stability of the females can be seen: crooked legs and uneven hooves.

- *Functional traits are important because they can make handling difficult or can lead to health problems.* The traits related to apparent musculature are usually unimportant from the productive point of view, because this appearance is usually unrelated to the quality of the carcass (e.g. expensive parts of the carcass, percentage of meat and similar traits). Conformation traits are often used to discard animals with problems, although they can sometimes be part of multi-trait BLUP.
- *Muscular conformation can be taken into account when pricing carcasses.* For example, in the southern half of Spain, some parts of Germany, Belgium and French Brittany, conformation is decisive in valuing the carcass, while in the northern half of Spain, the UK, the USA and most of Germany, conformation has little importance. Figure 8.11 shows one of the systems used to classify porcine carcasses according to their conformation.

This superiority can be largely illusory. Table 8.1 shows the results of an experiment by Blasco *et al.* (1994) comparing carcasses of high-conformation crosses with poor-conformation crosses. The Belgian Landrace terminal sire produced carcasses with much higher conformation than those that came from a cross with Large White or Duroc sire (for conformation score in pigs, a lower value indicates higher conformation). Comparing the data from the best carcasses (those with the Belgian Landrace terminal sire) with the worst (those with the Duroc terminal sire), their percentage of ham was very similar (25.1% and 24.5%, respectively), the percentage of meat in the ham was also similar (31.2% versus 30.7%) and the percentage of ham in the price of the carcass was practically the same (11.2% versus 11.0%). Thus, animals with better conformation can have some better carcass characteristics, but they may be less than what Fig. 8.11 seems to promise.

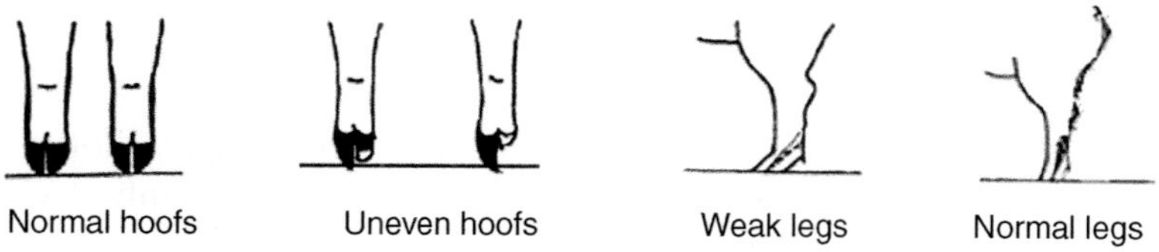

Fig. 8.10. Leg problems in pigs. Modified from Wallenbeck *et al.* (2020) (open access source).

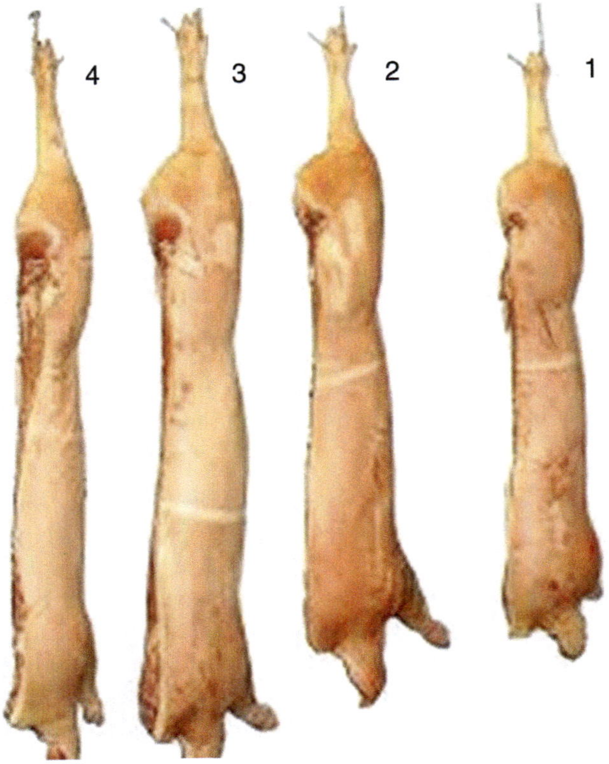

Fig. 8.11. Classification of pig carcasses according to their conformation. Modified from Masoumi (2020) (open access source).

Table 8.1. Composition of pig carcasses of Landrace × Large White females crossed with different terminal sires. From Blasco *et al.* (1994). Used with permission of Elsevier.

	Duroc	Large White	Belgian Landrace
Conformation	2.6	2.5	1.7
Ham (%)	24.5	23.8	25.1
Meat in ham (%)	30.7	30.0	31.2
Ham in the price of carcass (%)	11.0	10.9	11.2

- *Culling pigs for conformation in selection programmes before applying BLUP represents a problem.* This is because some animals with high economic additive values may be eliminated due to poor conformation. Around 30–40% of animals are slaughtered prior to application of the selection criteria (including those eliminated due to diseases and other causes).

A6. CARCASS QUALITY Income from the sale of animals to the slaughterhouse is affected by the *dressing out percentage* and by the quality of the carcass, mainly due to its meat content. In pigs, the *ham percentage* may also influence the returns.

- *Officially, pig carcasses are classified for their meat content into six groups*, according to a European classification system (SEUROP), although most carcasses fall into the central groups of the classification, and slaughterhouses pay a carcass price according to its meat content, estimated with precision by various systems independently of this classification.
- *The meat content of the candidate for selection is estimated in live pigs by ultrasound*, although nowadays it is common for large improvement companies to include data from meat content of relatives evaluated at the slaughterhouse, which makes the estimate more accurate.
- *As fat content is a highly heritable trait, selection for this trait has been successful.* As a consequence, nowadays carcasses are much leaner than they were when pig breeding programmes started, and the trait has lost the importance it once had. Moreover, pig fat is needed for processed products, which has led to increasing slaughter weight in some countries to produce pigs with higher fat content.

A7. MEAT QUALITY After the Second World War, consumers demanded cheap food, so genetic improvement was devoted essentially to traits related to efficiency in order to lower the price of meat. At present, when the price of pork meat is very low, there is an increasing demand for improving meat quality.

- *Genetic improvement has addressed meat quality mainly by creating specialized lines*, such as Iberian pork and Duroc pork with more intramuscular fat.
- *Meat quality is a difficult trait to be introduced in a breeding programme.* This is because it depends on many variables. Furthermore, the variables should be easy to measure, because they have to be recorded in many animals. Measuring many variables requires estimating many genetic parameters, with increasing probabilities of finding incoherent estimates of heritabilities and genetic correlations; moreover, when selecting for many traits, the response per generation in each trait is lower than when using few traits, and thus it is more difficult to monitoring the actual progress in many traits (see section 3.3). The most common proposal for improving meat quality is to use a single trait related to meat quality, such as the ultimate pH or the intramuscular fat, or to use a phenotypic 'global meat-quality index' of several traits weighed according to their importance for meat quality.
- *Selecting for a new trait implies applying less selection pressure on other traits.* A key issue for meat-quality improvement is how much it costs to measure it and how much profit is it expected to produce, because it should at least compensate the loss produced by applying less pressure in other traits. The problem is that it is difficult under current market conditions to assess the economic gains due to an improvement in meat quality.
- *Most meat-quality traits depend on many genes with small effect.* Genomic selection could be useful for selecting to improve meat quality, as it is not necessary to slaughter the animal to determine a genetic evaluation of its quality. However, the relationships between single-nucleotide polymorphisms (SNPs) and the genes responsible for the trait change from population to population, so the results in one population are not necessarily applicable to another (see section 6.3.1). In addition, the relationships between markers and genes change over the generations due to recombination, so that after two or three generations, the equations for genomic selection would have to be

re-estimated (see section 6.4.2), and meat-quality traits of several hundred animals have to be recorded, which is neither economical nor practical.

- *In some cases, a meat-quality trait depends on a single gene.* This is the case for the halothane gene that produces pale, soft and exudative meat (PSE), or the Rendement Napole (RN) gene, which affects the performance of cooked ham (see Purslow, 2017, Chapters 3 and 4, for a detailed description). In the first of these cases, the animals present a syndrome of sensitivity to stress, linked to a particularly lean carcass and a pronounced conformation (type 1 in Fig. 8.11), but also to a white meat with heavy losses of water after slaughter. PSE depends on a partially recessive gene (all the recessives and some of the heterozygotes show the syndrome) originally detected using halothane anaesthetic, and it is present in the Pietrain breed and the Belgian Landrace. RN is a dominant gene related to the water-retention capacity in cooked ham, producing a meat that is somewhat more acidic (with a lower final pH after slaughter), having some meat-quality traits that are better and others that are worse than standard meat; it is present in the Hampshire breed. Eliminating the homozygous does not solve the problem because the allele is still present in the heterozygote (see section 1.1.3), but the detection of carrier animals (heterozygotes) using molecular markers allows easy elimination of these problems.

B. Traits related to reproduction

Traits related to reproduction essentially affect fixed costs. If the pig litter size goes from 12 to 13 piglets, the costs of the facilities, feeding and handling of the parental stock, labour costs, etc. will be practically the same, so that, per pig sold, the fixed costs are distributed among more pigs.

B1. LITTER SIZE This is the most important reproductive trait, and is as important as feed efficiency. It is a trait of the female, although some traits of the male (e.g. semen quality) may have some influence on the trait. The genes of the male could influence litter size as well (e.g. some embryos could have advantages in implantation and development), but this influence is small, and litter size is essentially a trait of the female carrying the embryos. Litter size is easy and cheap to measure, but it is important to determine when it is done, as we will see below.

- *Data for litter size from multipliers are used in the prediction of the breeding values of the nucleus.* As females in multipliers are siblings of females in the nucleus, their data are useful for improving the accuracy of prediction of breeding values of the candidates for selection, so they are usually integrated in the evaluation when using BLUP.
- *Litter size at slaughter.* This is the most economically relevant trait, but it is not considered to be a female trait only, because it is affected by individual mortality from weaning to slaughter, over which the female has much less influence.
- *Litter size at weaning.* This is used in some swine programmes and is a trait that summarizes the ability of the female to produce a large number of young with the possibility of being able to keep them until weaning (essentially due to her milk capacity but also to her capacity for maternal care, behaviour, etc.). *Litter weaning weight* has sometimes been proposed as a trait that also relates to the litter size and the dairy

capacity of the female, but its heritability is near zero and it is difficult to clearly establish its economic weight.

- *Number of animals born alive.* This is a common trait in pigs for selection but has the drawback that due to crushing and mortality around birth (neonatal mortality), the value of the trait depends on when it is recorded. It does not have the advantage of considering the sow milk capacity, but in pigs, adoptions are common, so having litters with a high number of animals born alive may only be a problem in the long run.
- *Total litter size* (number of animals born alive plus stillbirths). This measure is more precise than the number of animals born alive and is common as a selection trait because it has a heritability greater than litter size at weaning, which is influenced by environmental aspects during lactation. When it has been selected, it has resulted in a higher number of animals born alive, but a proportionally higher number of stillbirths, and thus neonatal mortality seems to increase more with larger litters (Blasco *et al.*, 1995; Canario *et al.*, 2006).
- *Litter size 5 days after birth.* This measure has been successful in pigs because it is no longer influenced by environmental aspects that increase neonatal mortality. The response to selection for this trait has been the highest among all litter sizes.
- *Uniformity of litter weight at birth.* This has been considered experimentally, essentially because it favours survival of newborns. It is possible to select for it, and the problem of large litters is successfully addressed by carrying out adoptions and transferring young kits or piglets from large litters to smaller litters.

B2. LITTER SIZE COMPONENTS Litter size is a difficult trait to select. First, the female does not have any data of her own at the time of selection as she has not reached puberty yet, so she must be selected through the data of her relatives, or with the support of genomic selection. Second, heritability is low, so the response to selection is poor, despite the large coefficient of variation of the trait, and genomic selection requires large reference populations. The response to selection is usually lower than expected, indicating that the models used are not correct or are incomplete. This has led to various attempts to improve litter size from its components, with mixed success (see Blasco *et al.*, 1993, for a review):

- *Selection by ovulation rate.* This has been successful in increasing ovulation rate but not in increasing litter size, because there was an increase in prenatal mortality. *Homeostasis* phenomena frequently occur; the organism reacts against the modification of a trait by compensating for that improvement with the modification of another trait.
- *Selection for prenatal survival.* There is little experience in selection for prenatal survival. In pigs, after seven generations of selection, a modest response was obtained, which was less than a piglet and lower than what would have been obtained by direct selection by litter size; furthermore, the response was not transferred to the number weaned. Estimates of heritability of this trait have always been very low (Blasco *et al.*, 1998).
- *Selection by combination of ovulation rate and prenatal survival or litter size.* Greater success has been obtained in pigs by combining a prenatal survival index and ovulation rate, or selecting in pigs and rabbits by independent levels of ovulation rate and litter size. The response has been, in both cases, somewhat superior to the direct response for litter size, although it is doubtful that the cost of recording ovulation rate outweighs the benefit obtained.

- *Selection by uterine capacity*. Uterine capacity is defined as the capacity of the female to carry a litter when ovulation rate is not a limiting factor; that is, when having more ova does not produce larger litters. To measure it in a practical way, an ovary is removed; the other ovary, due to a compensatory phenomenon, doubles its ovulation rate; one uterine horn is removed, so the other horn is overloaded with embryos. It is assumed that under these conditions, females have larger litter sizes, have greater uterine capacity (understood as gestation capacity, not volume), and that with two uterine horns they will also have larger litter sizes. As ovulation rate is much greater than litter size in both species, it is presumed that litter size will increase due to selection for uterine capacity. Indeed, in the experiments that have been done, litter size increased but not more than what would be expected when selecting directly by litter size; moreover, the cost of estimating uterine capacity does not compensate for the progress achieved.

B3. LONGEVITY Longevity or productive life is a trait that has been drawn attention, mainly in pigs, rabbits and dairy cattle. In pigs and rabbits, it is usually measured as the number of parities that a female has before being culled due to diseases, problems with legs and feet, low fertility, etc. The inclusion of longevity in pig selection programmes is less evident than in dairy cattle, as we will see in section 9.1.2.B1. As in all selection objectives, selecting for longevity implies exerting less selection pressure on other traits, reducing the expected benefits, so the increase in longevity must compensate for these losses.

- *Longevity is not an economically important trait in itself.* This is because when dividing the replacement cost of a female by the number of piglets that she has had in her productive life (usually more than 50), the marginal cost is negligible. It has been argued that selection for longevity decreases the number of females eliminated in their first parities, but the cost of replacing females remains low, and the faster the females are replaced, the sooner they will benefit from the genetic progress of the nucleus.
- *Selection for longevity improves parental stock health, increasing their resistance to stress and disease.* The problem is that it is difficult measuring the benefits of this selection to calculate the economic weight of longevity. However, in a world in which the use of antibiotics in animal production is going to be increasingly restricted, it is convenient to develop lines that are more robust and less dependent on antibiotics, in preparation for the future.
- *Selection for longevity leads to lower pre-weaning mortality of the piglets.* This argument could have economic relevance but has been little studied.
- *Selection for longevity is difficult to carry out.* Direct measurement implies waiting for females to be culled, which lengthens the generational interval, and if sophisticated techniques such as analysis of censored data are used, their inclusion in multi-trait indices or BLUPs is very complex.

B4. OTHER REPRODUCTIVE TRAITS There are endless lists of interesting traits to select, many of them easy to measure and inexpensive; these include pre-weaning survival, age at puberty, fertility, individual or litter weight at birth, and traits related to the quality of semen. Traits related to the male have received some attention, but their cost per product (per pig sold to the slaughterhouse) is economically irrelevant; moreover, as with all objectives, selecting them implies reducing the pressure on other traits of economic interest.

Breeding companies either select a percentage of candidates that do not meet minimum requirements prior to applying BLUP, or they use phenotypic indices in which various reproductive traits are weighed and considered as if they were a single trait, either for the prior elimination of candidates or for inclusion of the indices in the selection programme. In general, the effectiveness of both procedures is very limited.

C. Disease resistance and resilience

Selection for disease resistance is rarely performed because it is not practical if animals have to be infected. In addition, after several years of selection, a vaccine may appear that readily solves the problem, as has already happened for some diseases.

- *Disease resistance is not generic*, and animals that show resistance to some diseases are sensitive to others.
- *Genomics does not offer a solution either*, unless the disease depends on a gene, something that rarely happens. However, there are some known cases in pigs where genetic modifications have led to resistant animals, as we will see in section 10.1.2.B. The reader interested in the genetics of disease resistance can consult the book by Bishop *et al.* (2010).
- *Currently, the focus is on resilience*. Resilience is the ability of an animal exposed to disturbances such as pathogens or stress to return to its original state and recover its original performances. It is a different concept from *resistance* (the ability of an animal to limit the invasion of pathogens), *tolerance* (the ability of an animal to withstand infection, maintaining performances) or *robustness* (the ability of an animal to maintain its performances in a wide range of environments). A resilient animal can be resistant, tolerant or both, and generally resilient animals are also robust. Measuring resilience does not need pathogen records as in resistance or tolerance, but some physiological or productive indicators are needed for a selection programme. Several indicators for resilience have been proposed in pigs (e.g. Laghouaouta *et al.*, 2021).

8.1.3 Genetic parameters and response to selection

A. Genetic parameters

- *Genetic parameters are specific to each line* and depend on genetic factors and environmental factors, because *better environmental control reduces environmental variability and increases the heritabilities*. However, it is common for the genetic parameters of a trait in different lines to be similar, as environmental control is extreme in all the nuclei.
- *Reproductive traits tend to have low heritabilities, less than 0.10, growth traits intermediate heritabilities, around 0.30, and traits related to carcass fat content high heritabilities, around 0.50* (higher heritabilities are very rare in farm animals). The genetic correlations between growth and reproductive traits are usually very low. The genetic correlations between FCR and growth rate are negative and high in pigs, and the genetic correlation between RFI and growth rate is near zero. The amount of fat is negatively correlated with the FCR and positively correlated with RFI. Feed intake is highly correlated with growth rate but also positively correlated with the amount of fat (see Fig. 8.9). Table 8.2 summarizes the main genetic parameters in pigs. These values

Table 8.2. Genetic parameters of productive traits in pigs. Heritabilities are shown in bold on the diagonal, and genetic correlations off the diagonal. The values shown are means of estimates from several publications. Data from Clutter (2011) and Saintilan *et al.* (2013).

	FI	FCR	RFI	GR	BF
Feed intake (FI)	**0.30**	0.65	0.60	0.67	0.40
Feed conversion ratio (FCR)		**0.30**	0.76	−0.56	−0.40
Residual feed intake (RFI)			**0.24**	0.10	0.44
Growth rate (GR)				**0.30**	0.14
Back fat (BF)					**0.50**

are merely indicative, as genetic parameters depend on the genetic variability of the line, the environmental conditions, and even the estimation method, which corrects for systematic environmental variance (e.g. season, farm).

- *The low heritability of litter size indicates that selection will be difficult*. However, the response to selection also depends on the phenotypic standard deviation of the trait. Litter size has a standard deviation relatively greater than growth traits (standard deviation relative to the mean; its coefficient of variation is around 30% compared with 10% for growth traits). Currently, pig breeding companies include information from the multipliers, where siblings of the candidates provide litter size data, increasing precision. It is also common in pigs to use genomic selection for predicting the litter size breeding value of the candidates to be selected, as it adds individual information to the information from relatives.
- *The high heritability of carcass fat has contributed to rapidly reducing fat content in the carcass*. The result of this is that its economic weight is not as high as before.
- In pigs, selection for FCR was done indirectly by selecting for growth rate. Nowadays electronic ear tags that allow the animal to be identified are used to record intake in individual feeders, so individual feed intake can be automatically recorded cheaply and efficiently. This individual measurement of consumption has given rise to a lengthy literature on the possible *genotype–environment interactions* that would appear when growing pigs are fed individually in the nucleus and collectively in commercial farms; obviously, the behaviour is not the same, and it may be that the best animals in individual feeding are not the best in collective feeding.

B. Response to selection

- *Feed efficiency*. Improvement in feed efficiency has been successful. In general, efficiency has been achieved mainly through a reduction in appetite. In pigs, FCR has gone from 3.3 in 1975 to 2.6 today (P.W. Knap, personal communication).
- *Carcass quality*. The improvement of carcass quality has been successful. Until the 1960s, the pig was a fatty animal, with more than 30 mm of backfat thickness, which has been reduced to less than half of that today. This has created problems in butcher shops due to loss of fat firmness and lack of sufficient fat to make sausages or cured products such as salami, which has led in some cases to the production of pigs with higher live weight for the slaughterhouse.

- *Litter size*. The low heritability of litter size, the fact that it is a trait expressed in only one sex and that fact that it is not expressed at the time the females are selected have led to very weak responses over the years. However, in recent years, there has been a notable increase in litter size, as shown by the genetic trends in Fig. 8.12. The causes are diverse: the information on siblings from the multipliers has been integrated in the evaluation; the selection criterion has moved to the litter size at 5 days after birth in pigs; and genomic selection is now used in pigs, allowing evaluation of the candidates for selection at the time of selection, even though they do not yet have their own data. The result is that it is now common to have litter sizes of around 13–14 piglets instead of the 11 piglets of some years ago.

8.1.4 Pig breeding programmes

Breeding programmes in pigs depend on the companies that develop them, but they tend to be quite similar in both their design and their results. In a market of near-perfect competition, products tend to look alike. Large companies can afford to have several lines to better adapt to demand in different areas, they can use more expensive methodologies such as genomic selection or index with data from slaughterhouses, and they can be more effective when using data from their nuclei and multipliers. However, local companies or large cooperatives may be able to better adapt their objectives to their customers' needs, which explains why, in pigs, large companies have not completely occupied the market. In the following sections, we will see a basic example of the breeding programmes of these companies that in practice allow several variants, which each company modifies according to its market shares, expectations, development possibilities, etc.

A. Three-way crosses

Intensive pig breeding programmes are generally based on three-way crosses (Fig. 8.2).

A1. OBJECTIVES The *lines for the crossbred commercial female* are selected mainly for litter size at birth or at 5 days, feed efficiency, growth rate and carcass meat content. The

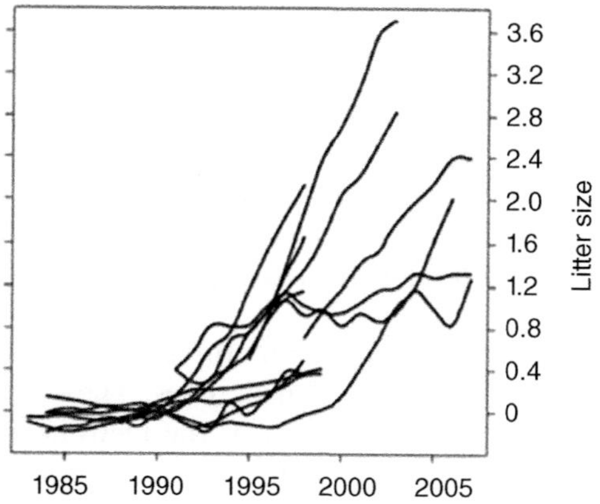

Fig. 8.12. Genetic trends for litter size in several lines. From Knap and Raw (2009).

terminal sire line is selected for feed efficiency, growth rate and carcass meat content, but not by litter size, because litter size is assumed to be essentially determined by the female. The terminal sire line can consider conformation if it is paid for in the market in which the final product is commercialized. In all lines, before selection, around 40% of the possible candidates are eliminated due to a long list of traits related to health, feet and legs, conformation and other commercial considerations.

A2. BREEDS As we indicated in section 7.1.1, *the term breed is not a technical term* and should be replaced by *line*, as within each breed there are lines with very different productive capacities. Having said that, it is very common to cross a Landrace breed line and a Large White breed line for the female, both lines of similar litter size, because as the lines come from different breeds, they have been in closed reproduction for a long time, guaranteeing a certain amount of heterosis (section 1.2.4), around 10% for litter size of the crossed female. These breeds have been selected for many years for litter size. The terminal sire is usually a Duroc breed line, traditionally with more intramuscular fat than other breeds, or, in cases where conformation is paid for, a Belgian Landrace line or a Pietrain line. These last lines are sensitive to stress, so the use of *synthetic lines* is often preferred, with marked conformation, if necessary, formed from several breeds and intensively selected for the objectives discussed above.

A3. METHODS The main objectives are entered in a multi-trait BLUP (see sections 4.3.2.C1 and 5.3), or in a single-step G-BLUP (section 6.2.3) with the aid of genomic selection. The economic weights are initially calculated as indicated in section 3.2.1 but in practice are modified using some version of the desired genetic gains method (section 3.2.1), or increasing or decreasing the pressure on the traits according to the response actually obtained. Genetic parameters are estimated within the line and are frequently re-estimated. Animals with signs of disease or inappropriate conformation, problems with feet and legs, or genetic defects are culled before applying BLUP.

B. Four-way crosses

Four-way programmes use four lines, two for the crossbred female and two for the terminal sire, which in this case is a crossed terminal sire (Fig. 8.5, right).

B1. OBJECTIVES The lines for the crossed female are selected as in the three-way crossing. For the terminal sire, one of the lines can produce animals with high conformation, which can be homozygous recessive for the PSE muscle gene, which means that they are also very sensitive to stress. These males are difficult to handle on commercial farms, so they are often crossed with a homozygous dominant male for that gene, so that the commercial farm receives heterozygotes that are much less sensitive to stress. Both lines are selected for the same objectives as in the three-way cross.

B2. BREEDS In four-way schemes, the most common terminal sires are crosses of Pietrain × Hampshire lines or Belgian Landrace × Duroc lines. The Pietrain × Hampshire lines cross is intended to preserve the conformation of the Pietrain line while still maintaining high meat content. This is useful when conformation is expensive, but in general the Hampshire lines do not have the same productive performances as the lines of other breeds selected by the breeding companies. Crossing lines of Duroc × Belgian Landrace or Duroc × Hampshire aims to keep the good traits of the Duroc lines for growth and meat-quality traits and the high meat content of the Belgian Landrace or Hampshire lines. Synthetic lines are also used sometimes in crosses.

B3. METHODS These are the same as in a three-way cross.

8.2 Poultry Breeding

8.2.1 Organization of genetic improvement and dissemination of genetic progress

Meat from birds comes from a variety of species including poultry, turkey, duck and some other minor species, but poultry represents 90% of the total meat from birds produced in the world, and thus we will focus on poultry breeding, noting that the schemes for other birds are quite similar:

- *Poultry breeding schemes are organized in nucleus–multiplier–commercial farms*, as in pigs (Fig. 8.1). The dissemination of genetic progress is similar to that of pigs, but is particularly impressive in poultry. Figure 8.13 represents a nucleus with two lines for laying hens, each with 250 females. Egg production per hen exceeds 300 eggs/year, of which half are female, and around one-third are culled, as in pig breeding, which means that approximately 100 females go to multiplication farms each year. As the cycle is approximately 1 year, each female in the multiplier gives rise to 100 crossbred hens ('hybrid' hens in popular language) for commercial farms each year, leading to a production of 1.5 billion eggs/year.
- Currently, *nine companies grouped in two holding companies dominate the world trade of genetically improved reproductive stock*: Aviagen and Cobb-Vantress.
- *The system of nuclei and multipliers dramatically lowers the cost of genetic improvement*, as in pigs (Fig. 8.4).
- *Crossing schemes* in poultry breeding programmes are based on three- or four-way crosses, as in pigs, as we will see in next sections.

8.2.2 Selection objectives

In poultry, the selection objectives are different for meat production than for egg production. As in pigs, the interests of farmers differ according to whether they are selling 1-day-old chicks, growing broilers or having hens for egg production.

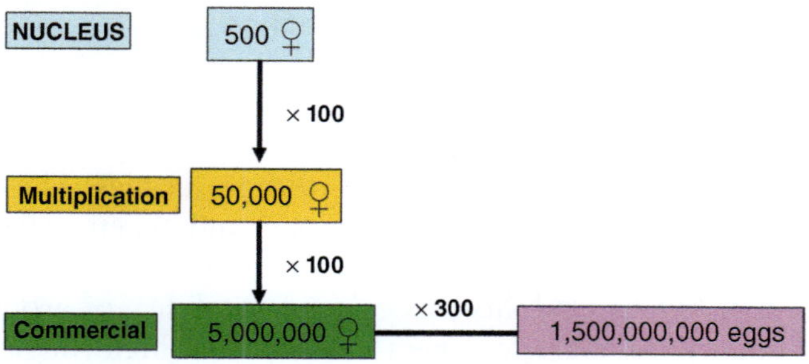

Fig. 8.13. Scheme of dissemination of genetic progress in poultry (example for layers).

A. Broilers

- *Feed efficiency* is one of the main objectives in poultry for meat but also in egg production. Feed efficiency is commonly measured as FCR, but also as RFI. The discussion about feed intake, growth rate, FCR and RFI in section 8.1.2.A can also be applied to poultry breeding.
- *Carcass quality* is a specific objective for broilers. *Carcass fat content*, which appears as a consequence of increased appetite when selecting for growth rate (Fig. 8.9), is difficult to evaluate because there are no simple methods to record it in the individual such as the ultrasounds used in pigs; however, selection for feed efficiency, negatively correlated to fat content, has allowed this trait to be controlled and currently it does not represent a problem. *Breast content* is an important trait, estimated by breast angle or subjective evaluation. *Conformation* is not used in poultry breeding; there is only interest in skeletal deformities and in breast content in broilers.
- *Meat quality* is difficult to be introduced in a selection index for the same reasons as in pigs: meat quality depends on many traits and should be measured on a large number of animals using non-expensive criteria. There are no single genes for meat quality, as in pigs, and meat quality appears to be under a multi-factorial genetic determination in poultry. As in pigs, the meat of younger animals is more tender and juicier, but has less flavour; these differences disappear when comparing them at the same age, which is an approximation to the comparison at the same degree of maturity (see review by Berri, 2000).
- *Health traits* related to some common problems in broilers, like ascites or leg health, have been addressed by including health and welfare traits in the selection indices (Avendaño and Ralph, 2018).

B. Layers

- *Egg production* is an important objective in layers but also in broilers, for the same reason as in pigs; hens in multipliers producing chicks are more efficient if they produce more eggs in the same period, or they have a longer laying period (called *persistency*).
- *Hatchability* is an important objective in both layers and female lines of broilers, for the same reason as the number of eggs.
- *Egg quality traits* are specific for layers. They include shell traits (mainly resistance to breakage) and inner traits such as albumen quality or organoleptic traits.
- *Feed efficiency* and *health traits* are common objectives with broilers, as discussed above.

8.2.3 Genetic parameters and response to selection

A. Genetic parameters

Genetic parameters in growth traits are similar to those in pigs. The heritabilities of feed intake, growth rate, RFI and FCR are moderate, and genetic correlations have a similar size and sign (Mebratie *et al.*, 2019). The heritability of egg production is also moderate, around 0.3, but hatchability has a low heritability of around 0.10 (Simm *et al.*, 2021). Egg quality traits are in general moderate to high (Blanco *et al.*, 2014); for example, breaking strength has a heritability of around 0.3. We should stress that genetic parameters vary among populations, and uncertainty, particularly in the case of genetic correlations, is high in prolific species, because the amount of available data for estimating genetic parameters is limited. Therefore, all comments about genetic parameters should be taken as merely indicative.

B. Response to selection

Due to the low generation interval, high selection pressures and moderately high heritabilities, responses to selection for growth traits have been spectacular. In broilers, in an experiment comparing animals from an unselected population from 1957 and a selected one from 2001 (Havenstein *et al.*, 2003), fed representative diets from 1957 and 2001, the FCR had gone from 2.14 to 1.63. There were no signs of reaching a limit, and genetic progress continues. Figure 8.14 compares, at various ages, a selected chicken line and a control line, both fed with a modern diet.

Egg production per year has changed from 230 to 300 over the period 1960–2005, and the number of eggs per kilogram of feed from five to nine in the same period (van der Steen *et al.*, 2005). The adult weight of laying hens has also been greatly reduced (e.g. in brown egg hens, from about 3 kg live weight to less than 2 kg). Nowadays, an increase in the number of eggs being produced is achieved by increasing the persistence in laying. The resistance to shell breakage has also increased.

8.2.4 Poultry breeding programmes

A. Broilers

Poultry companies usually have specific lines for specific market niches, such as slow-growing lines for markets where non-intensively growing animals are perceived to have better welfare (this is not necessarily the case, but this is the perception of a broad set of

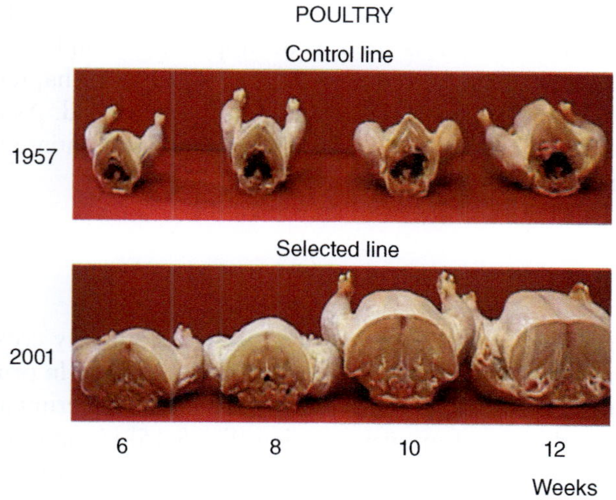

Fig. 8.14. Effect of selection on the quality of chicken carcasses. Both lines were fattened on the 2001 diet. From Havenstein *et al.* (2006) (open access source).

customers). Here, we will describe not the programmes for specific lines or local lines but rather the breeding programmes of the lines that provide chicken meat worldwide. The programmes for turkeys and other birds are similar.

Broiler chicken breeding programmes are based on three-way or four-way crosses. The most common is the three-way crossing (Fig. 8.2), although the volume of the industry, in which two large groups of companies dominate the world trade in poultry meat, has led to schemes of the nucleus × nucleus type (Fig. 8.6, left).

A1. OBJECTIVES The lines that give rise to the commercial crossed female are selected mainly for *egg production* (for the same reason that pigs are selected for litter size), *FCR*, *growth rate* and *breast content*. The sire line is selected for the same traits except for egg production, which is not relevant here. In addition, there is a prior elimination due to traits related to health and other traits difficult to include in an index. The *conformation* of relevance in poultry is the breast conformation and skeletal deformities. *Pathological problems associated with growth* (e.g. ascites) are corrected by selection, including the traits in the index, and by proper handling, and they do not pose a problem nowadays. Selection for feed efficiency, negatively correlated to fat content, has allowed this trait to be controlled and currently it does not represent a problem.

A2. BREEDS The female lines originally came from the White Rock Plymouth breed (called White Rock), created in the late 19th century as a dual-purpose egg–meat breed. White feather is important for western consumers because they prefer not to find black spots on the skin of the chicken, unlike eastern consumers. Sire lines come from the Cornish breed, created in the mid-19th century from fighting cocks, which have a particularly developed breast. For the same reasons as White Rock, the white-plumaged variety was created at the end of the 19th century. Note that 'breed' is not a technical term; current lines differ in productivity, size and appearance from the established breed standards, and we comment on the breed origins here only for historical context.

A3. METHODS As in pigs, the objectives and selection criteria are introduced in a multi-trait BLUP by a set of traits after culling, and are sometimes grouped in indices with biological significance. Genomic selection is used as an important complement allowing us to increase the selection response to around 20–30%, as we saw in Chapter 6, with genomic BLUP (G-BLUP) as the most common genomic selection method. As chick sexing is a complicated operation, it is common to include genes that differentiate male from female chicks (e.g. feathering speed).

B. Layers

Laying hen breeding programmes are generally based on four-way crosses to produce a laying hen, although three-way crosses are sometimes used as well. In principle, a two-way crossing would be enough, as the layer obviously does not need a terminal sire on the commercial farm; however, three- or four-way crossing (Fig. 8.15) is common so that multipliers can take advantage of heterosis in their reproductive stock. In the case of sire lines, using two lines in the multiplication instead of sending males and females of the same line prevents the possibility of appropriation of lines by multipliers. The use of several lines can also complement traits of interest that are more developed in one line than in another.

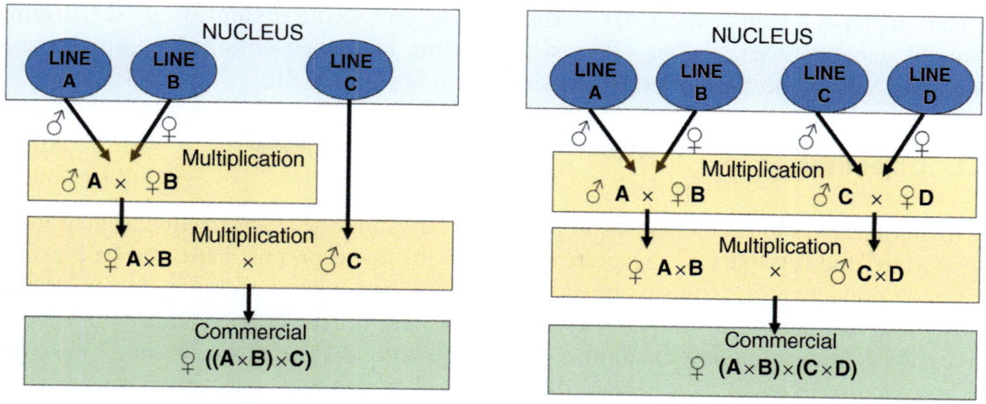

Fig. 8.15. Three- and four-way crossbreeding schemes for layers.

The original nucleus × nucleus type schemes, where the company owned not only the nucleus but also the nucleus × nucleus farms, have recently been replaced by large nuclei that allow selection with all the nucleus animals.

B1. OBJECTIVES The lines for the crossed female and the terminal sire are selected by number of eggs, egg quality (mainly resistance to breakage) and FCR to egg. There has also been selection to increase the weight of the hen at the beginning of laying (20 weeks of age) and reduce it at the end (90 weeks). Conformation is not used in layers.

B2. BREEDS White egg production is done by crossing White Leghorn breed lines, created in the mid-19th century from the white variety of local Tuscan chickens imported from Livorno (Italy). In brown eggs, the laying hen is the product of the cross of female Plymouth Rock Barred lines (a dual-purpose breed) with males of New Hampshire or Rhode Island Red lines, in order to facilitate sexing, as the male chick of these crosses has a whitish spot on the top of the head and can easily be removed.

B3. METHODS Multi-trait BLUP is used, with previous eliminations of candidates for selection, as in the previous programmes. G-BLUP is also used. As in meat poultry, sexing is done by plumage colour in brown egg layers or by feathering speed in white egg layers. Males are eliminated at birth, leading to a massive slaughter of animals, which is unpopular; sexing methods have recently been developed in the egg that avoid the slaughter of chicks, although their implementation is not widespread.

8.3 Rabbit Breeding

Rabbit production is a business with a much lower economic size than pigs, in which any additional fixed costs, such as the need for more hand labour for recording or management of breeding programmes, has a much greater impact on the price of the final product than on pigs. A consequence of the low size of the business is that small companies, cooperatives and large companies for rabbit breeding coexist.

Rabbits have a similar litter size to that of pigs, and so the organization and dissemination of genetic progress also follows the scheme of nucleus–multipliers–commercial farms (Fig. 8.1). Rabbit schemes are three-way crosses (Fig. 8.2).

8.3.1 Objectives

The main objectives in rabbits are, as in pigs, *litter size* and *feed efficiency* due to their influence on fixed and variable costs, respectively. Due to the low economic size of rabbit breeding mentioned previously, the traits to be recorded as selection criteria are limited to those that are easy and inexpensive to measure. In rabbits, the lines for the crossed female are selected only by litter size at weaning or number of animals born alive, and the sire line by growth rate, as an indirect way of increasing feed efficiency. There are some selection experiments for feed efficiency with the aim of developing new rabbit sire lines. Other experiments aim to develop resilient lines to improve resistance to diseases and stress by selecting for litter size homogeneity (Blasco *et al.*, 2017; Argente *et al.*, 2019). The rabbit is a very lean animal, so the meat content of the carcass is not a selection objective. In rabbit, conformation is not used.

8.3.2 Breeds

The breeds used in rabbit production are generally synthetic, either mixtures of many genetic origins or lines formed by crossing commercial crossbred females with crossbred males. As hybrids do not exist in animals, a line can be created by crossing two commercial 'hybrids' (see section 7.1.2) as a quick way to get enough genetic variability for selection. The origins of most maternal synthetic lines are from the California and White New Zealand breeds, and sire lines usually have some component from giant breeds.

8.3.3 Methods

BLUP is used for selection by litter size, and simple individual selection or BLUP are used for growth rate. Genomic selection is not currently used in rabbits due to the high marginal costs involved, but this may change soon with the decreasing price of sequencing and the use of low-density chips for imputation (see section 6.4.3).

Main Concepts

- ***Selection nucleus:*** The farm in which the genetic improvement is made. From there, the males of one line and females of another line are sent to the multiplier farms.
- ***Multiplier:*** The farm that crosses two 'pure lines'. It has males from one line and females from another line and produces crossbred females for commercial farms.
- ***Selection objectives:*** The economically relevant traits to be selected. These are not necessarily the measured traits, when objective traits are expensive or difficult to measure; they can be selected indirectly through other traits.
- ***Culling:*** Removing animals for poor health, bad conformation or undesirable traits before applying the selection indices or BLUP. In this pre-selection, 30–40% of the candidates for selection can be culled.

- *Residual feed intake:* Usually interpreted as the excess or lack of feed consumed by a growing animal, after correcting for the consumption necessary for maintenance and, in the case of pigs, for the amount of fat in the carcass. It represents the feed consumed above or below the theoretical maintenance and growth needs of the animal.
- *Conformation:* External traits of an animal relative to its apparent musculature. It is also used to refer to other functional traits, mainly legs and feet.
- *Three-way cross:* A production scheme in which two lines in closed reproduction are crossed and their female offspring are then mated with a male from a third line. The male offspring from the first cross are culled.
- *Four-way cross:* A production scheme in which two lines in closed reproduction are crossed and their female offspring are then mated with a male that is produced from the cross of another two lines kept in closed reproduction.

Problems

8.1. A pig cooperative that has 3000 sows goes to an official centre to request help to genetically improve their animals. What scheme can be proposed?
8.2. A pig company has a farm with 600 females and wants to become a breeding company with a nucleus that supplies 'pure line' animals to multipliers and terminal sires to commercial farms. What would be the number of sows of all commercial farms that the company can serve?
8.3. A selection nucleus for laying birds has two lines of 1000 females each. What will the egg production be in the commercial farms?
8.4. A group of farmers has 2000 sows. Can they set up a multiplier with 200 sows and two boars using artificial insemination?
8.5. A large rabbit farming company has 10,000 females. Can it set up its own nucleus?
8.6. Some rabbit breeding companies sell animals of their 'pure lines' for €90 each. How much does this affect the cost of the rabbit sold at the slaughterhouse? The average sale price is around €5 per rabbit at commercial weight.
8.7. What are the criteria for purchasing 'pure line' animals from a commercial company? How should you choose between several companies?
8.8. What are the criteria to start a multiplier or generate your own crossbred females?
8.9. What are the criteria for deciding to create a selection nucleus?

Self-assessment Questions (True or False)

1. Farmers can carry out genetic improvement on their own farm by self-reproducing offspring from their best animals, without the need to resort to genetic improvement companies.
2. Longevity is an economically important trait in prolific species due to the savings involved in delaying the purchase of new dams.
3. The cost of genetic improvement is very low; saving on the purchase of breeding stock by choosing poorer-quality animals is an inappropriate practice.
4. Establishing regional or local breeding programmes means freeing oneself from dependence on multinational companies; therefore, profits increase and it is a recommended practice.

5. Nucleus animals are raised in exceptionally good sanitary conditions, so commercial crossbred animals are likely to be particularly susceptible to disease.
6. Conformation in pigs is important because a better conformation indicates a significantly higher percentage of meat.
7. Conformation traits related to legs are important in pigs because they affect handling and it is advisable to eliminate animals with limb problems prior to selection.
8. Rabbit conformation is important because the carcass price is paid according to conformation.
9. Conforming to the traits of the breed standard is important for the sale of broilers.
10. Breeders of birds, rabbits and pigs must come from parents registered in the genealogical books of the breeds.

References

Argente, M.J., García, M.L., Zbynovska, M., Petruska, P., Capcarova, M. and Blasco, A. (2019) Correlated response to selection for litter size environmental variability in rabbits' resilience. *Animal* 13, 2348–2355. DOI: 10.1017/S1751731119000302

Avendaño, S. and Ralph, J. (2018) Economic challenges facing a global breeding company. In: *Proceedings of the 11th World Congress on Genetics Applied to Livestock Production, Auckland, New Zealand, 11–16 February*. WCGALP, Auckland, New Zealand, Vol. 23, p. 113.

Berg, R.T. and Butterfield, R.M. (1976) *New Concepts of Cattle Growth*. Wiley, Hoboken, New Jersey.

Berri, C. (2000) Variability of sensory and processing qualities of poultry meat. *World's Poultry Science Journal* 56, 209–224. DOI:10.1079/WPS20000016

Bishop, S.C., Axford, R.F.E., Nicholas, F.W. and Owen, J.B. (eds) (2010) *Breeding for Disease Resistance in Farm Animals*, 3rd edn. CAB International, Wallingford, UK.

Blanco, A.E., Icken, W., Ould-Ali, D., Cavero D. and Schmutz, M. (2014) Genetic parameters of egg quality traits on different pedigree layers with special focus on dynamic stiffness. *Poultry Science* 93, 2457–2463. DOI: 10.3382/ps.2014-04132

Blasco, A. (2017) *Bayesian Data Analysis for Animal Scientists*. Springer, Cham, Switzerland.

Blasco, A., Bidanel, J.P., Bolet, G., Haley, C. and Santacreu, M.A. (1993) The genetics of prenatal survival of pigs and rabbits: a review. *Livestock Production Science* 3, 1–21. DOI: 10.1016/0301-6226(93)90061-L

Blasco, A., Gou, P., Gispert, M., Estany, J., Soler, Q. *et al.* (1994) Comparison of five types of pig crosses. I. Growth and carcass traits. *Livestock Production Science* 40, 171–178. DOI: 10.1016/0301-6226(94)90046-9

Blasco, A., Bidanel, J.P. and Haley, C. (1995) Genetics and neonatal survival. In: Varley, M. (ed.) *The Neonatal Pig. Development and Survival*. CAB International, Wallingford, UK, pp. 17–34.

Blasco, A., Sorensen, D., and Bidanel, J.P. (1998) A Bayesian analysis of genetic parameters and selection response for litter size components in pigs. *Genetics* 149, 301–306. DOI: 10.1093/genetics/149.1.301

Blasco, A., Martinez-Alvaro, M., García, M.L., Ibañez-Escriche, N. and Argente, M.J. (2017) Selection for environmental variance of litter size in rabbits. *Genetics Selection Evolution* 49: 48. DOI: 10.1186/s12711-017-0323-4

Brody S. (1945) *Bioenegetics and Growth*. Reinhold, New York.

Canario, L., Roy, N., Gruand, J. and Bidanel, J.P. (2006) Genetic variation of farrowing kinetics traits and their relationships with litter size and perinatal mortality in French Large White sows. *Journal of Animal Science* 84, 1053–1058. DOI: 10.2527/2006.8451053x

Clutter A.C. (2011) Genetics of performance traits. In: Rothschild, M.F. and Ruvinsky, A. (eds) *The Genetics of the Pig*, 2nd edn. CAB International, Wallingford, UK, pp. 325–354.

Havenstein, G.B. (2006) Performance changes in poultry and livestock following 50 years of selection. *Lohmann Information* 41, 30–37. Available at: www.lohmann-information.com/content/l_i_41_2006-12_artikel5.pdf (accessed 15 July 2025).

Havenstein, G.B., Ferket, P.R. and Qureshi, M.A. (2003) Growth, livability, and feed conversion of 1957 versus 2001 broilers when fed representative 1957 and 2001 broiler diets. *Poultry Science* 82, 1500–1508. DOI: 10.1093/ps/82.10.1500.

Kennedy, B.W., van der Werf, J.H. and Meuwissen, T.H.E. (1993) Genetic and statistical properties of residual feed intake. *Journal of Animal Science* 71, 3239–3250. DOI: 10.2527/1993.71123239x

Kleiber, M. (1961) *The Fire of Life*. Wiley, New York.

Knap, P.W. and Rauw, W.M. (2009) Selection for high production in pigs. In: Rauw W.M. (ed.) *Resource Allocation Theory Applied to Farm Animal Production*. CAB International, Wallingford, UK, pp. 210–229.

Knap, P.W. and Wang, L. (2012) Pig breeding for improved feed efficiency. In: Patience, J.F. (ed). *Feed Efficiency in Swine*. Wageningen Academic, Wageningen, Netherlands. pp. 167–181.

Koch, R.M., Swiger, L.A., Chambers, D. and Gregory, K.E. (1963) Efficiency of feed use in beef cattle. *Journal of Animal Science* 22, 486–494. DOI: 10.2527/jas1963.222486x

Kyriazakis, I., and Whittemore, C. (eds) (2006) *Whittemore's Science and Practice of Pig Production*. Blackwell, Oxford, UK.

Laghouaouta, H., Pena, R.N., Ros-Freixedes, R., Reixach, J., Díaz, M., *et al.* (2021) A methodology to quantify resilience in growing pigs. *Animals* 11: 2970. DOI: 10.3390/ani11102970

Landgraf, S., Susenbeth, A., Knap P.W., Looft, H., Plastow, G.S. *et al.* (2006) Developments of carcass cuts, organs, body tissues and chemical body composition during growth of pigs. *Animal Science* 62, 889–899. DOI:10.1017/ASC2006097

Masoumi, M., Marcoux, M., Maignel, L. and Pomar, C. (2020) SpectralWeight: a spectral graph wavelet framework for weight prediction of pork cuts. *arXiv* 2005.05406 (preprint). Available at: https://doi.org/10.48550/arXiv.2005.05406 (accessed a5 July 2025).

Mebratie, W., Madsen P., Hawken, R., Romé, H., Marois, D. *et al.* (2019) Genetic parameters for body weight and different definitions of residual feed intake in broiler chickens. *Genetics Selection Evolution* 51: 53. DOI: 10.1186/s12711-019-0494-2

Noblet, J., Karege, C., Dubois, S. and van Milgen, J. (1999) Metabolic utilization of energy and maintenance requirements in growing pigs: effects of sex and genotype. *Journal of Animal Science* 77, 1208–1216. DOI: 10.2527/1999.7751208x

Purslow, P.P. (ed.) (2017) *New Aspects of Meat Quality*. Woodhead Publishing, Duxford, UK.

Saintilan, R., Mérour, I., Brossard, L., Tribout, T., Dourmad, J.Y. *et al.* (2013) Genetics of residual feed intake in growing pigs: relationships with production traits, and nitrogen and phosphorus excretion. *Journal of Animal Science* 91, 2542–2554. DOI: 10.2527/jas.2012-5687

Schwab, C.R., Baas, T.J., Stalder, K.J. and Mabry, J.W. (2006) Effect of long-term selection for increased leanness on meat. *Journal of Animal Science* 84, 1577–1583. DOI: 10.2527/2006.8461577x

Shirali, M., Varley, P.M. and Jensen, J. (2018) Bayesian estimation of direct and correlated responses to selection on linear or ratio expressions of feed efficiency in pigs. *Genetics Selection Evolution* 50: 33. DOI: 10.1186/s12711-018-0403-0

Simm, G., Pollott, G., Mrode, R.A. and Houston, R. (2021) *Genetic Improvement of Farmed Animals*. CAB International, Wallingford, UK.

van der Steen, H.A.M., Prall, G.F.W. and Plastow, G.S. (2005) Application of genomics to the pork industry. *Journal of Animal Science* 83 (Suppl. E), E1–E8. DOI: 10.2527/2005.8313_supplE1x

Wallenbeck, A., Eliasson, C., Lundeheim, N. and Nilsson, K. (2020) Leg health, growth and carcass characteristics in growing-finishing pigs of two different genotypes reared on Swedish organic farms. *Organic Agriculture* 10, 97–103. DOI: 10.1007/s13165-019-00260-8

Whittemore, C. (1986) An approach to pig growth modeling. *Journal of Animal Science*, 63, 615–621. DOI: 10.2527/jas1986.632615x

Wientjes, Y. C. J. and Calus, M. P. L. (2017) The purebred-crossbred correlation in pigs: A review of theory, estimates, and implications. *Journal of Animal Science*, 95, 3467–3478. DOI: 10.2527/jas.2017.1669

Solutions

Solutions to the problems

8.1. If they have 3000 commercial sows, they need approximately 300 dams in the multiplication farm, which allows them to set up a multiplier (see question 8.8). This multiplier could be supplied by a nucleus of 30 females, but this would be too few for a selection line and it would not be viable. The recommendation is to set up a multiplier or obtain crossbred females from a commercial company. For terminal males, a good recommendation is to set up an insemination centre with several males from one or several companies.

8.2. The company would have three lines to supply animals for three-way crossing. The two maternal lines would have 200 females each (400 in total), which means that they could supply to several multipliers with approximately 4000 female places. These multipliers would provide commercial farms with approximately 40,000 females. Before starting the business, you must be sure, therefore, that you will have a sufficient clientele, such as large integrations or cooperatives.

8.3. The 2000 females will produce approximately 200,000 hens for the multipliers, which will have to produce 20,000,000 laying birds for farms, and these will produce around 6000 million eggs/year. It is not, therefore, a small or medium-sized business. This capacity for multiplication is what has led to two large groups dominating the world market for egg laying.

8.4. In principle, they can set up the multiplier; there is no consanguinity because the males and females are not related as they are different breeds. However, using only two males implies taking a serious risk. As we saw in Fig. 2.2, section 2.3.1, the offspring of a male with good performance may not be as good as the male, so one way to reduce the risk would be to use semen from six or seven males.

8.5. The nucleus should have a minimum of 100 females, which is the limit to have one rabbit line instead of the three needed for the three-way crossing. However, it may be possible to have a single maternal line, to buy males from the other maternal line from a commercial company, and to buy terminal sires, as the money invested in males is much lower than in females. However, creating a selection line is a complex process and should not be the first recommendation to make (see question 9).

8.6. The cost of genetic improvement is in this case 90 − 4 = €86. Since a 'grandmother' (i.e. a dam from the nucleus) produces at least 15 crossbred females, and each of them will give rise to more than 40 rabbits for slaughter, a conservative estimate indicates that a 'grandmother' has produced 15 × 40 = 600 animals for slaughter. The genetic cost for each of these animals is 86/600 = €0.14 per animal. Actually, it is somewhat higher, because the multiplication is done with 'pure line' animals and not with crossbred females, which have poorer performances, but these figures show that the cost is not important.

8.7. In pigs, chickens and rabbits, the commercial companies offer similar products. In a world of almost perfect competition, if one company offered a clearly superior product, it would soon dominate the market. The criteria for choosing the company often depend on issues such as ease of service or technical advice, particularly in the case of poultry, in which the market is dominated by two large business groups. In the case of pigs, large companies can coexist with medium-sized companies, with cooperatives or integrated groups that have their own nucleus or with national programmes, and in the case of rabbits with much smaller companies. The advantage of large companies is their ability to invest in improvement programmes, but medium-sized companies can often better cover local needs and market niches, and can better adapt to their environment.

8.8. Setting up a multiplier is technically simple, as it consists of receiving reproducers from the nucleus lines, crossing them and supplying commercial farms with crossed animals. However, a multiplier has higher requirements than commercial farms. For example, the health of the animals has to be exemplary, as a multiplier introduces animals into other farms. At the slightest sign of illness, they have to interrupt the supply of reproducers and send the progeny they produce to the slaughterhouse. Moreover, animals from nucleus lines are less robust and less productive than crossbred females. The multiplier is usually set up in association with a large company, cooperative or integration, which ensures that there will be no surplus of crossbred females.

8.9. Creating a selection nucleus is a complex and risky operation, which is why it is generally not recommended. The lines must be kept in a closed population, and it is necessary to identify the animals, record several traits and avoid inbreeding when mating. A geneticist should implement and monitor selection methods and culling. All this generates heavy expenses. The health of the animals must be impeccable, so animals are slaughtered at the slightest sign of illness, which generates losses. Animals that have reached the age of sale but have not been sold due to lack of orders are also slaughtered, although their cost is much higher than the cost of a commercial animal due to the genetic process (e.g. records, evaluations). In addition, a considerable number of clients from commercial farms are needed to put the entire production of the nucleus in multipliers. In pigs and rabbits, small nuclei have been created in large cooperatives, or they have been associated with national programmes or with public research centres that manage their genetic programmes. Thus, creating a nucleus is a business initiative that must be approached with special caution.

Solutions to self-assessment questions

1. F, 2. F, 3. T, 4. F, 5. F, 6. F, 7. T, 8. F, 9. F, 10. F.

9 Genetic Improvement in Ruminants

Abstract

This chapter deals with the genetic improvement of ruminants, divided into two main groups, milk and meat production, although their selection schemes are similar. Cattle are predominant, followed by sheep and goats. The chapter starts by describing the organization of breeding, registration and dissemination of genetic progress in dairy cattle. Selection objectives are discussed, focusing mainly on milk and its components, but other traits of interest such as longevity and fertility are also examined, as well as 'new traits' such as feed efficiency, heat tolerance and methane production. Genetic parameters and the selection response are examined, with emphasis on the use of genomics in dairy cattle. In beef cattle, selection objectives are examined, focusing on feed efficiency and fertility, and selection criteria are analysed, particularly growth rate and conformation. The chapter ends with a brief discussion of sheep and goat programmes, which are similar to those for cattle but on a smaller scale.

9.1 Dairy Cattle

The main animal production in ruminants, both in cattle and in goats and sheep, is milk production. Meat production is less developed than milk production, not only because a large part of beef production is a by-product of dairy cattle programmes but also because sheep and goat meat production, and part of beef cattle production, are frequently associated with the conservation of rural areas in danger of desertification, and often represent a low-tech part of genetic improvement.

9.1.1 Organization of dairy cattle breeding and dissemination of genetic progress

A. Characteristics of ruminants affecting the organization of genetic improvement

Animal breeding in ruminants cannot be organized using the nucleus–multiplier–commercial farm schemes that we saw in the previous chapter, as ruminants are not prolific species and cannot benefit from the price reduction that multiplication allows. Smith (1988) proposed turning the cow into a prolific animal through multiple ovulations of the best cows and embryo transfer schemes, but in practice, attempts to implement this scheme for genetic selection have been unsuccessful, although these techniques are tools currently used in breeding programmes. The following characteristics of ruminants are relevant for genetic improvement schemes:

- *Ruminants are not prolific species.* In beef cattle, which are often non-intensively raised, there is an average of one calf per parity and five calves throughout the life of the animal, but in intensive dairy cattle, this number has dropped to an average of three calves per cow and sometimes even lower, depending on the country and the genetic programme (de Vries, 2020). This implies that 1.5 females are produced per

DOI: 10.1079/9781836990994.0009

cow on average, so almost all females of the progeny are engaged in replacement. Currently, sexed semen can be used to produce more females per cow, which allows some selection of the best cows for replacement, with the worst cows being used to produce calves for slaughter.

- *Milk production is expressed only in one sex.* However, due to the low prolificacy of the species, genetic improvement has been concentrated in males.
- *The generational interval is high,* around 6 years (van Tassell and van Vleck, 1991). This is because data from the daughters are needed to accurately evaluate a male, and the following must be taken into account:
    - The bull must reach 2 years of age to produce semen effectively.
    - Gestation lasts 9 months. Insemination is not 100% successful, with a success rate of around 40% in cows and 60% in heifers, so cows that do not get pregnant have to be re-inseminated 3 weeks later, when they are in heat again. As we will see later, success after three inseminations in cows is 80%.
    - Daughters must be 2 years old to be inseminated and then have a gestation and a lactation. Again, the insemination success is 60%, so it may be necessary to re-inseminate, and on average the process can be delayed several weeks.
    - After gestation, lactation (10 months) should be measured. After measuring lactation of around 100 daughters, the bull being tested is genetically evaluated and it is decided whether it will be selected or not. Genomics allows us to skip this process and reduce the generation interval, as we see later.
- *Artificial insemination is efficient.* Unlike in pigs, in which the number of possible dilutions per semen ejaculate is low and freezing of semen is not very effective, artificial insemination in cattle with frozen semen has a high success rate of fertilization. In cows, it has dropped from 70% in the 1970s to 40% today, but after three attempts, the success rate is 80% (Dobson *et al.*, 2008). Each ejaculate allows hundreds of dilutions, making it possible to produce hundreds of doses for insemination. The ease of freezing also allows a global semen trade, leading to an easy dissemination of genetic progress.
- *The sex of the embryo can be determined by semen sexing.* This is useful for producing candidate bulls to be tested and for producing replacement females from the best cows. The use of sexed semen is becoming more and more frequent.
- *Frozen embryos have a high viability*, similar to artificial insemination (Hansen, 2020). This allows use of the most productive cows to produce embryos through hyperovulation. Males go to the artificial insemination centre to be evaluated and to produce semen, and female embryos are sold as highly productive females.
- *The value of the animals is high.* Compared with prolific species, a cow costs much more than a sow, hen or rabbit, and due to the success in artificial insemination, the number of males required in cattle is very low, so the value of each selected male is much higher than that of equivalent males of prolific species. This allows the genotyping costs for genomic selection to be relatively much more affordable than in the case of other livestock species.
- *Data are collected from production farms rather than from selection nuclei.* This makes it possible to have a large amount of data available, which facilitates having large reference populations for genomic selection, and also allows improvement of the precision of the estimates of selection parameters. There is an exception for new traits, which we will discuss later (see section 9.1.2.C).

These characteristics mean that the organization of genetic improvement of cattle is different from that of prolific species. Genetic progress is obtained mainly through the males. *Genomics* is used for the first evaluation of the male's breeding value. Once they have had daughters, the males are re-evaluated, and both their predicted breeding value and the precision of the prediction increases (see section 2.3.3.D). Information from relatives is also used in the prediction of breeding values.

Using data from daughters dramatically slows down the male's evaluation process, greatly increasing the generation interval. The arrival of genomic evaluation in 2009 made it possible to evaluate males without data from their daughters (although with lower precision), reducing the generational interval and significantly increasing genetic progress per unit of time (see section 9.1.3.B) (García-Ruiz *et al.*, 2016).

B. Organization of genetic improvement in dairy cattle

Starting in the 1950s, when artificial insemination with frozen semen began to develop, livestock associations created artificial insemination centres to provide semen from selected bulls. The basis of the selection was the progeny test, in which each tested bull had at least 100 daughters who had had at least one lactation, which were used to evaluated the sire. Later, at the end of the 1970s, information from all the relatives was also integrated by means of best linear unbiased prediction (BLUP) with the relationship matrix, as we saw in section 4.3, but the fundamental part of the information still came from the daughters.

This scheme was used until recently. When genomic evaluation became available, it was no longer essential to have the offspring of a bull to evaluate it, although it is still necessary to continue collecting data, as described below.

B1. DAIRY CONTROL Unlike prolific species, in which the data are collected only in a selection nucleus, here the data is collected on the farms of the farmers who agree to have their cows under dairy control (more than 50% of cows are under dairy control in developed countries). There are no specific farms for dairy cattle genetic selection, with a few exceptions (see section 9.1.2.C); rather, the data come from commercial farms to genetically evaluate both bulls and cows. Semen from bulls that still do not have offspring is used for inseminating cows, and when these cows have daughters, they are also inseminated so that they become pregnant and have a lactation (in the insemination of daughters, the semen is irrelevant for genetics, its objective is that the daughter becomes pregnant to produce milk to be recorded).

- *An official controller of the association* visits the farms often (around once a month) and takes production data from all cows. Farmers bear the cost of dairy control, although they often receive subsidies to carry it out, generally covering part of the cost.
- *The most important data recorded are quantity of milk and its composition.* However, other important data are also taken, such as the somatic cell count (an indicator of mastitis), fertility, calving ease and pathological problems. To these data is added the morphological qualification of the bulls and cows, which is assessed once a lifetime by specialized qualifiers, as we will see later. These data are used in the genetic evaluation of bulls, which are in the artificial insemination station producing semen or have already been eliminated after freezing a sufficient number of their semen doses. Bulls are continuously re-evaluated as the data accumulates on their daughters and relatives.

- *In practice, there may be particular problems.* For example, it is necessary to avoid biases in inseminations with semen from bulls that do not yet have offspring (e.g. avoiding a bull inseminating the worst cows). Moreover, it is not uncommon for the farmer to treat daughters of very good bulls better, because their semen is more expensive, without communicating this to the controller; some mathematically complex solutions have been proposed to mitigate these undeclared preferential treatments.
- *The need to continue with dairy control as it is currently applied has been questioned.* This is because it is already possible to evaluate a bull through genomics without having to wait for data from his daughters. However, dairy control is necessary for a number of reasons. As discussed in section 6.4.2, the prediction equations must be recalculated because recombination causes genetic associations to be lost. Phenotypic records are also necessary so that the farms are well connected when applying a BLUP in order to separate environmental effects from genetic ones. The dairy controls also have other benefits: they serve to integrate farmers, relating them to common objectives, and dairy control is also used to advise them on how to improve their practices (e.g. better management), which may be especially useful in developing countries.

B2. SELECTION SCHEME BASED ON A PROGENY TEST Figure 9.1 shows the scheme based on the progeny test.

- *The selection process has two parts*, one is the choice of the best bulls according to their breeding value in terms of the traits of interest, but the other, which is even more important, is to decide *which bulls are to be tested*. The most effective way is to produce sexed embryos with the best possible genes. For this, cows with high breeding value are inseminated with semen from the best bulls at a global level. The best cows in the world do not produce milk but embryos; their ovulation is induced and embryos are produced with the best available semen, to give rise to the bulls that will be tested.
- *Genomic evaluation is carried out on each bull before they are tested*, and the semen from the best bulls is made available to farmers before they have daughters; they are

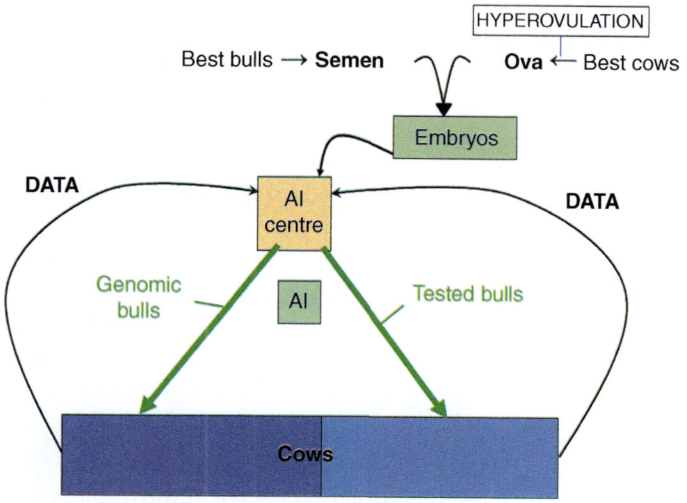

Fig. 9.1. Selection scheme in dairy cattle based on a progeny test. AI: artificial insemination.

popularly called *genomic bulls*. They are evaluated with the genomic data and the data from all available relatives (e.g. mother, siblings). Normally, at least ten bulls undergo genomic evaluation and the best one is selected, and semen from this bull is then made available to farmers, although the number may be higher depending on the programme.

- As the bulls have daughters and they pass to dairy control, their breeding value is re-evaluated and the precision increases. In cattle, *reliability* (the square of precision; see section 2.3.3.A) is used as one of the criteria for choosing semen doses. A bull with 100 daughters has a reliability of around 0.95 (it is usually expressed as a percentage, 95%), but a bull without daughters, evaluated only by siblings, the mother and other relatives using BLUP, only achieves a reliability of 20%, which results in an excessive risk when buying semen to substitute cows when they have finished their productive life. With the advent of genomic selection, genomic bulls without progeny testing can achieve reliabilities of 60–70%. This reliability is enough for farmers to use semen from these bulls before they have offspring, as young bulls have on average better breeding values due to the general genetic progress (see section 9.1.3.B).
- *In this system, there are no separate generations.* Instead, it is a continuous process in which the semen of the selected males is distributed, and catalogues are published with the genetic characteristics of these males, allowing farmers to acquire the semen they want, whether from genomic bulls without offspring or from bulls with higher reliability because they have data from their daughters.
- *Farmers can make their own genetic improvements.* Genetic improvement is not carried out in a nucleus and distributed later, as in prolific species; instead, it is up to the farmers to make the decisions for improving their herd. This is an important characteristic of the breeding scheme, as farmers can make their own genetic improvement by buying semen from bulls that have high breeding values for the traits that are important to them on their specific farm. For example, the price of milk can be penalized for a farmer due to a lower percentage of protein, while another farmer might have a problem with teat conformation, if the teats do not fit well with the milking machine (this is an intermediate optimum trait: neither long nor short nipples are desired). In these cases, each farmer can choose semen that is appropriate for their farm to improve these particular traits.
- *The use of sexed semen* makes it possible to choose the best cows as mothers and to use the worst cows to produce calves for sale to feedlots, as we will see in section 9.2.2.A.
- The percentage of bulls with offspring and genomic bulls varies according to the country or to genetic programme being used; in general, *more than half of inseminations are from genomic bulls*.
- As in any estimation process, some bulls have an overestimated breeding value, but as the confidence intervals of genomic bulls are greater than those of bulls with data from daughters, the overestimation can be worse, so it is common for genomic bulls with higher breeding values to see their breeding value reduced when data becomes available from his daughters. This also happens with highly reliable bulls, but as their confidence interval is very small, the effect is hardly noticeable.

B3. SIZE OF THE ORGANIZATION The organization of a genetic improvement programme based on progeny testing requires a large number of cows in production. Half of the offspring will be males unless sexed semen is used. Moreover, the fertilization rate is around 60% in heifers and 40% in cows; this implies that many cows will need to be re-inseminated 3 weeks later if they did not become pregnant in the first insemination, and again 3 weeks later if they did not become pregnant in the second insemination (if they do not become pregnant after three inseminations, the cow is usually culled). With this fertilization rate, 80% of cows and 94% of heifers will become pregnant with one to three inseminations. Figure 9.2 shows that, on average, 250 cows would be needed to produce 100 daughters, resulting in 94 lactations. These are average data, but to ensure that most of the bulls will have data from 100 lactations, it is necessary to inseminate many more cows per bull. When genomic evaluations were not available, it was common to inseminate 1000 cows per bull so that most of them had data from 100 lactations. If 10% of the best bulls were selected, this meant that for each bull in the catalogue, controls of 10,000 cows had been used, so the cost of the test was very high, around €30,000 per bull and €300,000 per bull that definitely entered the catalogue (numbers adapted from Funk, 2006).

At present, the possibility of carrying out genetic evaluations with moderately high reliabilities without the need for progeny testing has made the process considerably cheaper, as the selection of bulls is carried out on the basis of genomic tests. Furthermore, the increasingly frequent use of sexed semen makes it possible to directly produce males to be genotyped. Dairy control is still maintained for the reasons that we explained above, but data from daughters are no longer necessary for the selection of bulls, so the costs of each bull in the catalogue have been greatly reduced and this has led to a greater number of available bulls; today, around 4000 Friesian cattle bulls are listed annually in the world. As in the previous chapter, these numbers are approximate and they only intend to give an idea of the size of an organization dedicated to genetic improvement in dairy cattle.

Due to the efficiency of artificial insemination and the large number of dilutions that can be extracted from each ejaculate, a popular bull can produce 100,000 doses of semen per year (Funk, 2006). The availability of semen doses from these bulls means that in dairy cattle, farmers around the world are supplied with semen from a reduced number of bulls, so that the population of cows (notably those of the Friesian breed) is effectively a 'world nucleus'. This situation has led to an increase in inbreeding, which in global terms has gone from 2.7% in the 1970s to values close to the current 10%, and therefore to a reduction in the effective number of the population (see section 1.2.2). Nevertheless,

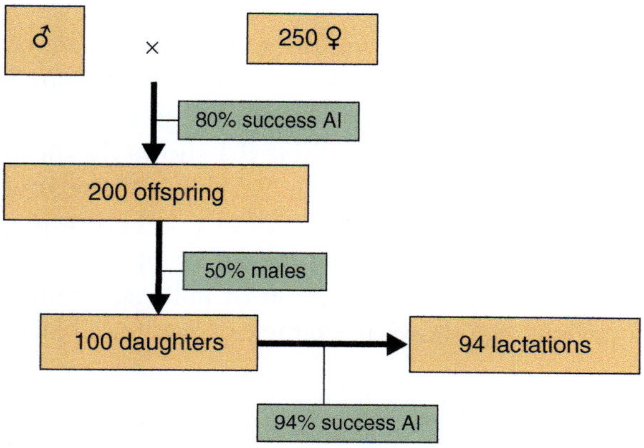

Fig. 9.2. Inseminations needed for a bull to have data of 100 daughters. AI: artificial insemination.

the increase of inbreeding has been slow, having less consequences in terms of fertility than inbreeding produced by mating close relatives, as we will discuss in section 9.1.2.B (see also section 1.2.2, and González-Recio *et al.*, 2007).

C. Dissemination of genetic progress

As mentioned earlier, farmers can make their own genetic improvements by choosing semen from particular bulls in a catalogue. These catalogues can come from private companies or associations of farmers, and provide genetic information on a group of males whose semen is available to the farmer:

- *Bull catalogues provide estimates of the breeding value of bulls and their reliability* for productive traits and also for the type, a concept similar to conformation (see section 8.1.2.A5), which we will discuss in more detail in section 9.1.2.B. All the catalogues indicate the estimate of the breeding value of the bull in terms of: (i) quantity of milk (measured in kilograms or pounds); (ii) milk fat content (measured in kilograms or pounds of production in a lactation or as a percentage); (iii) milk protein content (measured in kilograms or pounds or as a percentage); and (iv) traits relative to type. These values are given with reference to a mean, normally the mean performance of a specific year, which is updated every year. Information on longevity, calving ease, fertility, somatic cells and other traits is also frequently offered.
- The breeding value of the daughters is half the additive value of the father and half that of the mother. Assuming that medium-type mothers are inseminated, *the production expected from the daughters over the mean is half the breeding value of the male.*
- *The data come from official milk controls* that are carried out on the farms, and from official evaluations of the type carried out by expert judges who visit the farms and evaluate cows in their first lactation.
- *National evaluations are forwarded to an international association, Interbull*, based in Sweden, which combines evaluations from all national associations to calculate breeding values for bulls with data from all countries where they have daughters, and also their genomic values calculated using all the information available globally. Bull owners can use this information for their own catalogues and for the advertising they provide to sellers of frozen semen. Bulls are registered in the breed's genealogical book, providing their ancestors.
- *The catalogues often offer economic indices* giving an estimate of the profit that can be expected by using doses of these bulls. They are versions similar to the economic additive value that we saw in section 3.2.2, although their justification is different depending on the economic index used. Table 9.1 shows the mean production values for Holstein Friesian cattle in Spain, and Fig. 9.3 shows the relative weights of a number of traits included in the economic indices of several European programmes. Figure 9.4 shows a fragment of a catalogue of a private company in which some bulls are offered, and Fig. 9.5 shows the classification of bulls according to the criteria of a national association. Figure 9.6 shows an example of data for a bull offered by a company selling its semen.
- *Often, the best bulls have lower reliabilities.* Young bulls recently entered in the catalogue have low reliability because they do not yet have data from daughters; however, due to the genetic progress of the population, they often have on average the best

Table 9.1. Mean production values for Holstein Friesian cattle in Spain. Adapted from Chafferdine and Pérez-Cabal (2019).

Characteristic	Mean value
Milk at 305 days (kg)	10,174
Fat (%)	3.67
Protein (%)	3.22
Number of somatic cells	261
Culling (%)	27
Number of calvings per cow	3.4
Age at the first calving (months)	25.9
Interval between calvings (days)	416
Average inseminations per heifer	1.65
Average inseminations per cow	2.42

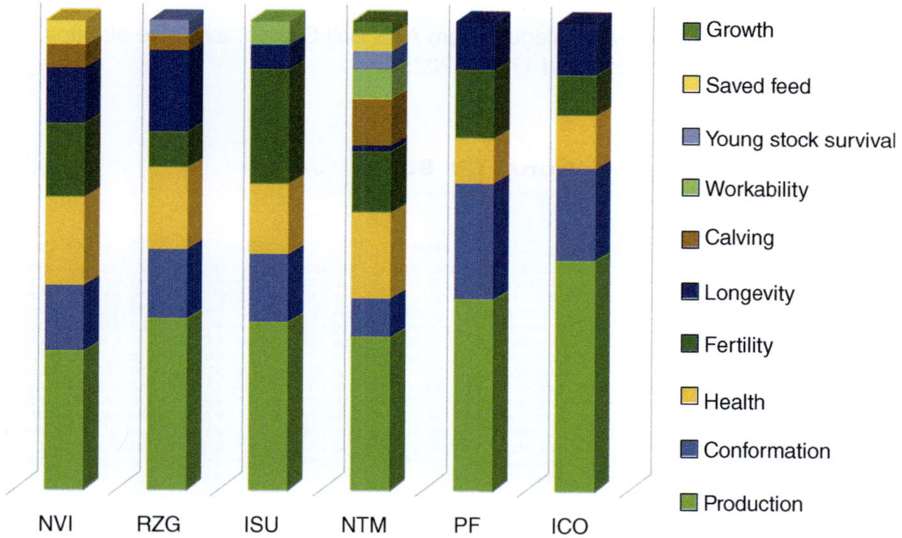

Fig. 9.3. Relative weights of several traits included in the indices of various European countries. From Eurogenomics, available at: www.eurogenomics.com/genomic-breeding-values/6-total-merit-indexes.html (accessed 17 July 2025).

breeding values. How to diversify the investment is a personal matter for the farmer; for example, a farm with many cows can diversify the investment, buying some of their semen from safe proven bulls with high reliability and some from less reliable bulls but with a higher estimated breeding value. A dose of semen from a good bull with high reliability does not guarantee obtaining a good cow, but the average of several daughters will be good; however, this good average is not guaranteed if the reliability of the bull is low (see section 2.3.1), so *it is usually recommended to diversify the acquisition of doses from bulls of low reliability*. If semen from several low-reliability bulls is used, the joint risk is lower than if only one or two low-reliability bulls are used (see Problem 2 at the end of this chapter).

⚥	♀	Code Name	NAAB	NM$	TPI®	CM$	Milk	CFP	Pro	Fat	Pro%	Fat%
⚥	♀	AMECHE HEISMAN x ALONSO	29HO21522	1083	3249	1125	273	146	38	108	0.11	0.36
	♀	HARMONY LEEDS x MODELO	29HO21534	1057	3227	1084	1096	150	52	98	0.06	0.19
⚥	♀	ENERGIZE LEEDS x ALTAPLINKO	29HO21551	1052	3222	1052	1885	146	55	91	-0.02	0.05
	♀	MYSTERY MONTANA x DUFFY	29HO21641	1048	3324	1072	705	124	34	90	0.04	0.23
⚥		MOTIF EXPRESS-P x LIONEL	29HO21803	1030	3294	1078	314	172	39	133	0.11	0.45
⚥	♀	HOOVER LEEDS x MENDEL-P	29HO21699	1027	3309	1052	1194	147	52	95	0.05	0.17
⚥	♀	ENCOURAGE LEEDS x ALTAPLINKO	29HO21552	1019	3318	1055	957	153	50	103	0.07	0.23
⚥		MANSON LEEDS x GAMEDAY	29HO21850	1017	3193	1039	1164	136	48	88	0.04	0.15
⚥		ETHAN EASTON x DELUXE	29HO21886	1011	3351	1053	621	148	45	103	0.09	0.29

Fig. 9.4. Example of company sire catalogue. From ABS Bull Search, available at: https://absbullsearch.absglobal.com/ (accessed 17 July 2025).

U.S. Registered Holsteins FOR MAXIMUM PROFIT
HOLSTEIN ASSOCIATION USA, INC.

Top 100 International TPI® Bulls APRIL 2025

Reliability of at least 80% traditional OR 85% genomic for production and Reliability of at least 80% for type in one of the countries included in MACE evaluations

						PRODUCTION				HEALTH						CONFORMATION				
Rank	Name	%RHA	Origin	NAAB	Pro	Fat	Milk	FE	%R	SCS	PL	LIV	FI	PTAT	%R	UDC	FLC	BWC	TPI®
1	GENOSOURCE CAPTAIN-ET TR TE	99I	USA	551HO04119	64	123	1872	371	99	3.05	3.7	-0.7	0.5	0.28	99	0.83	-0.26	-2.36	3441G
2	SDG CAP GARZA-ET MW TC	99I	USA	551HO04474	54	141	1686	378	97	2.91	3.4	0.0	-0.1	-0.18	90	0.30	-0.48	-1.49	3418G
3	STGEN COWEN THORSON-ET TR TE	99I	USA	551HO04520	60	137	1398	378	94	3.09	3.3	-0.2	0.6	-0.15	89	0.02	-1.43	-1.30	3389G
4	TERRA-CALROY ZURI-ET TR TP	99I	USA	097HO42585	56	108	1105	307	96	3.01	1.0	-2.0	0.0	1.37	85	0.88	0.39	-0.60	3348G
5	C-HAVEN POSITIVE DELUXE-ET TR	99I	USA	551HO04641	54	131	1581	327	99	3.02	0.2	-2.6	-1.7	1.16	97	0.36	0.15	0.49	3306G
6	S-S-I EISAKU PAYLOAD-ET TR TP	99I	USA	007HO15348	67	86	1654	293	99	3.13	2.2	0.7	-0.9	1.18	98	0.77	0.94	-0.18	3297G
7	STGEN DUBAI ON-DUTY-ET TE TC	99I	USA	551HO04250	38	51	490	171	98	2.90	2.6	1.3	3.7	1.92	97	2.29	-0.29	-0.03	3286G
8	PLAIN-KNOLL RENEGAD TROOPER TR	99I	USA	014HO15179	48	84	483	194	99	3.00	2.9	-0.6	1.4	1.16	96	0.60	2.13	1.02	3271G
9	GENOSOURCE CAP LEVOY-ET TE TC	99I	USA	551HO04581	42	116	1088	316	90	3.00	3.0	0.7	0.0	0.10	85	0.53	-0.45	-1.66	3257G
10	RMD-DOTTERER SSI GAMEDAY-ET TR	99I	USA	007HO15167	38	94	700	253	99	2.71	2.6	-1.4	-0.7	1.36	99	1.60	0.02	-0.50	3253G
11	DENOVO 16333 ENVY-ET TR TE	99I	USA	029HO19510	56	120	1430	295	98	2.96	1.3	2.1	-2.3	0.75	91	0.11	1.07	0.60	3251G
12	WINSTAR CONWAY CUMULUS-ET TR	99I	USA	014HO15932	61	106	1317	305	92	2.96	1.8	-0.1	-0.7	0.38	85	-0.65	0.53	-0.10	3248G
13	PLAIN-KNOLL CRUSHER-ET TR TP	99I	USA	007HO15465	41	115	156	290	99	2.86	3.5	3.5	0.1	0.21	96	-0.55	-0.43	-0.05	3238G
14	BACON-HILL COLDPLAY-ET TR TP	99I	CAN	200HO12085	48	78	1003	213	90	2.82	2.7	0.2	0.3	0.88	87	1.50	-0.32	0.37	3216G
	WINSTAR ALTAREGULUS-ET TR TP	99I	USA	011HO15783	35	95	-294	221	93	2.71	3.1	2.5	1.3	0.42	88	0.20	-0.66	0.64	3216G
16	WESTCOAST LAMBEAU-ET TR TP	99I	CAN	200HO11862	43	90	818	216	99	2.98	2.1	0.0	0.6	1.25	98	0.57	0.20	1.48	3208G
17	MR T-SPRUCE FRAZZ LIONEL-ET TR	98I	USA	007HO14454	68	111	2575	332	99	2.95	0.9	-2.5	-0.9	0.07	99	-0.09	-1.14	-0.41	3204G
18	PEAK ALTAZEMINI-ET TR TP	99I	USA	011HO15624	60	78	1051	253	99	2.81	1.6	0.7	0.0	0.41	94	0.10	-0.31	0.23	3203G
	PINE-TREE LIONEL COWEN-ET TR	99I	USA	551HO04138	60	116	1668	322	93	3.18	0.6	-2.4	-1.2	0.52	87	0.11	-0.22	-0.52	3203G
20	FLY-HIGHER SUGARHIGH-ET TR TP	99I	CAN	200HO12090	51	57	1451	192	94	2.82	3.1	0.3	0.7	1.46	91	1.55	-0.13	0.15	3195G

Fig. 9.5. Example of a sire list from the US Holstein Association. Available at: https://www.holsteinusa.com/genetic_evaluations/bull_lists.html (accessed 17 July 2025), used with permission.

9.1.2 Objectives and selection criteria in dairy cattle

In the case of dairy cattle, as its structure is based on collecting data from commercial farms, the phenomenon of stratification of interests that we saw in the case of prolific species does not occur. The returns of the farmer come mainly from the sale of milk, which is valued according to its fat and protein content, and is penalized by the concentration of somatic cells found in the milk, which are related to mastitis. Other supplementary income comes from the sale of calves for fattening, which we will discuss in section 9.2.2.A. The costs, apart from the fixed cost of labour, buildings and management, come

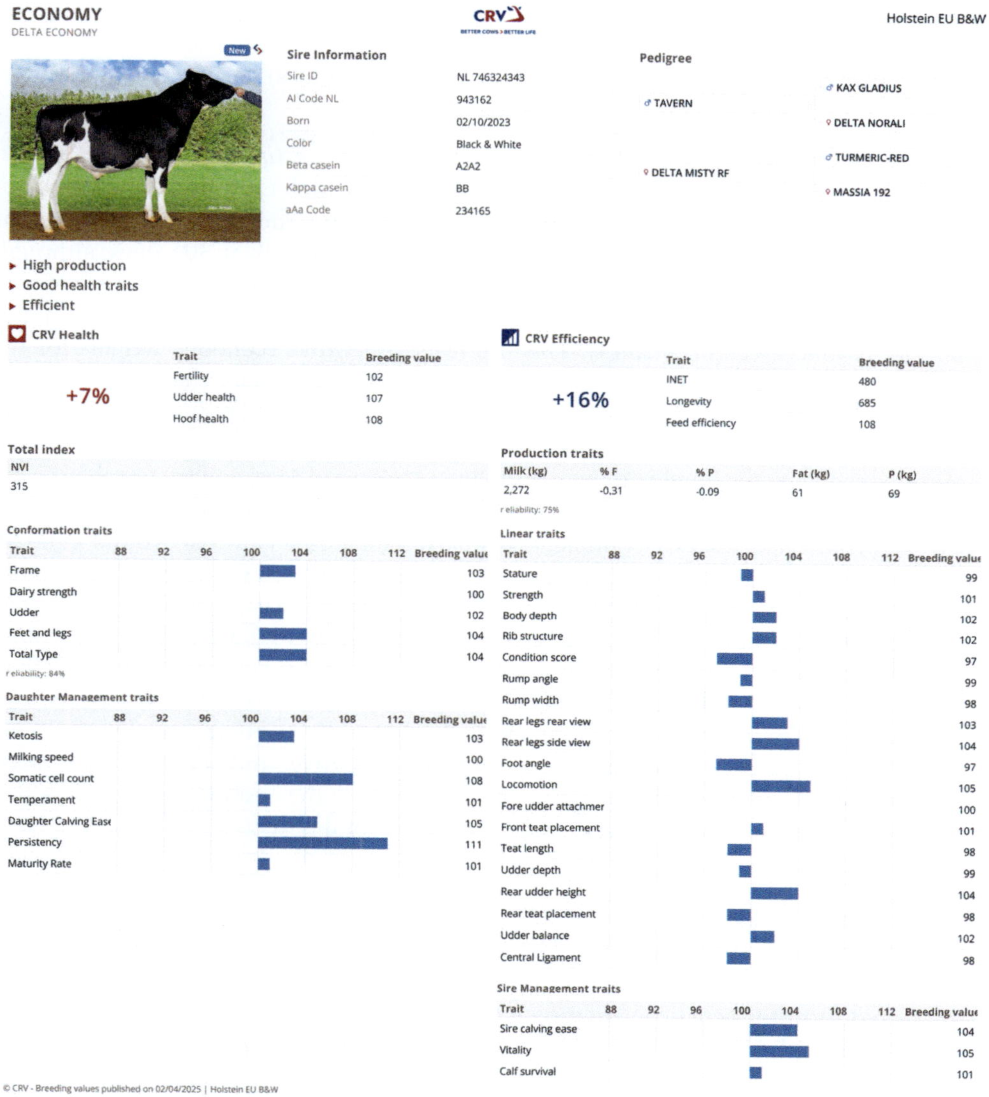

ECONOMY
DELTA ECONOMY

New

Holstein EU B&W

Sire Information

Sire ID	NL 746324343
AI Code NL	943162
Born	02/10/2023
Color	Black & White
Beta casein	A2A2
Kappa casein	BB
aAa Code	234165

Pedigree

♂ TAVERN — ♂ KAX GLADIUS, ♀ DELTA NORALI
♀ DELTA MISTY RF — ♂ TURMERIC-RED, ♀ MASSIA 192

- High production
- Good health traits
- Efficient

CRV Health +7%

Trait	Breeding value
Fertility	102
Udder health	107
Hoof health	108

CRV Efficiency +16%

Trait	Breeding value
INET	480
Longevity	685
Feed efficiency	108

Total index

NVI
315

Production traits

Milk (kg)	% F	% P	Fat (kg)	P (kg)
2,272	-0.31	-0.09	61	69

reliability: 75%

Conformation traits

Trait	Breeding value
Frame	103
Dairy strength	100
Udder	102
Feet and legs	104
Total Type	104

reliability: 84%

Daughter Management traits

Trait	Breeding value
Ketosis	103
Milking speed	100
Somatic cell count	108
Temperament	101
Daughter Calving Ease	105
Persistency	111
Maturity Rate	101

Linear traits

Trait	Breeding value
Stature	99
Strength	101
Body depth	102
Rib structure	102
Condition score	97
Rump angle	99
Rump width	98
Rear legs rear view	103
Rear legs side view	104
Foot angle	97
Locomotion	105
Fore udder attachmer	100
Front teat placement	101
Teat length	98
Udder depth	99
Rear udder height	104
Rear teat placement	98
Udder balance	102
Central Ligament	98

Sire Management traits

Trait	Breeding value
Sire calving ease	104
Vitality	105
Calf survival	101

© CRV - Breeding values published on 02/04/2025 | Holstein EU B&W

Fig. 9.6. Information on a bull from a sire catalogue. From the CVR Holstein catalogue, available at: https://shop.crv4all.com/en/sires/holstein/nl-bw-holstein (accessed 17 July 2025), used with permission from CRV International.

from feeding (which represents around 40–60% of the costs), and culling of the animals, usually due to fertility problems or type problems related to the legs (lameness) or udder (udder health, teats not well adapted to the milking machine). As the average number of lactations of a cow is already less than three, *longevity* is an important trait to reduce the cost of the cow replacement. *Fertility* has also become an important trait because it has been decreasing over time, with consequences on the improvement schemes that we saw in the previous section. Furthermore, some traits that were not possible to include in a breeding programme, such as milk feed efficiency or the reduction of methane emissions, are now feasible through genomic evaluation.

A. Traits related to milk production

A1. MILK PRODUCTION The amount of milk produced in a lactation, measured in kg, is corrected to standardize it at 305 days of production:

- *It is enough to take samples from the first lactation for estimating breeding values*, because the genetic correlations between lactations are high (>0.80) for milk production and composition. Monthly samples can be taken from the first 3 months of lactation to provide a reasonable assessment of the amount of milk at 305 days, although in practice samples are taken every month and from all lactations.
- Due to surplus milk production, production quotas can be established so that milk cannot be sold beyond a certain quantity. This can lead to negative economic weights for the quantity of milk. In principle, production can be adjusted by reducing the number of cows to produce a given quantity of milk, but in practice, it is difficult because the production cycles are long, the cows are expensive and in general the system is much less flexible in terms of changing the number of animals than in the case of pigs, poultry or rabbits.
- *The cost of genetic improvement is a very small part of the production cost*, as was the case in prolific species. For example, a standard semen dose can cost around €10–20, and twice this if it is sexed semen. The farmer receives around €0.33 per litre of milk, depending on the country and the evolution of the milk prices, which for an average production of around 10,000 litres of milk per lactation and an average of around three lactations per cow implies an average total return per cow of around €10,000, from which costs should be deducted to calculate the profit. Costs are close to returns, as in every farming scenario; in our example the cost can be around €9000. Even though the numbers in the example are merely indicative, and several semen doses may be needed for a cow in production, it is evident that the cost of semen (i.e. the cost of genetic improvement) is a very small part of the total costs.

A2. MILK FAT CONTENT Fat content in milk can be expressed as the total amount in kilograms after 305 days of lactation, or as a percentage of fat in milk. In the first case, the genetic correlation with the amount of milk is positive, while in the second it is negative. The preferred form for genetic evaluation is the amount in kilograms in one lactation, to avoid problems with ratios similar to the problems with feed conversion ratio (FCR) that we saw in section 8.1.2.A2.

- *Milk fat content is valued by the dairy industry*, on the basis of a reference percentage of fat, and they pay more or less depending on whether the milk is over or under this reference percentage.
- *Milk fat content was the main quality factor in the past.* With the reduction in the consumption of animal fats for dietary reasons, fat has lost part of its economic value, but its economic weight continues to be positive, and there has even been some re-evaluation, in part due to the decrease in consumption of other fats more harmful to health such as palm oil. Although the consumption of saturated fat is clearly related to heart disease (see review by Zong *et al.*, 2016), milk and dairy products have a large number of fatty acids and other components, so there is no evidence of an association between the consumption of non-skimmed dairy products, including cheese, and heart disease, which is causing many of the dietary recommendations regarding dairy products to be revised (see reviews by Drouin-Chartier et al., 2016a,b).

A3. MILK PROTEIN CONTENT Milk protein content can also be expressed as the total amount in kilograms after 305 days of lactation, or as a percentage in milk, but the amount in kilograms is the preferred form for genetic evaluation for the same reason as in milk fat content.

- *Milk protein content is valued by the dairy industry* as milk fat content, with bonuses or penalties if the content is over or under a reference value.
- *Protein content is the main economic component of milk* due to the decrease of the economic weight of the quantity of milk and fat content.

A4. CELL COUNT Mastitis is a bacterial infection of the udder that can lead to milk having to be treated before consumption, depreciating the value of this milk. The presence of somatic cells in milk is an indication of mastitis, and the price of milk is penalized according to the cell count in the samples.

B. Other traits

B1. LONGEVITY *Longevity* is defined as the number of days of a cow's life from the first calving to its death.

- Cow replacement usually occurs on the farm, taking advantage of some of the calvings of the cows, and although the cost of semen is not important, the cost of rearing until the new cow enters production can be expensive (e.g. around €1800), so longevity is an economically relevant trait, especially as the number of calvings per cow has dropped from five in the 1970s to the current average of three. In the current improvement programmes, the main aim is for this number not to continue to decrease.
- *Selection for longevity is not easy*. To record this trait, it would be necessary to wait for the death of each cow. However, if a cow has reached, for example, her third parity, we know that at least she has got to this point, and we have more information than if we only had the first parity. These incomplete data are known as *censored data*, and there are complex mathematical techniques to include them in breeding programmes.

B2. TYPE *Type* is a set of external morphological characteristics of the cow and bull that are assumed to have direct or indirect economic relevance. Morphological qualification is mandatory for all animals included in the herd book.

- Type is a selection criterion to improve objectives related to longevity, health status or productivity, although, as we will see below, these relationships are not very evident.
- Qualification is carried out by people specially trained for this.
- *Many traits (usually more than 20) are evaluated*, but not all type traits are of equal importance. Some udder traits are important because they are related to the cow's ability to adapt well to milking machines and to the propensity for mastitis. Legs traits are important because they are related to lameness and culling animals from these causes (Fig. 9.7).
- *In economic indices, type traits* (mainly legs and udder traits) *have a small weight*, between 7% and 12%, compared with productive traits (see Fig. 9.3).

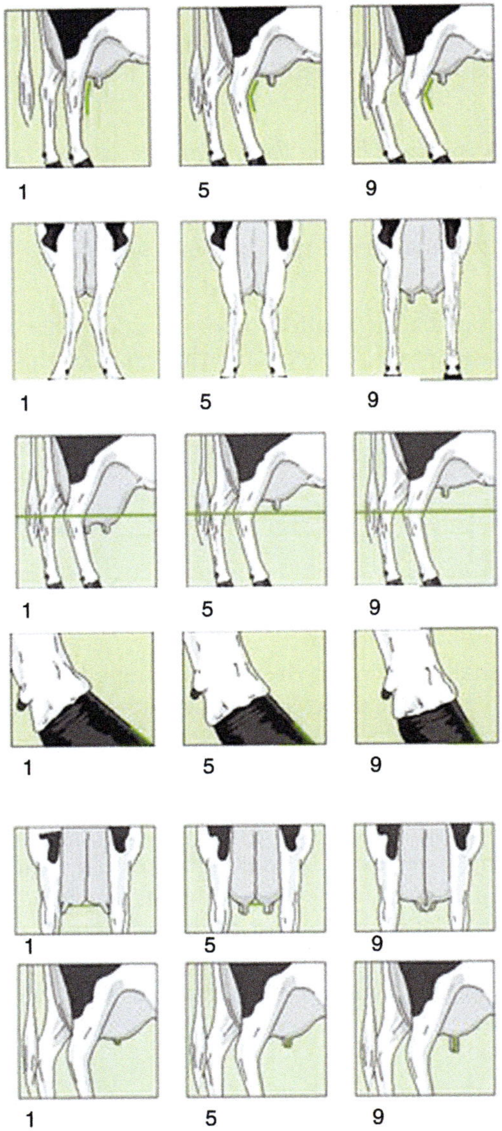

Fig. 9.7. Some type traits scored for the international type evaluation of dairy cattle. From World Holstein Friesian Federation (2022), used with permission.

- *The numerical classification used in type traits can come from measurements* (each point corresponds to a certain number of centimetres), which is the most common, *or by subjective scoring*. In this second case, as in all subjective scores, a score of 4 does not indicate that the animal gives rise to twice as many problems as one of 8. In general, in all cases the first two or three classes usually indicate problems, and the remainder indicate acceptable situations, except for some traits with intermediate optimum.
- *The genetic correlations between type and longevity, and between type and production, are low*, as we will see later. Phenotypic relationships may be better, particularly

non-linear relationships, but genetic correlations are always linear because under the infinitesimal model, traits are normally distributed and the relationships between traits are linear under normality.

B3. FERTILITY The reduction in fertility in recent decades has led to consideration of this trait as a selection objective. The decrease in fertility in Friesian cattle could be attributed to the increase in inbreeding, but as inbreeding increased slowly, it seems that unfavourable alleles related to fertility should have been purged (lost due to genetic drift, see section 1.1.3), and *only recent inbreeding* (mating close relatives) *produces serious problems of fertility decline* (González-Recio *et al.*, 2007), but farmers avoid crossing between close relatives. Thus, it seems that the decrease in fertility is mainly due to its negative genetic correlation with milk production (see section 9.1.3.A) more than to the overall inbreeding.

C. New traits

Genomic selection has opened up new breeding opportunities for traits that are difficult or expensive to measure. Among the new traits we will consider are the FCR of feed into milk (or feed efficiency, more generally), methane production and heat tolerance:

C1. FCR In intensive milk production, the cost of feeding the lactating cow can represent more than 50% of the total cost. The problem is that data are collected on farms, which do not have facilities or staff to collect feed data.

- At present, there are several institutions that are collaborating to analyse the feed efficiency of milk production, although with fewer than 10,000 cows analysed at the time of writing, it is too early to have reliable estimations of breeding values for this trait.
- Genomic selection would allow us to evaluate any animal on any farm, measuring the conversion efficiency only on a few farms (Li *et al.*, 2020). A problem, however, can arise if there are genotype–environment interactions, as the farms in which feed efficiency is measured are not necessarily standard farms – they may have a better environment than the that of standard farms.
- The heritability of feed efficiency is around 20% (Houlahan *et al.*, 2022). This is notably lower than the heritabilities of milk production and composition, which means that more individuals are needed to produce a reasonable reliability. Moreover, with low heritabilities, large reference populations are needed to establish the equations for genomic selection. This does not imply that the necessary data will not be available in the future, but it is unlikely to be in the immediate future.

C2. METHANE PRODUCTION The growing concern about global warming has highlighted that the production of methane by cattle is one of the main factors responsible for greenhouse emissions, as we will see in section 10.3.1. Reducing production decreases the amount of methane produced; the problem is how to maintain productivity while also reducing the amount of methane.

- The first problem is finding a simple way to measure methane so that cheap measurements can be made on a large number of animals; it would not be practical, for example, to use breathing chambers. The fatty acid composition of milk has a moderate ability to predict methane. The same occurs with milk composition evaluated by mid-infrared spectroscopy. A third method is measuring methane and carbon dioxide directly from the cow's respiration during milking (Lassen and Difford, 2020).

- The heritability of methane production is around 0.2, and its genetic correlation with milk yield is around 0.3 (Manzanilla-Pech *et al.*, 2020), thus it would be possible to include in a selection index.
- There has been intense research in the genetics of rumen bacteria (using metagenomics) to detect genes or bacteria that can reduce methane production without affecting production (McAllister and Newbold, 2008) (see section 10.1.3).

C3. HEAT TOLERANCE With Friesian cattle, and in general with breeds from temperate climates, milk production suffers when the ambient temperature is above 25°C (although this depends on physiological state, diet, etc.).

- In tropical climates, and generally in hot climates, breeds that are more resistant to heat are used, but in temperate climates, the number of hot days throughout the year may be economically relevant due to the lower production of breeds currently used in temperate climates.
- The trait is difficult to define and record. Hot days productivity is commonly used as an approximation, but it is not easy to calculate the economic weight of the trait, nor is it easy to integrate it into a selection programme (Carabaño *et al.*, 2019).
- The use of genomic selection allows us, in principle, to genetically evaluate bulls and cows that do not have these data. Garner *et al.* (2016), using genotyped cows that had climatic information, prediction equations were developed using over 600,000 single-nucleotide polymorphisms (SNPs), and this information was validated using a reduced sample of cows in climatic control chambers in which heat waves were simulated. The conclusion of the experiment was that the better cows that had been genetically evaluated for heat tolerance were more productive during heat waves, which, in principle, would allow this trait to be included in genomic selection relatively easily.

9.1.3 Genetic parameters and response to selection

A. Genetic parameters

The genetic parameters of milk production and its composition are shown in Table 9.2, integrating results from various publications, while the genetic correlations between type traits and production are shown in Table 9.3. These values are the average of several publications and should be taken as a guide only.

Table 9.2. Genetic parameters of milk production traits. Heritabilities are on the diagonal in bold. Genetic correlations are above the diagonal. From Simm (1998).

	Milk (kg)	Fat (kg)	Protein (kg)	Fat (%)	Protein (%)
Milk (kg)	**0.39**	0.75	0.91	–0.42	–0.50
Fat (kg)		**0.36**	0.82	0.28	–0.09
Protein (kg)			**0.36**	–0.20	–0.10
Fat (%)				**0.58**	0.60
Protein (%)					**0.62**

Table 9.3. Genetic correlations between linear type traits, longevity and milk production traits. From Simm (1998).

	Longevity	Milk (kg)	Fat (kg)	Protein (kg)
Stature	0.00	0.22	0.16	0.25
Chest width	–0.13	–0.14	–0.10	–0.13
Body depth	–0.13	0.24	0.25	0.24
Angularity	0.13	0.44	0.42	0.43
Rump angle	0.03	–0.11	–0.08	–0.11
Rump width	–0.11	–0.01	0.01	0.01
Rear legs, side view	–0.05	0.07	0.06	0.07
Foot angle	0.09	0.02	0.05	0.07
Fore-udder attachment	0.06	–0.29	–0.23	–0.28
Udder depth	0.14	–0.48	–0.40	–0.44
Central ligament	0.02	0.10	0.16	0.15
Front teat placement, rear view	0.03	–0.05	–0.03	–0.01
Teat placement, side view	–0.02	0.38	0.35	0.36
Teat length	–0.19	0.18	0.12	0.17

- Note that heritabilities are high, higher even than those published in the 1950s. Given that the selection process has been in operation for many years, this may seem surprising, but a large part of this increase in heritability is due to the methods used for better correcting environmental effects (REML, Bayesian methods), as discussed in Chapters 4 and 5.
- The genetic correlations between milk, fat and protein are positive when fat and protein are expressed in kilograms, because the greater the quantity of milk, the greater the quantity of its components; however, note that they are negative in terms of their composition in percentages, which implies that selection tends to dilute the milk components.
- The genetic correlations between fat and protein are high, whether they are expressed as a percentage or in kilograms, which favours the selection by milk composition.
- *Genetic correlations between type traits and traits of milk production and composition are, in general, very low*, as can be seen in Table 9.3, which integrates data from various publications. Particularly noteworthy are the low genetic correlations with longevity. This is important, because some farmers can give an importance to type traits that they do not have. Nevertheless, farmers can choose animals in the catalogues with the characteristics that they want to improve in their farms, helped by economic indices, which include an evaluation of the type, as we saw in Figs 9.4 and 9.5.
- *Fertility* (which is not included in Table 9.3) *has a very low heritability* of around 0.03, and a negative genetic correlation with the amount of milk, fat and protein of around –0.5 (Veerkamp *et al.*, 2001), which explains the decline in fertility observed in recent decades.

B. Response to selection

The selection method used in dairy cattle is BLUP and some of its variants, usually with a *repeatability animal model* (see section 4.3). The generalized use of genomic selection has imposed the *single-step genomic BLUP (G-BLUP)* (see section 6.2.3) as the preferred method.

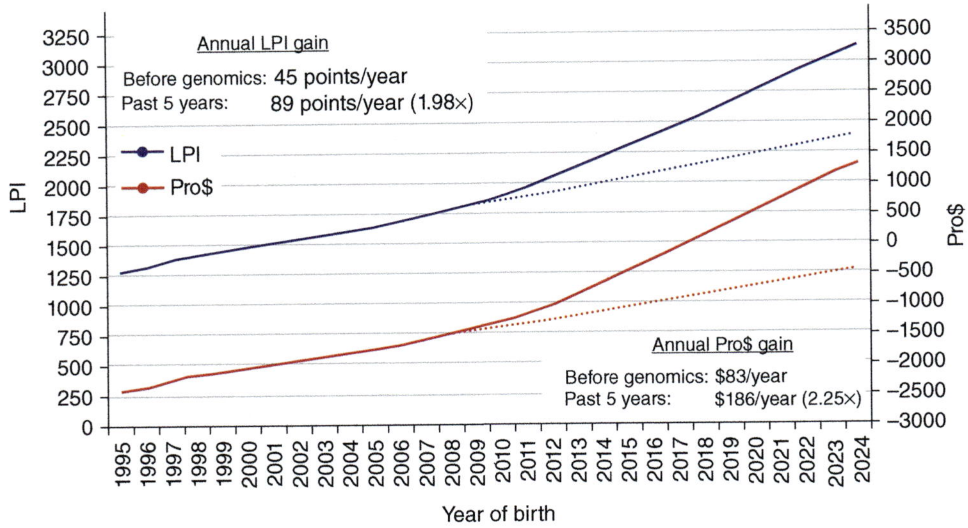

Fig. 9.8. Response to selection in Canadian Holstein dairy cattle measured with two economic indicators, LPI (life performance index, in points) and Pro$ (profit in Canadian $). Genomic selection started in 2009. The dashed line indicates the genetic trend that would have occurred if genomic selection had not been applied. From Sweett (2024).

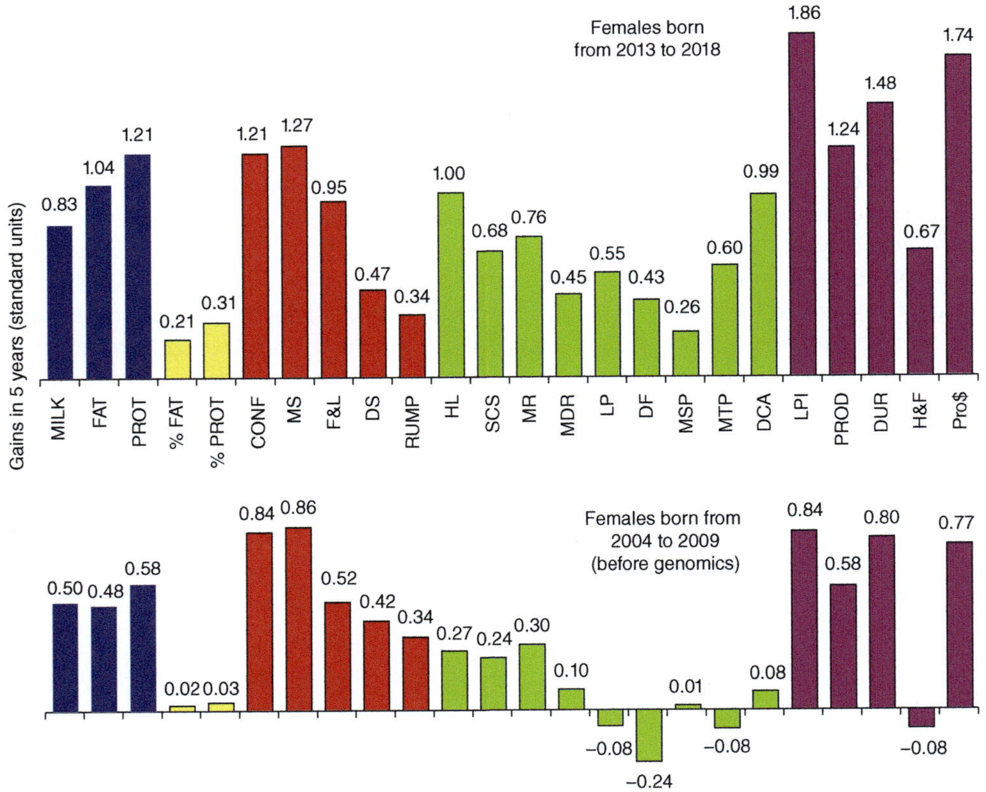

Fig. 9.9. Relative genetic gain before the use of genomics (2004–2009) and after the use of genomics (20012016) for Canadian Holstein cattle. From van Doormal (2019). PROT: protein,

Continued

The response to selection in dairy cattle has been consistently successful for as long as there have been records, with no sign of declining. Moreover, since genomic selection has been applied, the response has increased notably due to the reduction of the generational interval by half, which greatly compensates for the lower precision of the genomic bulls. Figure 9.8 shows, for Holstein cattle in Canada, what the global genetic progress has been, using two indices of economic type, one in points and the other in Canadian dollars, taking as the origin the time at which genomic selection began to be applied (2009). It can be seen that, since 2009, global economic progress has been much higher than before.

The response to selection has not been the same for all traits. Figure 9.9 compares the response before and after genomic selection for a number of traits, also for Canadian Holstein cattle. As can be seen, the effect of genomic selection is particularly noticeable when used for traits that are expressed in later life, such as longevity.

Fertility, despite being a trait with low heritability and a negative genetic correlation with productivity, also shows positive trends. In addition, by having genomic information, combinations of alleles harmful to fertility have been detected and avoided through targeted mating.

9.2 Beef Cattle

In contrast to dairy cattle, beef cattle present a wide variety of situations in production. In Europe, a large part of beef production is a by-product of dairy cattle farming. The dairy farmer only needs one parity to replace the cows using sexed semen, or two parities on average if not. For other parities of the cows, it is common to inseminate with beef cattle with a high growth rate, so many calves sent to the feedlot are crossed animals. The use of sexed semen favours this practice, using only well-evaluated cows for self-replacement, and leaving the others to produce fattening animals. Beef cattle breeding programmes are carried out on breeds with high growth rates or on breeds that are adapted to poor environments.

9.2.1 Objectives and selection criteria in beef cattle

A. Selection objectives

A1. FEED EFFICIENCY In beef cattle, the main selection objective is *feed efficiency*, as feed represents more than 60% of the production cost in beef cattle. However, as feed efficiency is expensive to measure, the selection criteria (measured traits) are not necessarily the same as the objectives.

Fig. 9.9. Continued.
CONF: conformation; MS: mammary system; F&L: feet and legs; DS: dairy strength; HL: herd life; SCS: somatic cell score; MR: mastitis resistance; MDR: metabolic disease resistance; LP: lactation persistency; DF: daughter fertility; MSP: milking speed; MTP: milking temperament; DCA: daughter calving ability; LPI: life performance index; PROD: production; DUR: daughter durability; H&F: health and fertility; Pro$: profit in Canadian $.

A2. CARCASS MEAT CONTENT AND PROPORTION OF EXPENSIVE RETAIL CUTS These are also important objectives, but obtaining the appropriate carcass records is expensive, and thus the selection criteria are not necessarily the same as the objectives, as with feed efficiency.

A3. EASE OF CALVING Some breeds have special problems in terms of ease of calving, particularly those with the myostatin gene (section 9.2.2.C).

A4. FERTILITY *Fertility* is also an objective, but its heritability is very close to zero, so it is often not considered. As in monogastrics (see section 8.1.2), the objectives of the farmer are not always the same as those of the slaughterhouse, butcher or consumer; for example, the killing-out percentage matters little to the butcher, while the proportion of expensive parts of the carcass does not concern the consumer.

B. Selection criteria

B1. BODY WEIGHT AND GROWTH RATE The most frequent selection criterion is weight at a given age, usually *weaning weight*. In more developed programmes, weight is also taken at a later age and *growth rate* is computed.

- *Genetic parameters and selection criteria.* All feed efficiency traits are highly genetically correlated, as expected. Weight and growth rate have fairly high heritabilities (~0.3–0.4) and the genetic correlation of FCR with growth rate is relatively high (around –0.6) (Berry *et al.*, 2015). *The usual way of improving feed efficiency is by selecting for growth rate.* These numbers are merely indicative, as genetic parameters depend on the genetic variability of different populations, and on the diverse influence of the different environments, as we have discussed elsewhere.
- *Increase in birth weight.* A frequent problem derived from selection for weight at a particular age or for growth rate is an increase in birth weight, which can end up causing problems at delivery. For this reason, it has been suggested using programmes that include restrictions on birth weight.
- *Increase in adult size.* Heavier parents are more expensive to maintain, particularly in beef cattle when compared with prolific species, because the cost of the parents' maintenance is distributed only over five calves on average. Selection for weight at a particular age or for growth rate leads to an increase in adult weight, so it has been suggested to select for growth rate relative to adult weight, or to use an index with restrictions for adult weight. However, when applying restrictions, the heritability of growth rate is much lower and it does not economically compensate for this type of selection. Theoretically speaking, an index (or BLUP) including growth rate and adult weight, with their respective economic weights and genetic parameters, should be the optimal solution, but the consequences of still increasing adult weight will probably create management problems in the long term, particularly in extensive systems, so it is a problem that does not yet have a clear solution. If the selected breed is used in a cross as a terminal sire by artificial insemination, the maintenance is less of a problem because terminal sires are a small proportion of the parental population.

B2. CONFORMATION Conformation assessment in living animals and in carcasses is frequent, with criteria related to muscularity and the skeleton:

- *Degree of maturity.* It is common to argue that beef cattle breeds have a better shape than dairy breeds, and that they have a higher percentage of meat:bone ratio or a higher proportion of expensive retail cuts of the carcass. A large part of these differences is due to the fact that *meat breeds grow rapidly and are generally slaughtered younger than dairy breeds*. When comparing carcasses at market weight, different breeds are usually slaughtered at a different stage of maturity. The stage of maturity (also called degree of maturity) is defined as the weight at the time of measurement, divided by the adult weight. A breed slaughtered at 40% of its adult weight will have a different carcass composition than if it is slaughtered at 70% of its adult weight, as we saw in Fig. 8.8, because the relative development of tissues and the parts of the carcass change with age (see Berg and Butterfield, 1976, for a discussion on cattle, and Taylor, 2009, for a general discussion on comparisons at the same degree of maturity).

  If breeds or crosses are compared at the same degree of maturity, despite differences in conformation, the similarities are striking. Table 9.4 shows the composition of Holstein carcasses (dairy cattle) and the frequent crossing of Holstein × Charolais (a typical beef breed) evaluated at the same degree of maturity; note the similarity of results. This does not mean that crossbreeding with meat breeds should not be used; as we saw in Fig. 8.7, they grow faster so they have better feed efficiency, and also tend to have a more favourable carcass composition, as they are slaughtered younger than Holsteins at the same commercial weight. As a whole, although they have a higher proportion of bone and worse dressing percentage, having a better FCR compensates the inconveniences (see section 8.1.2.A4).

  To determine the degree of maturity, it is necessary to know the adult weight, which is not always feasible, but as fat is a late tissue, it is often assumed that the breeds or treatments being compared are at the same degree of maturity when they have the same fat content; an approximate way of looking at it is to add a fat covariate to the breed or treatment comparison model, to correct for differences in fat content. As age and stage of maturity are highly correlated, an approximate alternative is to perform comparisons at the same age.

  Kempster *et al.* (1982) compared crosses of Holstein cows with males of various breeds, including Charolais and Holstein, and found similar results. There is an abundant bibliography comparing bovine carcasses, although they are not always compared at the same degree of maturity. Using cows of beef cattle breeds, Wheeler *et al.*

Table 9.4. Carcass composition of Holstein Friesian and Charolais × Holstein Friesian steers measured at the same degree of maturity. Adapted from Berg and Butterfield (1976).

	Holstein	Charolais × Holstein
Lean (%)	63.8	63.6
Lean:bone ratio	4.0	4.3
Wholesale cuts:		
Rump (%)	7.1	7.2
Top side (%)	22.5	22.9
Fore rib (%)	5.0	5.1
Loin (%)	9.3	9.5
Lean in high-priced cuts (%)	47.8	47.7

(1997) compared the carcasses of a three-way cross between crossed mothers (Hereford × Angus) and males of the cattle breeds Charolés, Gelbvieh, Pinzgauer, Shorthorn, Galloway, Longhorn, Piedmontese and Salers, and a zebu (Nelore), all at the same point of maturity, and found that the differences between them were small, with the notable exception of the Piedmontese breed, a breed with a muscle hypertrophy gene that we will talk about later in section 9.2.2.C. Discounting the Piedmontese, the percentage of bone ranged from 13.7% to 14.2% (95% confidence interval ± 0.4%) (Piedmontese 11.6%), and the percentage of cuts from 62.6% to 63.4% (95% CI ± 1.5%); the same similarity was found in the remaining characteristics of the carcass.

- *Carcass conformation* is important due to its possible relationship with the percentages of meat and expensive cuts of the carcass, but this relationship is not strong; it is more apparent than real. The problem with the percentages is that the numerator (weight of a retail cut or weight of the meat content of the carcass) and the denominator (weight of the carcass) are highly correlated, so the variability of the ratio is small and it is difficult to make accurate predictions for this ratio; the same happens with the meat content of the retail cuts with respect to the weight of the retail cut. A measure of the reliability of a prediction is, as in dairy cattle, the coefficient of determination, R^2. This coefficient indicates the proportion of the observed variability explained by the prediction equation; that is, the part that does not depend on the residual. In a prediction equation:

$$y = b_0 + b_1x_1 + b_2x_2 + \cdots + b_nx_n + e$$

where the variables x_i are external measurements in a live animal or in the carcass (e.g. conformation, fat scan, longitudinal measurements), and the variable y is what is intended to be predicted (e.g. meat content, meat/bone, fat content, proportion of expensive cuts), so the coefficient of determination is:

$$R^2 = 1 - \frac{\sigma_e^2}{\sigma_y^2}$$

When there is only one variable x, the coefficient of determination is the same as the square of the correlation coefficient; it is similar to the prediction reliability that we saw in section 2.3.3.A.

Table 9.5 shows the coefficients of determination of conformation measurements on a live animal, and those of equations that include scans taken on the live animal for

Table 9.5. Coefficients of determination (R^2) for the prediction of percentages of meat, bone fat and expensive cuts in the carcass, based on several muscle measurements or skeletal measurements taken from live animals at slaughter. Prediction equations include muscle scores, and muscle content and fat content scans. Adapted from Conroy *et al*. (2009).

	Meat (%)	Fat (%)	Bone (%)	High-priced cuts (%)
Muscle score	0.27–0.41	0.00–0.06	0.00–0.09	0.04–0.19
Skeletal score	0.02–0.12	0.02–0.09	0.49–0.67	0.01–0.14
Prediction equation	0.53	0.48	0.68	0.80

predicting meat and fat content. As can be seen, the scoring measures for morphological traits are poor predictors, especially of the percentage of expensive cuts of the carcass but also of the meat and fat content of the carcass.

The external appearance of the animal suggests that its relationship with these traits is going to be much closer, but the shape has a strong dependence on the covering fat, leading to 'optical illusions' in which visual estimates of muscle development have little significance, as noted by Berg and Butterfield (1976). This implies that conformation should not be overemphasized. The results in Table 9.5 are common to many experiments; in fact, the ratios tend to be lower if the conformation assessors are poorly trained.

9.2.2 Genetic improvement programmes

A. Breeds and crosses

- *Most of the beef consumed in Europe comes from Holstein dams*, either pure Holstein or crossed with a beef breed. When a gestation of a dairy cow is not intended for replacement, the cow is often inseminated with males of beef cattle breeds (e.g. Charolais, Limousin, Belgian Blue), which have better growth and feed efficiency and are therefore more valuable. Despite this, pure Holstein cattle are also destined for meat, either because the result of insemination for replacement with Holstein is a male, or simply because available Holstein semen was used.
- *Industrial crossing* is also common. This is done with native breeds with low growth performances but with good maternal characteristics (e.g. fertility, calving ease) with a high growth performance beef cattle breed. There are other types of crosses, but they are less common.
- Outside Europe, the production systems are different. In America (North and South), it is common to breed productive beef cattle such as Angus or Hereford, without crosses.
- *In difficult environments, it is common to use breeds adapted to the environment*, even if they are foreign breeds. A notorious example is the Nelore zebu, which is widespread in Brazil although its origins are in India, or the synthetic Santa Gertrudis breed, which is a mixture of zebu and Beef Shorthorn.

B. Beef breeding programmes

Unlike dairy cattle with the Friesian breed, in beef cattle there is no global nucleus of a specific breed. There are breeds with larger population numbers and more developed programmes (e.g. Angus, Hereford), but in general, beef breeding, particularly in Europe, depends on small associations, often with insufficient development of artificial insemination.

- *Preselection is usually made by conformation*, not only taking note of legs problems but also the external conformation to which we referred in the previous section.
- *The selection method is normally BLUP*, or G-BLUP when genomic selection is used. One problem is having sufficient connections between herds, which is not always the

case, particularly in small breed associations with underdeveloped artificial insemination programmes. BLUP evaluations with low herd connections produce predictions with low reliability.

- *Other interesting objectives, such as calving ease or intramuscular fat content are often achieved through crosses* (the latter is important in North America and East Asia); for example, cows of breeds without calving difficulties are crossed with bulls of breeds with high growth rate.
- *Individual performance and progeny test*: As weight and other traits can be measured in the individual itself, and the heritability is high, it is not really necessary to carry out progeny testing. In practice, however, progeny data are often included in several programmes for a number of reasons. First, the males are first selected by their own data (e.g. for weaning weight or growth rate), but when measuring weight at a later age for growth rate, they may have early data from their offspring (e.g. weaning weight). Furthermore, the use of artificial insemination allows connecting farms to correctly apply the BLUP and separate environmental effects from the additive values of the evaluated animals. Finally, as there is often no official control of weaning weight, and it is up to each farmer to provide their numbers, the progeny test allows farmers to verify that, indeed, the semen they are going to buy comes from a well-evaluated bull.

C. Genomic selection and the search for major genes

Genomic selection has been proposed in beef cattle for reasons similar to those in dairy cattle: it makes it possible to shorten the generational interval if progeny tests are carried out and it is possible to increase the precision together with the data recorded on the individual itself.

- The use of genomics in beef cattle has not had, however, the spectacular result that it has today in dairy cattle for two reasons: the first is that a *reduction in the generational interval is not an objective,* as bulls can be evaluated with their own data.
- The second reason is that *the heritability of weight or growth rate is around 0.4 in cattle*, which is also the value of reliability in individual selection (see Example in section 2.3.3.B), which is 40% when expressed as a percentage, or higher when including information from relatives, so that in the end the reliability is not very far from the 60% of genomic bulls. Although the joint use of genomics and the individual's data increases precision, the use of genomics does not have the same consequences as in dairy cattle, in which the reliability of a bull without information from his daughters is around 20%.
- There is a third reason that affects beef programmes in general and that is *the lack of organized programmes* with official controls and good connections between farms by artificial insemination that allow the application of a BLUP in good conditions.
- Genomic selection could be used for traits that are not evaluated in the selection candidate, such as carcass traits, or for traits that are measured in few individuals, such as feeding efficiency; however, the number of measurements carried out today is insufficient to establish and update the prediction equations.
- Beef is a special case in terms of the existence of major genes: the myostatin gene has several mutations that produce a noticeable increase in the proportion of meat in the carcass (~20%), giving rise to animals that called 'double muscled or 'double rumped'.

These animals have calving difficulties and higher nutritional requirements, so they are not usually an interesting alternative to other breeds (de Smet, 2014). In section 10.1.2.B, we will discuss genetic modifications to the myostatin gene.

9.3 Genetic Improvement of Sheep and Goats

We will divide this section into sheep for milk production, sheep for meat, and goats, although in dairy sheep there is a by-product from the sale of lambs for meat. The production of wool has fallen in most countries, and nowadays it is dominated by China and Australia, each producing around 20% of world production, although it is possible that in the future it will be of interest again, as wool is biodegradable while synthetic fibre is not.

9.3.1 Dairy sheep

Sheep's milk is produced for the manufacture of cheese. Often, a variety of cheese is associated with a specific breed and a production area; for example, Rochefort cheese must come from Lacaune sheep, Manchego cheese from Manchega sheep, and Idiazábal from Latxa and Carranzana sheep. This does not imply that these breeds have any historical cultural component; for example, the Lacaune breed is a synthetic breed created in the mid-20th century. In addition, sheep are often raised extensively in relatively difficult environments to which it is necessary for the breed to be well adapted. Both the cheese specificity and the different environments result in a wide variety of breeds, which limits the size of the breeding programme and its possibilities.

- In dairy sheep, the *objectives* are similar to those of dairy cattle, but with the peculiarity that the milk is dedicated to cheese, so the protein concentration in the milk is particularly important:
  - The main objectives are milk production and its composition, as in dairy cattle. In sheep, lactation is assessed at 120 days (150 days in highly productive breeds).
  - Traits related to type are considered, mainly the mammary morphology.
  - The relative importance of the objectives depends on each programme; for example, in the Churra breed it is 60% for the amount of milk, 20% for protein concentration, 5% for the number of live births, 10% for mammary morphology traits and 5% for body morphology traits.
- *The organization of genetic improvement* is similar to that of dairy cattle, but the populations are much smaller and are less well connected by artificial insemination, given their extensive management and the fact that they are inseminated with induced heat, which affects to fertility. This leads to lower reliability estimates of additive values than for dairy cattle. Refrigerated semen is frequently used, which is cheaper and easier to handle, although it has a lower success in fertilization (50%) than frozen semen when it is applied by intrauterine insemination. The doses are cheaper than those of dairy cattle (~€8), and thus genetic improvement represents a very low proportion of the production cost. It is common to use mixed systems of controlled natural breeding and artificial insemination.
- *Catalogues* for dairy sheep are a simplified version of dairy cattle catalogues. As the evaluation is carried out within breeds, there is no national evaluation, nor a joint

evaluation similar to Interbull that we saw in section 9.1.1.C. A catalogue contains the males of a breed and offers the additive value of milk and its composition, and sometimes the economic additive value and reliability.

- The *genetic parameters* of the traits related to milk production are similar to those we saw in dairy cattle (Table 9.2), and the response to selection is around 3%/year, also similar to that of dairy cattle before the use of genomic selection. The response has been variable, depending on the associations. For example, for the Spanish breed Latxa, the response to selection has been spectacular for milk production, which has almost doubled its production from when the programme began in the 1990s until today.
- *Scrapie* is a neurodegenerative disease that affects sheep and goats, caused by a prion, as in the case of bovine spongiform encephalopathy (mad cow disease), but is not transmissible to humans. Resistance to scrapie can be detected: the gene that gives rise to the prion has an allele called ARR that gives rise to animals with a very low risk of scrapie if it is homozygous, and with variable risks if it occurs in other combinations. Catalogues often report the scrapie alleles of their males; in Europe, it is mandatory by law to have a plan to eradicate this gene.
- *Genomic selection* is implemented in France in the Lacaune and Manech breeds and is starting in New Zealand and in the Latxa breed in Spain. The reasons why it has not been applied to date are economic; but as in monogastrics, imputation allows us to solve a large part of this problem.
- As in all species, BLUP is the preferred selection method, and single-step G-BLUP if genomic selection is used.
- Sheep genetic programmes are often subsidized, and in many cases, they are linked to conservation programmes. In section 7.3.1, we saw that when *in situ* conservation is the solution adopted, a selection programme can be established to improve the breed's performance and make it more competitive, reducing the need for it to be kept in the long run.

9.3.2 Meat sheep

The consumption of sheep meat is much lower than meat from other livestock species (see section 10.3.1 and Fig. 10.8). Meat sheep are usually under extensive production and frequently in difficult places where breeds adapted to the environment are used. This gives rise, as in the case of dairy sheep, to various genetic programmes associated with the different breeds. Some dairy sheep breeds are also being used to develop meat sheep programmes.

- *Types of meat lamb.* Several types of lamb are produced in meat sheep – milk-fed lamb (unweaned, 4–6 weeks old), lamb (<12 months old), mouton (mature sheep, normally <20 months old) and other types – so the objectives are not necessarily the same. Suckling lambs whose diet is based on mother's milk give rise to light carcasses (~6 kg), while older grazing lambs have carcasses of 10–14 kg; there is also semi-intensive production of fattened lambs with heavier carcasses. Another characteristic of both meat and milk sheep is the rare presence of multiple births; the average litter size is around 1.3.

- *Breeds and crosses.* As in beef cattle, sheep crosses are frequent, particularly what is known as an industrial cross, which consists of crossing a poorly productive local breed being crossed with a breed with high growth rate to produce lambs that are fed intensively. There are many autochthonous breeds that have a genetic improvement programme. Industrial crosses are made with high-growth breeds such as Ile de France, Landschaf, Fleischschaf, Texel and Suffolk.
- *Objectives and criteria.* Common criteria are prolificacy (number of lambs weaned) and growth (to improve efficiency), with some type traits. The genetic programme of the milk-fed lambs of the Churra breed in Spain gives the following weights to its economic index: 50% for the number of live births, 30% for age at commercial weight, 10% for sheep score, 5% for calving interval and 5% for success in artificial insemination. In the Merina breed used for meat, the traits commonly included in a selection programme are weight at weaning (a way of evaluating the mother's milk capacity), growth rate and prolificacy, and some type traits.
- *The organization of genetic improvement* has problems similar to those of dairy sheep; populations are small and are less well connected by artificial insemination (which is much less used in meat sheep than in dairy sheep), given its extensive management. As in beef cattle, growth-related traits can be measured in the individuals themselves, which questions the progeny test due to the lengthening of the generational interval, although the increase in the number of multiple births, given its low heritability, could justify the progeny test. As in the case of dairy sheep, the programmes are often subsidized.
- *Genomic selection* is not generally implemented. Some markers close to major genes that increase prolificacy are used in some breeds. As in dairy sheep, there is genetic information about scrapie and attempts are being made to eliminate it.

9.3.3 Goats

Goat genetic improvement focuses on milk production; the consumption of goat meat is marginal.

- *Objectives.* As goat's milk is destined for cheese, its objectives are similar to those of sheep's milk: milk production and composition, particularly protein content. The type is also evaluated, particularly the mammary morphology. As in dairy cattle, an economic index is also made.
- *Genetic parameters and selection method.* Genetic parameters are similar to those of dairy sheep. The evaluation system used is BLUP or G-BLUP, as in previous cases.
- *The organization of genetic improvement* also has problems similar to those of dairy sheep; populations are small and poorly connected by artificial insemination. The programmes are often subsidized.
- *Genomic selection.* In general, there are no genomic selection programmes, but genomic evaluations have recently begun in the Alpine and Saanen breeds in France, Australia and New Zealand. Genes with favourable alleles for cheese making are known in goats, and males have this information in the catalogues.

Main Concepts

- *Progeny test:* A selection scheme in which males are evaluated for traits related to milk production, mainly using information from their daughters (and also from other relatives), in order to obtain a high reliability for the breeding values. The use of genomic selection has made it possible to reach reasonable reliabilities without the need to use data from daughters, but the scheme is maintained because it is necessary to have data in order to establish the genomic equations. In addition, dairy control on farms allows connection of farm data and correcting for farm environmental effects, allowing data from different farms to be used together.
- *Sire catalogue:* A list of males whose semen is available so that farmers can self-replace their cows. It includes the breeding value for the type and for the traits of economic interest as well as the reliability. Often, it includes complementary information, such as scrapie or casein alleles, information on the parents of sires, number of daughters, number of herds in which the sire has daughters.
- *Longevity:* A trait defined as the number of days of life of a cow from the first calving until its elimination. Culling is usually due to udder health problems, lameness, difficulties in adapting to the milking machine, illness or other causes.
- *Type:* A set of external morphological traits of the cow, sheep or goat that are assumed to have direct or indirect economic relevance. Genetic correlations with productive traits, including longevity, are generally low. In practice, certain udder and leg traits are the most relevant because they influence health, adaptability to the milking machine, and handling.
- *Conformation:* A set of external morphological traits that are supposed to have direct or indirect economic relevance. The measured traits are intended to predict the composition of the carcass (i.e. percentage of fat, meat and bone, and expensive parts of the carcass). In general, its predictive capacity is low, even when using equations that combine several conformation traits.

Problems

9.1. In dairy cattle, proven bulls with 100 daughters have a reliability of around 0.95, but the generation interval is 6 years. Genomic sires have an accuracy of around 0.60 and their generation interval is 3 years because they do not need to have daughters evaluated. Based on these numbers alone, and assuming that the intensity of selection is the same, which is the most efficient scheme?

9.2. As we saw in the last point of section 9.1.1.C, genomic bulls are younger, have been selected more recently and, due to continuous genetic progress, tend to be better than the most proven bulls, but they have lower reliability. A dairy farm wants to use unrelated genomic sires with an average reliability of 70%, with higher average additive values than a proven bull with 95% reliability. How many genomic bulls should be used so that the whole batch of genomic bulls equals the reliability of the proven bull?

9.3. Compare the data of these two catalogue bulls, Carmen Mixer and Balisto. Based on productive data only, which sire would be more appropriate? In this case, milk is sold at €0.32 per kg of milk, and the price increases by €0.003 for each 0.1% extra fat content and by €0.004 for each 0.1% extra protein content.

	Reliability	Milk (kg)	Fat (%)	Protein (%)
Carmen Mixer	96%	1217	–0.07	–0.01
Balisto	98%	655	0.49	0.40

9.4. In a dairy cattle artificial insemination station, they have the mean values of many daughters of each bull for milk production and milk fat percentage. Are these values a sufficient guarantee to ensure that a daughter of one of these good sires will be a good milk producer? Are they a guarantee that the average of several daughters of one of the sires will be as good as the average of daughters that have already been evaluated?
9.5. The size of the udder is obviously related to the amount of mammary tissue and therefore milk production. Would it not be easier to improve the volume of the udder rather than to use dairy controls?
9.6. If the type is not related to production, why measure it?
9.7. Why should you import semen for use in dairy cattle? Is it not cheaper to import a bull for a local insemination centre?

Self-assessment Questions (True or False)

1. Farmers can do genetic improvement on their own farm by self-replacing daughters of their best animals, buying semen from sire catalogues.
2. Longevity is an economically important trait in dairy cattle because the average number of offspring per cow is around three and should not be further reduced.
3. The cost of genetic improvement is very low. Saving on the purchase of semen is generally an inappropriate practice.
4. Establishing regional or local genetic improvement programmes in dairy cattle means freeing oneself from dependence on multinational companies and substantially increasing profits, so it is therefore a recommended practice.
5. Establishing genetic improvement programmes in native sheep and goat breeds is inefficient due to their low productivity, and it would be always preferable to use imported breeds selected for productivity.
6. Conformation in beef cattle is important because a better conformation in live animals indicates a higher percentage of meat.
7. Type traits in dairy cattle, particularly legs and udders, are highly related to longevity.
8. An expensive dose of semen from a highly reliable bull (99%) always guarantees a single highly productive daughter if inseminated into a highly productive cow.
9. In the previous question, if there were several doses of semen from that bull inseminating highly productive cows, the average of the daughters would have a high productivity.
10. Inseminating high-productivity cows with semen from bulls whose daughters on average have relatively low reliability (70–80%) but high breeding value would give rise to daughters whose average productivity should be high.

References

Berg, R.T. and Butterfield, R.M. (1976) *New Concepts of Cattle Growth*. Sidney University Press, Sidney, Australia

Berry, D.P., Kennedy E. and Crowley J.J. (2015) Genetics of feed intake and efficiency. In: Garrick, D.J. and Ruvinsky, A. (eds) *The Genetics of Cattle*, 2nd edn. CAB International, Wallingford, UK.

Carabaño, M.J., Ramón, M., Menéndez-Buxadera, A., Molina A. and Díaz, C. (2019) Selecting for heat tolerance. *Animal Frontiers* 9, 62–68. DOI: 10.1093/af/vfy033

Chafferdine, N. and Pérez-Cabal, M.A. (2019) Estudio económico para la actualización del ICO y definición de nuevos índices de mérito económico total. *Frisona Española* 230, 48–49.

Conroy, S.B., Drennan M.J., Kenny D.A. and McGee M. (2009) The relationship of live animal muscular and skeletal scores, ultrasound measurements and carcass classification scores with carcass composition and value in steers. *Animal* 3, 1613–1624. DOI: 10.1017/S1751731109990395

de Smet, S. (2014) Double muscled-animals. In: Jensen, W.K. (ed.) *Encyclopedia of Meat Sciences*. Elsevier, Amsterdam, Netherlands, pp. 397–402.

de Vries, A. (2020) Symposium review: why revisit dairy cattle productive lifespan? *Journal of Dairy Science* 103, 3838–3845. DOI 10.3168/jds.2019-17361

Dobson, H., Walker, S.L., Morris, M.J., Routly J.E. and Smith R.F. (2008) Why is it getting more difficult to successfully artificially inseminate cows? *Animal* 2, 1104–1111. DOI: 10.1017/S175173110800236X

Drouin-Chartier, J.P., Brassard, D., Tessier-Grenier, M., Côté, J.A., Labonté, M.È. *et al.* (2016a) Systematic review of the association between dairy product consumption and risk of cardiovascular-related clinical outcomes. *Advances in Nutrition* 7, 1026–1040. DOI: 10.3945/an.115.011403

Drouin-Chartier, J.P., Côté, J.A., Labonté, M.È., Brassard, D., Tessier-Grenier, M. *et al.* (2016b) Comprehensive review of the impact of dairy foods and dairy fat on cardiometabolic risk. *Advances in Nutrition* 7, 1041–1051. DOI: 10.3945/an.115.011619

Funk, D.A. (2006) Major advances in globalization and consolidation of the artificial insemination industry. *Journal of Dairy Science* 89, 1362–1368. DOI: 10.3168/jds.S0022-0302(06)72203-2

García-Ruiz, A. Cole, J.B., VanRaden, P.M., Wiggans, G.R., Ruiz-López, F.J. and van Tassell, C.P. (2016) Changes in genetic selection differentials and generation intervals in US Holstein dairy cattle as a result of genomic selection. *Proceedings of the National Academy of Sciences USA* 113, E3995–E4004. DOI: 10.1073/pnas.1519061113

Garner J.B., Douglas, M.L., Williams, S.R.O., Wales, W.J., Marett, L.C. *et al.* (2016) Genomic selection improves heat tolerance in dairy cattle. *Scientific Reports* 6: 34114. DOI: 10.1038/srep34114

González-Recio, O., López de Maturana, E. and Gutiérrez, J.P. (2007) Inbreeding depression on female fertility and calving ease in Spanish dairy cattle. *Journal of Dairy Science* 90, 5744–5752. DOI: 10.3168/jds.2007-0203

Hansen, P.J. (2020) The incompletely fulfilled promise of embryo transfer in cattle – why aren't pregnancy rates greater and what can we do about it? *Journal of Animal Science* 98: skaa288. DOI: 10.1093/jas/skaa288

Houlahan, K., Schenkel, F.S, Miglior, F., Jamrozik, J., Stephansen, R.B. *et al.* (2022) Estimation of genetic parameters for feed efficiency traits using random regression models in dairy cattle, *Journal of Dairy Science* 107, 1523–1534 DOI: 10.3168/jds.2022-23124

Kempster, A.J., Cook, G.L. and Southgate, J.R. (1982) A comparison of the progeny of British Friesian dams and different sire breeds in 16- and 24-month beef production systems. 2. Carcass characteristics, and rate and efficiency of meat gain. *Animal Production* 34, 167–178. DOI: 10.1017/S0003356100000647

Lassen, J. and Difford, G.F. (2020) Genetic and genomic selection as a methane mitigation strategy in dairy cattle. *Animal* 14 (Suppl. 3), s473–s483. DOI: 10.1017/S1751731120001561

Li, B., VanRaden P.M., Guduk, E., O'Connell, J.R., Null, D.J. *et al.* (2020) Genomic prediction of residual feed intake in US Holstein dairy cattle. *Journal of Dairy Science* 103, 2477–2486. DOI: 10.3168/jds.2019-17332.

Manzanilla-Pech, C.I.V., Løvendahl, P., Mansan Gordo, D., Difford, G.F., Pryce, J.E. *et al.* (2020) Breeding for reduced methane emission and feed-efficient Holstein cows: an international response. *Journal of Dairy Science* 104, 8983–9001. DOI: 10.3168/jds.2020-19889

McAllister T.A. and Newbold, C.J. (2008) Redirecting rumen fermentation to reduce methanogenesis. *Australian Journal of Experimental Agriculture* 48, 7–13. DOI:10.1071/EA07218

Simm, G. (1998) *Genetic Improvement of Cattle and Sheep*. CAB International. Wallingford, UK.

Smith, C. (1988) Applications of embryo transfer in animal breeding. *Theriogenology* 29, 203–212. DOI: 10.1016/0093-691x(88)90040-4

Sweett, H. (2024) Celebrating 15 years of genomic selection! *Lactanet*. Available at: https://lactanet.ca/en/celebrating-15-years-of-genomic-selection/ (accesed 22 July 2025).

Taylor, St C.S. (2009) Genetic size-scaling. In: Rauw, W. (ed.) *Resource Allocation Theory Applied to Farm Animal Production*. CAB International, Wallingford, UK, pp. 147–168.

van Doormal, B. (2019) 10 years of genomic selection: what's next? *Lactanet*. Available at: www.cdn.ca/document.php?id=530 (accessed 18 July 2025).

van Tassell, C.P. and van Vleck, L.D. (1991) Estimates of genetic selection differentials and generation intervals for four paths of selection. *Journal of Dairy Science* 74, 1078–1086. DOI:10.3168/jds.s0022-0302(91)78258-1

Veerkamp, R.F., Koenen, E.P.C. and de Jong G. (2001) Genetic correlations among body condition score, yield, and fertility in first-parity cows estimated by random regression models. *Journal of Dairy Science* 84, 2327–2335. DOI: 10.3168/jds.S0022-0302(01)74681-4

Wheeler, T.L., Cundiff, L.V., Koch, R.M., Dikeman, M.E. and Crouse, J.D. (1997) Characterization of different biological types of steers (cycle IV): wholesale, subprimal, and retail product yields. *Journal of Animal Science* 75, 2389–2403. DOI: 10.2527/1997.7592389x

World Holstein Friesian Federation (2022) *Type Harmonisation Report and Recommendations*. WHFF, Leeds, UK. Available at: https://whff.info/wp-content/uploads/2023/04/WHFF-progress-of-type-harmonisation-2022.pdf (accessed 18 July 2025).

Zong, G., Li, Y., Wanders, A.J., Alssema, M., Zock, P.L. *et al.* (2016) Intake of individual saturated fatty acids and risk of coronary heart disease in US men and women: two prospective longitudinal cohort studies. *British Medical Journal* 355: i5796. DOI: 10.1136/bmj.i5796

Solutions

Solutions to the problems

9.1. The response to the selection is $R = \rho_{A,C} \cdot i \cdot \sigma_A$ and the reliability is $\rho^2_{A,C}$ (see section 2.3.3.A). The annual response in the case of progeny testing will be:

$$R_{TEST} = \frac{1}{6}\sqrt{0.95} \cdot i \cdot \sigma_A = 0.16 \cdot i \cdot \sigma_A$$

and the annual response in the case of genomic selection will be:

$$R_{GENOMIC} = \frac{1}{3}\sqrt{0.60} \cdot i \cdot \sigma_A = 0.26 \cdot i \cdot \sigma_A.$$

In both cases the additive standard deviation is the same because it is a population parameter, and we assume that the intensity of selection is the same in both cases, so genomic selection is more efficient. In practice, it is also necessary to continue doing progeny tests for the reasons stated in section 9.1.1.B.

9.2. To obtain the same reliability, they must have the same error variance. The variance of the individual prediction error is, as we saw in section 2.3.3.A, $\sigma_e^2 = \sigma_A^2\left(1-\rho_{A,C}^2\right)$. If we call $\rho_{\bar{A},C}^2$ the reliability of the mean of the additive values of the group, its error variance will be:

$$\frac{\sigma_A^2\left(1-\rho_{\bar{A},C}^2\right)}{n}$$

because the variance of a mean of the residuals is the mean of the variances when they are uncorrelated (see Appendix 2.3 for the formula for the variance of a residual):

$$\left(1-\rho_{A,C}^2\right) = \frac{\left(1-\rho_{\bar{A},C}^2\right)}{n} \rightarrow n = \frac{\left(1-\rho_{\bar{A},C}^2\right)}{\left(1-\rho_{A,C}^2\right)} = \frac{1-0.70}{1-0.95} = 6$$

Therefore, to obtain the same error variance, six genomic bulls are needed.

9.3. Carmen Mixer has an additive value for milk of 1217 kg, but Carmen Mixer decreases the fat content (–0.07%) and the protein content (–0.01%), so it has a penalty for fat and protein. Therefore, its economic additive value would be:

$$0.32 - 0.7 \cdot 0.003 - 0.1 \cdot 0.004 = 0.3225 \text{ €/kg}$$
$$1217 \text{ kg} \cdot 0.3225 \text{ €/kg} = \text{€}392$$

Balisto has an additive value for milk of 655 kg and two bonuses, one for fat (0.49%) and another for protein (0.40%). Therefore, its economic additive value would be:

$$0.32 + 4.9 \cdot 0.003 + 4 \cdot 0.004 = 0.3532 \text{ €/kg}$$
$$655 \cdot 0.3532 \text{ €/kg} = \text{€}231$$

The reliability is practically the same. Based on this data, we would choose Carmen Mixer.

9.4. There are never any guarantees that a particular daughter will be good, for the reasons stated in section 1.1.1 and Fig. 1.1; certain crosses can give rise to combinations of alleles that give particular bad daughters, even though the parents are good. This means that it can be ensured with certain reliability that if the average number of daughters of a sire is high, the average number of new daughters should also be high, as we explained in section 2.3.1 and Fig. 2.2 (relying, obviously, on the fact that the cows to be inseminated have similar yields to the mothers of the daughters already tested; the expected productive value of a daughter depends on the father and the mother).

9.5. The size of the udder is determined both by milking capacity and by connective tissue, so the correlation between both traits is not high and the coefficient of determination (the square of the correlation coefficient), which indicates the goodness of the prediction of milk quantity from udder volume, is even lower. In general, all type traits are poorly correlated with production traits and the coefficients of determination are very low, as can be seen if the correlation coefficients in Table 9.3 are squared.

9.6. The correlations are not zero and can provide some useful information, but the main reason is that certain traits of the type, particularly those related to legs and the morphology of the udder and teats, are useful to eliminate animals with problems with lameness, udder health and adaptation to the milking machine. However, these case types (feet and udders) only have a small weight (between 7% and 12%) in the economic indices, compared with the productive traits, as we saw in section 9.1.2.C and in Fig. 9.3.

9.7. There are many reasons for avoid importation of a specific bull to serve a group of farms. First, it is rare for a company to sell a proven bull, but if it does, given how expensive it is to prove the genetic value of a bull, the price will be extremely high. To compensate for the investment, the bull has to carry out many inseminations over a long time period. If we consider that genetic improvement is progressing at a constant rate, young bulls' semen with much better characteristics will be available soon, making the semen of imported bulls obsolete. In addition, the massive use of doses from a single bull turns a large part of the population into half-sisters, so care must be taken when making new inseminations so as not to increase inbreeding. Finally, the farmer is encouraged to buy doses of this bull, losing the options provided by a semen catalogue.

Solutions to self-assessment questions

1. T, 2. T, 3. T, 4. F, 5. F, 6. F, 7. F, 8. F, 9. T, 10. T.

10 The Future of Animal Genetic Improvement

Abstract
This chapter deals with a number of issues affecting animal breeding that are likely to become increasingly important in the future. It begins by examining new technologies: cloning, genetic engineering and metagenome analysis. It is argued that cloning is of no interest in animal breeding, as traits are generally determined much more by the environment than by genes. The achievements of genetic engineering and gene editing are presented, highlighting the difficulties in their use and dissemination in animal breeding, and ending with a brief comment on the possible use of the microbiome in animal breeding programmes. The chapter then continues with a discussion on whether limits to selection will be reached and stresses that, for the time being, there is no indication that this is the case. The chapter ends by discussing the use of breeding to combat global warming.

The use of livestock gives rise to a number of problems, such as pollution, increased global warming of the planet, ethical problems regarding animal welfare and problems derived from the use of genetically modified animals. In this chapter, we will examine the impact of animal breeding on these problems and its contribution to solving or alleviating them.

10.1 New Technologies

10.1.1 Somatic cloning

Cloning consists of obtaining individuals with the same genotype. Somatic cloning is achieved by replacing the nucleus of an embryo with the donor nucleus of a suitably reprogrammed somatic cell, so that the embryo gives rise to an animal with the same genotype as the donor. The technique was developed in 1997 for sheep (the famous 'Dolly' the sheep), and although it was very ineffective in its early days (277 embryos were needed to clone Dolly), the technique has notably improved in cost and efficiency, although today only 10–20% of embryos reach adulthood. It was originally thought that it could be used for cloning transgenic (often sterile) animals to produce pharmaceuticals that are expensive to obtain by other means, such as transgenic sheep to produce human blood-clotting factors in milk. It was soon suggested that somatic cloning could also be useful for animal breeding; it is the dream of any farmer who has a highly productive animal to repeat it, and cloning seemed to guarantee this repetition. If cloning becomes economical and efficient, we can suppose that all the animals of a farm would have the same genotype, and only a few varieties of cows or pigs will be used, in the same way that apples and oranges are divided into a few genotypes that are repeated to produce the commercial varieties that we know. This, however, is unlikely to happen, even if cloning is cheap and efficient, for reasons discussed below.

DOI: 10.1079/9781836990994.0010

A. The genotype is cloned, not the phenotype

First, it is essential to realize that it is the genotype and not the phenotype that is being cloned. The value observed in an individual has a genetic and an environmental component, and we have already seen that most of the variability that we observe among individuals is due to environmental causes. The heritability of most of the productive traits varies between 0 and 0.40 (e.g. Tables 8.2 and 9.2). The consequence is that the observed excellence of a single animal compared with the other animals in the population is more likely to be due to environmental reasons than genetic reasons, and the risk of having a much less productive clone than the original animal is very high.

However, it seems intuitively reasonable that all the animals of the same clone should produce as much as the animal from which they were cloned, so we will give an example that will clarify this matter. Let us take the case of pig litter size, a trait whose repeatability (see section 2.2.1) is around 0.15; that is, the phenotypic correlation coefficient between the litter size of two parities is 0.15, although we know that the genes determining litter size in all parities are almost the same. With such an extremely low phenotypic correlation, no one expects that if a sow has 20 piglets at one farrowing, she will have 20 at her next farrowing; in fact, she could have almost any number, as we know that the value of 20 was due essentially to environmental causes. However, a sow has the same genotype at one farrowing as the next; the sow is (informally) a 'clone of itself'. If we obtain a clone of this sow, the clone would have a productivity the same as at any other parity of the sow, far from the 20 piglets that the original sow had, so we should not expect this clone to have a litter size equal to that of this extraordinary farrowing.

B. Using cloning in a genetic programme

Let us examine what would happen if a genetic programme was based on cloning. We will give an example in pigs. To detect outstanding animals, many animals need to be measured. In pigs, many animals are measured only for litter size, as growth rate, feed efficiency and meat content are measured only in the selection nucleus, while litter size is measured in multipliers and commercial farms. In this example, the heritability of the trait is low, but even if it were higher, the same phenomenon would occur, as the productive traits have heritabilities very far from 1, and only when $h^2 = 1$ is the value of a trait determined solely by genetic causes. Here, we will only examine the example of possible somatic cloning in pigs, but the same arguments for dairy cattle can be found in van Vleck (1999). Further details of our example can be found in Blasco (2008).

A genetic programme based on cloning should have the following steps:

- *Outstanding animals should be detected for a trait of economic interest.* Cloning is only worth it for exceptional clones. Even animals with an outstanding phenotype generally have low genetic values. Suppose that in a pig population, with a mean litter size of ten, an outstanding individual has a litter size of 20; the heritability of the trait is 0.10, and the phenotypic standard deviation is 3. The estimated additive value of the clone, adding the mean of the population, and its 95% confidence interval are (see sections 2.3.3.A and B):

$$m + \hat{A} = 10 + 0.10(20 - 10) \pm 2\sqrt{0.1} \cdot 3 \cdot \sqrt{1 - 0.1} = 11 \pm 1.8$$

so the prediction will be that the clone has an additive value between 9.2 and 12.8; therefore, it could even happen that the clone had an average litter size smaller than the animal from which it was cloned.

- *The clone should be produced and tested for this trait and all other traits of economic interest.* This is important to avoid having a clone with a high value for one trait but a low overall economic value. It can be argued that we should look for clones of animals that are economically outstanding, considering all traits of economic interest, but, as we have said, in pig production few traits are recorded in a large number of animals. Testing a clone requires a considerable number of copies of the clone for the same reasons as in the previous section; the causes for which a specimen of a clone is exceptional may be environmental.
- *The clone should also be tested for other traits that may have undesirable consequences.* These might include fertility (sterility of a clone is frequent), genetic sensitivity to diseases and traits that can cause management problems.
- *The clone should be implemented in the commercial population*, reserving genetic variability only for nuclei in which the traditional selection process will continue. Suppose that we have succeeded and have found (by chance, as it is highly unlikely) an animal that is genetically far superior to all the animals in the selection nuclei, not only for a specific trait but also for its total economic value. Let us also suppose that we have an economically viable cloning method capable of producing clones in a functional way and at low cost. In this case, we would have the problem that diffusion of the clone would end the genetic variability of the traits of economic interest. There would only be genetic variability in the selection nuclei, as farmers would want to use only this highly productive clone. The lines of these nuclei, whose animals would be genetically on average well below the clone, would be economically unviable.

It is also important to note that *successful cloning would represent a serious loss of flexibility needed for constantly evolving markets.* If the selection objectives change, the cloning process could not be repeated because we need to measure many animals to find one extraordinary one, and commercial farms would be occupied by clones, with no genetic variability to cope with the change of objectives. The clone would therefore be a solution in the short and medium term but not in the long term.

C. Possible applications of cloning

Cloning can be useful to replicate animals producing goods that are particularly expensive and difficult to produce under other circumstances, as human blood factors from genetically modified animals. Some traits cannot be collected in a selection candidate and a clone can be used instead (e.g. disease resistance, carcass traits). Estimation of genotype–environment interaction is difficult, but it can be done straightforward by evaluating clones in several environments. Cloning can be useful for epigenetic studies, research requiring no genetic variability, and other uses. For all of these uses, cloning needs to be functional and economical, which is not yet the case.

10.1.2 Genetically modified animals

Genetically modified animals (GMOs) refer to animals whose DNA has been manipulated by genetic engineering procedures. These interventions can consist of inactivation of a gene (*knockout*) by simple deletions (suppressing some bases), or in more complex cases, insertion of a gene (*transgenesis*), which can be from another species, or insertion only of base pairs. In Europe, the consumption of genetically modified animals is prohibited, but in the USA, there is a single authorized case, transgenic salmon. These prohibitions do not have a clear scientific basis, as there do not appear to be human health problems caused by the consumption of transgenic animals, nor are there problems associated with animal welfare or the environment. Due to these prohibitions, since the appearance of *genetic editing* as discussed below, it has been argued that genetically edited animals should not be classified as transgenic animals, because, in most cases, genetic editing does not introduce a gene from another species; rather, the DNA of the species itself is modified. However, these arguments have not yet been successful, and in Europe both types of animals are considered GMOs.

A. Techniques for the genetic modification of animals

Techniques for the genetic modification of animals have greatly changed since they were introduced in the 1980s. The first transgenic animals developed for a productive trait were mice with a gene for rat growth hormone, which grew twice as large as normal (Palmiter *et al.*, 1982). Mice with the human growth hormone gene were then produced, with equally spectacular results (Palmiter *et al.*, 1983), and in 1985, the first transgenic animals in livestock species were produced (rabbits and pigs; Hammer *et al.*, 1985).

- At first, mechanisms were used allowing *random insertion* of a DNA fragment into the genome of the animal, by microinjection into the pronucleus of the fertilized ovum. This technique is very ineffective; Nieman and Kues (2003) reported that 36,500 embryos were needed to obtain 18 transgenic calves expressing the trait of the transferred gene, so that a transgenic cow could cost at that time more than US$500,000. Subsequently, a type of retrovirus, the lentiviruses, were used for random insertion into the DNA, and the process became more efficient, although it did not work in all species. Today, these methods are scarcely used.
- Recently, *targeted insertion* mechanisms have started to be used, which allow decisions to be made about where to insert the gene. These mechanisms also allow the deletion of bases or the addition of new bases to the genome, and so are authentic mechanisms of gene editing. These new mechanisms are based on restriction enzymes that recognize specific DNA sequences and can cut at specific points. The first enzymes used were yeast meganucleases, but these were impractical because they could only cut specific DNA sequences; the first useful genetic editing tools for use in animals appeared in 2009 (Fig. 10.1):
    - *ZFNs (zinc-finger nucleases).* These are artificial proteins whose structure is determined in part by zinc ions. These proteins have the ability to bind to DNA at different sites, depending on their amino acid sequence. In association with the restriction enzyme *Fok*I, they allow DNA to be cut at points determined by their amino acid sequence. In 2009, the first rats with several deletion-inactivated genes (knockout rats) were produced using ZFNs, and they were subsequently used for various livestock applications. Today, the system is not used due to the costs involved in designing a protein for each cutting point, and due to problems related to the intellectual property of the technique.

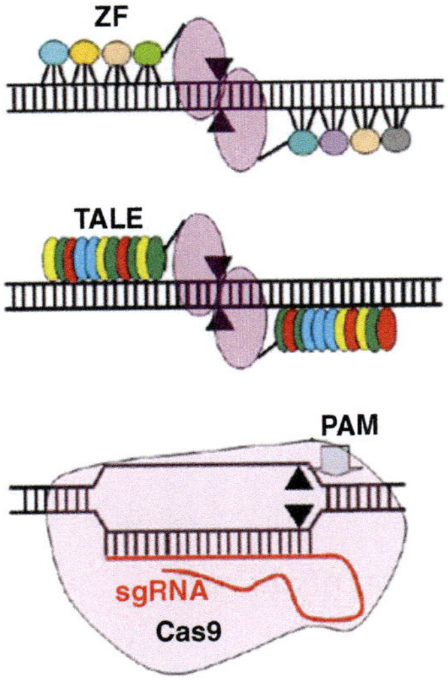

Fig. 10.1. Gene editing tools. CRISPR-Cas: clustered regularly interspaced short palindromic repeats/CRISPR-associated protein; PAM: protospacer adjacent motif; sgRNA: small guide RNA. TALEN: transcription activator-like effector nuclease; ZFN: zinc finger nuclease. Adapted from Yu *et al.* (2016) (open access source).

- *TALENs (transcription activator-like effector nucleases).* The TALEN system is similar to that of ZFNs, but here the proteins are bacterial (not artificial, like ZFNs) and bind to DNA based on their amino acid sequences, but in a different way from ZFNs. As they are natural proteins, there are no intellectual property problems, which has made this technique popular. This method was developed in 2011 to generate knockout rats, and was subsequently applied to livestock species. The technique is very labour intensive and is rapidly being displaced by the CRISPR tool.
- *CRISPR (clustered regularly interspaced short palindromic repeats).* This more recent system was first developed in 2012 and verified in 2013; it has been widely used for its simplicity, lower cost and greater versatility. It consists of an RNA guide chain designed with 20 specific nucleotides that join to the desired place in the genome. The *leader sequence* is attached at one end to the Cas9 endonuclease, which cuts the DNA just after the designed 20 nucleotide sequence. It is also necessary that the genome to be edited has a short DNA sequence that activates the Cas9 endonuclease; this sequence is called PAM (protospacer adjacent motif). As cleavage depends on the genome having this PAM sequence, this limits the possible cleavage sites, but as the sequence is short (2–6 base pairs), there are many occurrences in the genome, and with bioinformatics programmes, it is not difficult to find many sites for possible cutting. The described system is known as CRISPR-Cas9, and is currently the most widely used for genetic modifications, but there are many variants with both Cas9 and other endonucleases (e.g. Cas12a, Cas13; see Burgess, 2020). Finding the most appropriate CRISPRs is an area of intense research today.

B. Examples of genetically modified animals for possible use in animal production

The first application of transgenic animals was to produce human metabolites. In the late 1980s, the sheep 'Lucy' was used produced human blood-clotting factors in its milk. Apart from their possible use in producing useful metabolites in human medicine, which remains an important field of research, a number of genetically modified animals have been produced for animal production (Tait-Burkard *et al.*, 2018):

- *Transgenic pigs with the enzyme phytase in saliva.* The main source of phosphorus storage in plants is phytic acid, but monogastric animals cannot digest it, so dietary phosphorus supplements are common to avoid phosphorus deficiency problems. However, this creates a serious environmental problem due to the high phosphorous levels in the manure that is produced, so a common practice is to add phytase to the feed, which reduces the amount of phosphorus in the manure. Salivary phytase transgenic pigs excrete 75% less phosphorus in manure than non-transgenic pigs, which avoids adding phytase in feed.
- *Transgenic pigs with the bovine α-lactalbumin gene.* α-Lactalbumin is involved in the lactose production mechanism. Sows with this gene have a higher litter weight at weaning. Individual weaning weight represents approximately 20% of the economic weight of litter size (Quinton *et al.*, 2006). Litter size is today around 50% of the total economic weight, so the economic weight of the weaning weight is rather small.
- *Transgenic pigs with human growth hormone.* Pigs with human growth hormone or another similar hormone (IGF-1) had a 12% faster growth rate and an 18% better conversion rate, but also had health problems and were particularly sensitive to stress, so the project was abandoned.
- *Pigs resistant to porcine reproductive and respiratory syndrome virus.* This syndrome, also known as 'blue ear', is a serious and common disease in pigs; inactivation of the *CD163* gene produces animals resistant to the virus. There are other investigations of resistance to viruses, particularly in chickens. In swine, there is a study in progress on resistance to African swine fever, a disease for which there are no vaccines yet, although there is also a study to produce a vaccine against this disease. All these studies on resistance to diseases of viral origin have to evaluate whether it is worth investing in genetic modifications compared with the use of vaccines. As in the case of vaccines, they also need to be able to cope with the frequent mutations of many of the viruses.
- *Sterile pigs.* Inactivation of the *Nanos2* gene produces males without germ cells but with a fully functional reproductive system, which allows the transplantation of germ cells from other males so that the pig can act as a 'surrogate father'. This has led to imaginative proposals to change selection schemes in pigs and make them more similar to those used in plants (Gottardo *et al.*, 2019), and these are currently under study.
- *Transgenic cows with greater resistance to mastitis. Staphylococcus aureus* is responsible for approximately 30% of mastitis cases in dairy cattle. Staphylococci are lysed in the presence of lysostaphin, so transgenic cows that produce lysostaphin in milk have better resistance to mastitis. Stott *et al.* (2006) proposed a relative economic weight of 17% for mastitis for a selection index, so the final economic value of transgenic animals is not particularly high.

- *Transgenic cows with supplemental copies of the genes encoding bovine κ- and β-casein.* Transgenic cows produced up to 13% more protein in milk, although with highly variable results in terms of composition. The economic weight of the protein represents approximately 50% of the total economic value of a selection index. Again, the final economic value of transgenic animals is not particularly high.
- *Atlantic salmon with the growth hormone gene of the king salmon species.* King salmon is three times heavier than Atlantic salmon. In transgenic Atlantic salmon with the growth hormone gene of the king salmon, individuals have a growth rate more than twice that of normal Atlantic salmon. This is the only genetically modified animal species authorized for human consumption in the USA, but is not authorized in Europe.
- *Pigs rich in ω-3.* Pigs with a gene from the nematode *Caenorhabditis elegans* produce meat enriched with ω-3 acids, with a notably lower n-6:n-3 fatty acid ratio than normal pork. Although the experiment was successful, the added value of these pigs is not very relevant because the same effect can be obtained with feed rich in ω-3, and also because there are many ways to enrich the human diet with ω-3.
- *Animals with muscular hypertrophy.* In section 9.2.2.C, we described muscular hypertrophy in cattle. These animals with hypertrophy have around 20% more meat on the carcass but also have calving difficulties and higher nutritional requirements, so this is not a clear alternative to common beef cattle. This trait is regulated by the myostatin gene, also called the hereditary muscle hypertrophy gene, which has several mutations producing this effect. These mutations comprised deletions of 1 to 11 base pairs, depending on the breed of cattle. Using genetic editing techniques, these deletions can be reproduced in other species, with variable results. In pigs and rabbits, animals with a higher percentage of meat have been obtained, but in both cases, they tend to have problems with their legs and their health in general. In cattle, the results are variable, depending on the breed. In sheep, the technique seems to have been successful.
- *Polled cows.* The practice of cauterizing the buttons through which the horns will come out is frequent in dairy cattle, and causes pain to the calf unless anaesthetics and anti-inflammatories are used (mandatory only in certain countries). Some breeds of cows lack horns because they have the dominant polled allele. This allele can be implemented by introgression (see Fig. 10.2), but this is a very slow process in cattle, so this could be compensated for by using gene editing mechanisms, which have already been successful.

C. Difficulties in the use of genetically modified animals in animal production

The appearance of the first transgenic animals created great expectations about what could be done with genetically modified animals for animal production. However, few genetically modified animals are now available in animal production, and in fact only one (salmon) is used for commercial purposes. Genetic modification has a clear role in human health, such as in the fight against cancer, in rare disorders dependent on a gene, in the production of animals capable of generating organs that can be used by humans (xenotransplantation), in the production of difficult-to-obtain human metabolites, and in other health-related fields. However, genetic modification is unlikely to play a major role in animal breeding programmes, for the following reasons:

- *Very few productive traits are determined by major genes.* This is the main limitation of transgenic animals for animal production. The purpose is to find genes with important effects on traits of productive interest and transfer them to farm animals. Unfortunately, the most economically interesting traits are determined by many genes with small effects, so the infinitesimal model works well (section 2.1.2.A). This is also the main problem of genome-wide association studies (GWAS) for detecting major genes, as we saw in section 6.3.1. In humans, the situation is the same – the main diseases depend on many genes of small effect; however, if a genetic disorder dependent on a single gene causes a seriously disease in one person in every 1000, any effort that is invested in solving it through genetic editing is an important achievement; in animals, it is enough to eliminate the animal with the disorder.
- *Some researchers have proposed modifying dozens of genes in the same embryo* when many genes determine a trait, but this has two major difficulties: knowing which genes to modify and how to do it. Moreover, there are interactions between genes, and modifying one can have effects on others. In the current state of gene editing, many modifications markedly increase the risks of mosaicism, which we will discuss later in this section. Another added difficulty is the possible pleiotropic effects of the modifications affecting other traits in an unfavourable way. Finally, it is unlikely that editing a few dozen genes will modify a trait in a relevant way; currently, it is considered that productive traits are influenced by a large number of genes, more than 1000 in many cases, and it has even been proposed that a considerable part of the entire genome has an influence on quantitative traits (Boyle *et al.*, 2020).
- *Genetically modified animals should be outstanding for a trait of economic interest,* to make it worthwhile carrying out the transgenesis process. In section 10.1.2.B, we saw several examples of genetically modified animals. Except for transgenic salmon, which has spectacular growth, it is doubtful that in the other cases the effect will compensate the work required for the genetic modification and its diffusion. As in cloning, and for the same reasons, genetically modified animals, should be tested for all productive traits; an outstanding animal in one trait can have low performances in other economically relevant traits. In addition, genetically modified animals should compensate for the genetic delay involved in their creation and dissemination; while the genetic modification process takes place, the commercial nuclei should continue with selection.
- *Genetic modification should be more effective than classical breeding methods.* Even if a particular trait is determined by a major gene, genetic modification may be unnecessary:
  - *Classical selection is very effective in increasing the frequency of major genes.* Genes with important effects are selected directly by the classical methods discussed in Chapters 2–4; for example, commercial pig populations have fixed several genes with a major effect on fat content. Moreover, nowadays it is possible to utilize markers for selection of major genes.
  - *Some genes can be easily captured by introgression.* The objective is to implement a favourable allele for a trait in a population that lacks it, while preserving the rest of the alleles for the other traits. For example, there is an allele in the Australian Merino Boroola (let us call it A), which produces a high ovulation rate and an average litter size of 2.4 times the normal one. Crossing with a local breed that only has the unfavourable allele, allele a, we will obtain heterozygous Aa sheep for that gene. By crossing this population with the local breed several times, and selecting animals with high litter size or using genetic markers, we would obtain

sheep with genotypes Aa and aa, but the rest of the genes would be more and more similar to that of the local breed (Fig. 10.2). Finally, by crossing this population with itself, we would obtain the three possible genotypes (AA, Aa and aa), and by means of genetic markers we could detect those animals that are AA, generating an AA homozygous population for the gene of interest but having a genetic background similar to the local breed. This process is not very effective in the case of species with a long generation interval, such as cattle.

  - *The number of genetically modified animals obtained is small.* Due to the technical difficulties of genetic modification, normally few useful genetically modified animals would be obtained, which means that their diffusion can be slow. In a selection nucleus, we will need several generations, and high inbreeding can occur during the process; in dairy cattle, the diffusion of the gene, so that the bulls available in the catalogues of the companies and their daughters have it, will also be a very slow process.
  - *Genetically modified animals should compensate for the genetic lag involved in their creation and dissemination.* During the process of creation and dissemination of a genetically modified animal, the industry continues selecting its animals, and therefore the expected productivity of a genetically modified animal should not be compared with that of the current productions of its competitors but with the productions expected when the transgenic gene is diffused into the population.
  - *Genetically modified animals should also be tested for traits that may impair their functionality* (e.g. resistance to disease, fertility). These problems appear frequently, so these tests are necessary.

- *Technical difficulties.* When genetic modification was first developed, random insertion was very ineffective, but now *gene editing* has solved many of the technical problems of obtaining animals with transferred genes, or simply with genes modified with insertions, deletions or substitutions of some pairs of bases. However, there are still technical problems that need to be resolved. In gene transfer, regulatory mechanisms do not always work well, and there may be transferred genes that are not expressed, or that are expressed in one species but not in another, or sometimes that are only transmitted in one or two generations. In gene editing, even using CRISPR, editing errors are made, as DNA fragments may be cut but not at the desired cut points. In addition, although DNA can be cut in a specific place, the DNA repair mechanisms are not under control, and unwanted variants may be generated (this is what is called

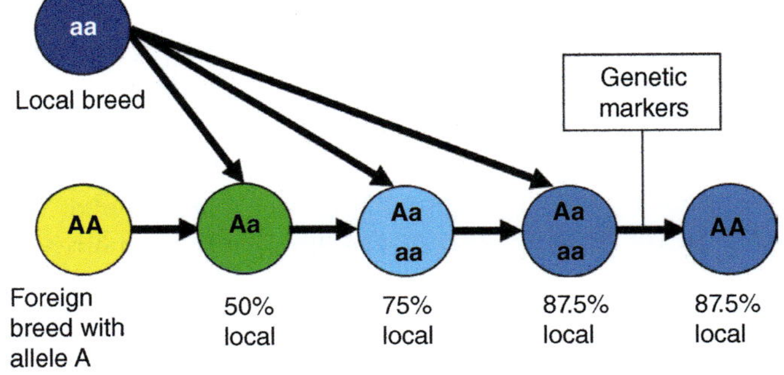

Fig. 10.2. Introgression of allele A of a major gene into a local breed.

genetic mosaicism). Some of these variants may modify other genes in addition to the desired one, or may alter the expression of other genes. The advantage of gene editing is that these variants can be detected at the cellular level, so there is no need to wait for the development of the animal to express them.

10.1.3 The metagenome and the integration of omics

The genome of gastrointestinal microorganisms of the reproductive system, or of other parts of the body (known as the *metagenome*), is considered the second genome of animals and people. The gastrointestinal metagenome participates in the development of the immune system, in the production of bioactive compounds, in the collection of energy and in other processes such as carbohydrate metabolism. In recent years, the availability of sequencing methods at low cost has allowed the identification of microbes (the *microbiota*) by identifying a characteristic gene of each species (the 16S rRNA gene), and complete sequencing of the bacterial genome is also available at a reasonable cost. Recent studies also show that the composition of the microbiota can be influenced by the genome of the individual in which it occurs, indicating interactions between genes of the individual and its microbiota. The genetic determination of the traits seems to depend on an extended genome composed of the genes of the animal plus the genes of the gut microbiome and the microbiome of other parts of the organism.

Studies of the microbiome have opened up new possibilities for research on the metabolic pathways that determine the expression of traits, and have also opened the door to new procedures to act on them. Furthermore, research on the metabolites present in the gut has been boosted by the recent development of kits allowing us to analyse from a few hundred to more than 1000 metabolites simultaneously, at a reasonable price. This makes it possible to analyse not only the set of metabolites in the blood of an individual (the *metabolome*) but also the metabolome of their bacteria (i.e. the metabolites they produce). The set of microbes, their genes and the metabolites they produce is called the *microbiome* (Fig. 10.3).

If we add the *proteome* (the proteins produced by genes) and the *transcriptome* (the genes being expressed at a given time), the relationship between phenotype and genotype can be examined using several intermediate steps, as indicated in Fig. 10.4.

All this technical development has generated an immense amount of data that not only has to be analysed separately but is also important to determine relationships in order to explain the biological mechanisms that are acting at a given moment. The challenge of investigating the relationships between variables when there are thousands of variables and only a few dozen animals requires statistical techniques different from those we have been studying, and this is a very active and recent field of research. It is important to realize that when many variables are considered, anomalous results can occur by chance; for example, correlations much higher than those existing in the population, or extreme values of some variables, can appear by chance because a large number of variables have been measured in a small number of individuals. This is a problem similar to the multi-test that appears in GWAS analyses. Multivariate analysis techniques (partial least squares), classification techniques (partial least squares-discriminant analysis, random forest), techniques based on artificial intelligence (machine learning, deep learning) and a wide spectrum of mathematical techniques are being investigated to handle these large numbers of variables.

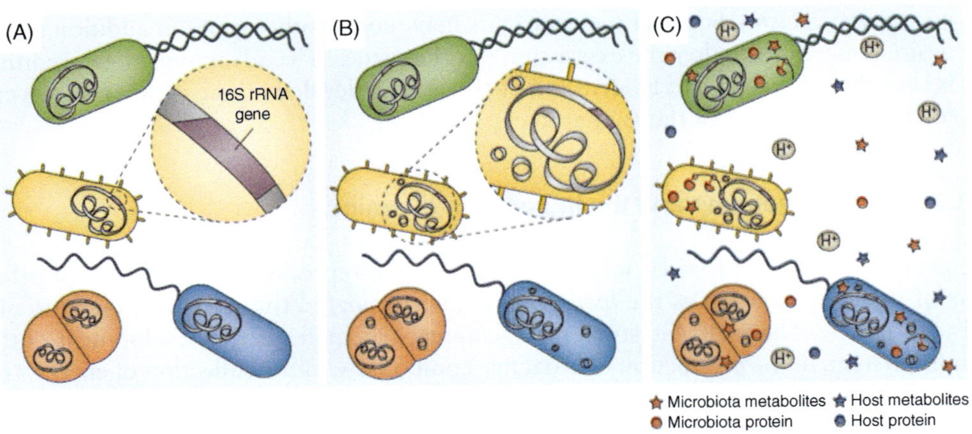

Fig. 10.3. (A) Microbiota: all microorganisms present, classified taxonomically with the 16S rRNA gene or from the metagenome. (B) Metagenome: genes and genomes of the microbiota, obtained by complete sequencing. (C) Microbiome: the microbiota, metagenome and metabolites produced by microorganisms and those existing in the environment. From Whiteside *et al.* (2015), used with the permission of Springer Nature.

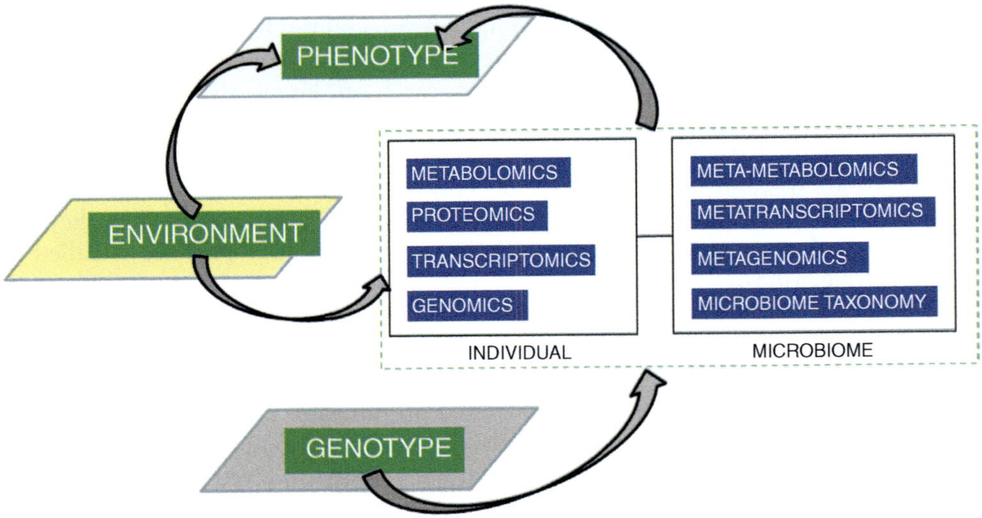

Fig. 10.4. Levels of study between phenotype and genotype. Adapted from Fontanesi (2016).

Modification of the microbial content of individuals has been carried out mainly by changes in diet, and some successes have been obtained in ruminants, particularly to control methane production. In metagenomics, there is intense investigation with two main objectives: (i) to understand the metabolic routes that govern the traits of economic interest; and (ii) to examine the possible use of the genes of bacteria in the prediction of breeding values. At the time of writing, it is too early for conclusive results, but in the near future, we will see how these techniques can find feasible applications in animal breeding.

10.2 Limits to Selection

The question of whether there are limits to selection needs to be further clarified. Obviously, there are certain traits that have selection limits by definition; for example, survival to weaning cannot be greater than 100%, or oil content reduction in maize seeds cannot be less than zero. However, natural selection has shown that traits can be selected to the extreme of generating new species. We are now going to examine whether there is evidence of limits to selection for traits in current breeding programmes, such as milk production and composition, growth rate, feed conversion rate, litter size and number of eggs. Is there evidence that genetic variability is being depleted due to selection? We already anticipated in section 7.2.3 that there seems to be no reason for being pessimistic on this point; we will examine the situation in more detail below.

10.2.1 The theory of limits to selection

Selection modifies the frequencies of genes: the frequencies of the favourable alleles increase (section 1.1.3). In the infinitesimal model, there is no fixation or loss of genes, and it is assumed that there are infinite genes that gradually modify their frequency, as we saw in section 2.1.2.B1, but we know that this is a simplification of the reality; eventually, the selected allele or alleles will be fixed. This would stop genetic progress, but we know that mutations occur, which, if they have a favourable effect, will be selected. Mutations are a rare event; they occur around one in 10^{-4}–10^{-6} times per generation at a particular locus, and most are unfavourable or neutral and disappear. However, as animals have billions of bases in their DNA, a favourable mutation is not such a rare occurrence. The proportion of new additive genetic variability per generation due to mutations (the fraction of the heritability due to mutations) is around 0.001 in a wide range of species. The maintenance of genetic variability depends on this new variation from mutations, and on the effective number of the population. For example, to maintain additive genetic variability in a trait with a heritability of 0.3, the effective number of the population should be 150 (Hill, 2016), which would correspond to a population with approximately 38 males (see section 1.2.2), something perfectly admissible in the selection nuclei. Hill (1982) concluded that, after 20 generations of selection, the new mutations contribute substantially to genetic variability, allowing the selection process to continue.

In addition, epistasis generates additive genetic variability. If certain combinations of alleles are favourable, selection acts by increasing the frequencies of these alleles, so that the favourable combination is more and more frequent and the trait responds to selection. In section 1.3.1, we discussed this issue, because confusion between the additive, dominant or *epistatic effect* and the additive, dominant or *epistatic value* is frequent; the latter depend on the allelic frequencies. Some sets of genes may have an epistatic effect, but in one population they generate additive values and in another they do not, depending on the frequencies of their alleles. A discussion of the theoretical limits to selection can be found in Caballero (2020).

10.2.2 Response to long-term selection in practice

With some few exceptions, such as speed in horse racing, limits to selection in productive traits of farm animals have not appeared (Hill, 2008); on the contrary, with the recent use of genomic selection in breeding programmes, the response in these traits has increased more rapidly. In Table 10.1, we can see the global progress (genetic and environmental) in several productive traits, before the implementation of genomic selection programmes. In some cases, particularly in broilers (Havenstein *et al.*, 2003), it has been possible to accurately estimate, with control populations, the proportion of this change due environmental aspects, mainly nutrition, and the proportion of change due to genetic improvement, the latter being around 80% of the total response achieved. In dairy cattle and other species, the environmental effect can be separated from the genetic one by means of the mixed model, as we saw in section 4.3.2.B (see Fig. 9.8). In general, a large part of the improvement in production is due to genetic improvement (see sections 8.1.4 and 9.1.3).

Table 10.1. Change in several productive traits between 1960 and 2005. From van der Steen *et al.* (2005), with the permission of Oxford University Press.

		1960	2005	Change (%)
Pigs	Pigs weaned/(sow·year)	14	21	50
	Feed conversion ratio	3.0	2.2	27
	Lean %	40	55	37
Poultry	Days to 2 kg	100	40	60
	Breast meat %	12	20	67
	Feed conversion ratio	3.0	1.7	43
Laying hens	Eggs/year	230	300	30
	Eggs/kg feed	5	9	80
Dairy cattle	Milk (kg)/(cow·lactation)	6000	10,000	67

Genomic selection has improved these values, increasing the response by around 20–30% in monogastric animals, beef cattle, sheep and goats, and by more than 50% in dairy cattle, as we saw in Fig. 9.8, without a decrease in response over time in any of the cases. Furthermore, long-term selection experiments do not show any signs of stagnation. Figure 10.5 shows the genetic progress in body weight of selected Japanese quail over 97 generations. Both the P and T lines increased their differences with respect to the corresponding PC and TC control lines, without apparently stopping genetic progress. In laboratory mice, there are also long-term selection experiments for live weight that do not find signs of reaching a plateau (Fig. 10.6). Sometimes, plateaus in response to selection have been found, but if selection continus, in general the response also continues. The same seems to be the case in plants, the best example being the famous Illinois selection experiment for oil content in maize, which has more than 114 generations of selection with no plateau in response (http://mooselab.cropsci.illinois.edu/longterm.html; accessed 21 July 2025)).

Selection can lead, in the long term, to problems of a pathological nature or to lack of adaptation of the animals to the facilities. For the latter, a solution would be to adapt the facilities to the new production conditions, but pathological problems are usually induced by the physiological changes produced by selection. The most frequently cited example is the increase in ascites in chickens due to their high growth rate and high percentage of breast. Ascites, or pulmonary hypertension syndrome, is a disease that can cause the death of chicken; it is the result of the lack of the necessary supply of oxygen to maintain the high growth of these birds. Resistance to ascites has a relatively high heritability, and although it is negatively correlated with weight, genetic improvement companies often apply some selection pressure against ascites. The optimal system is a selection

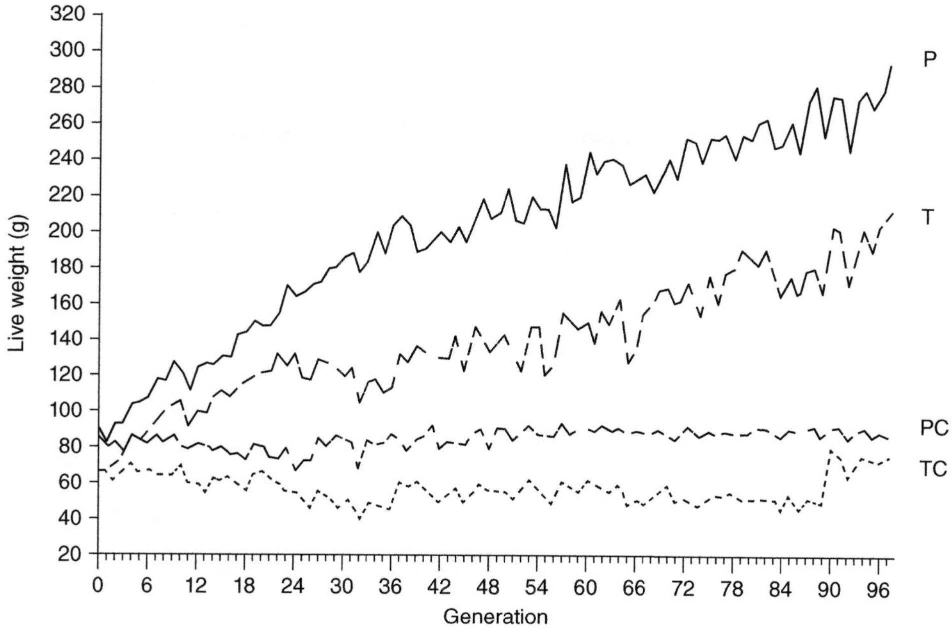

Fig. 10.5. Selection for body weight in Japanese quails. P and T: selected lines; PC and TC: control lines. From Marks (1996) (open access source).

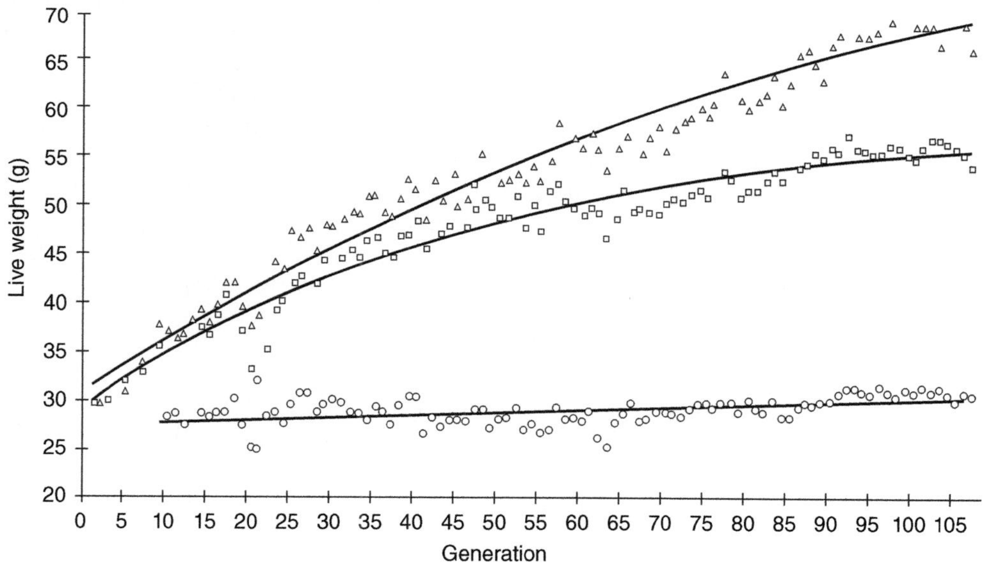

Fig. 10.6. Body weight selection in two lines of mice compared with an unselected control line. From Renne *et al.* (2003), used with the permission of Elsevier.

index with its economic weights, but as the incidence is not high (~2%), it is not imperative to introduce the trait in the index. Selection against ascites is an example where the unfavourable consequences of selection for certain traits can also be counteracted by selection.

At present, there is a debate on the sustainability of highly productive animals; for example, when the back fat of pigs is excessively reduced, problems appear in the processing of the carcass (fillet cuts are irregular). It does not currently seem possible to draw general conclusions, and it is necessary to examine each case. Again, changes in selection indexes can move to a new economic optimal point.

10.3 Sustainable Animal Production and Animal Breeding

10.3.1 Livestock and global warming

Livestock is responsible for nearly 12% of greenhouse gas emissions produced by human activities, according to estimates by the Food and Agriculture Organization of the United Nations (FAO, 2023). Among the species, the most polluting in absolute terms is cattle, both dairy and beef, which are much higher than other species (Fig. 10.7). Buffalo, sheep and goat are low polluting due to their small number; they are minor animal productions. Pork and chicken are very numerous, but they are nevertheless less polluting both in absolute and relative terms, because they are monogastric, usually raised in intensive production systems, and intensive production reduces global pollution as we will see below. Figure 10.8 shows the world production of meat; note that pigs and poultry account for more than two-thirds of the world's meat production.

In general, intensive production produces less pollution than extensive production *per kg of product*, including greenhouse gases. This is particularly noticeable in monogastrics and in particular in broilers. The reason is apparent; if a chicken in organic farming takes 12 weeks to reach commercial weight and a chicken in intensive farming takes 6 weeks, then during the time that elapses between slaughter of the intensively reared chicken and the slaughter of the chicken from organic farming, the later continues to produce CO_2 and other pollutants, so that *per kg of meat produced*, chicken from intensive livestock is less polluting. Table 10.2 compares the production of pollutants in intensive chicken farms with organic production, from a report commissioned by the British government to Cranfield University (Williams *et al.*, 2006). It can be seen that intensive systems are notoriously less polluting, despite the fact that food generation and pollution products from transport trucks and auxiliary machinery are included in the calculations.

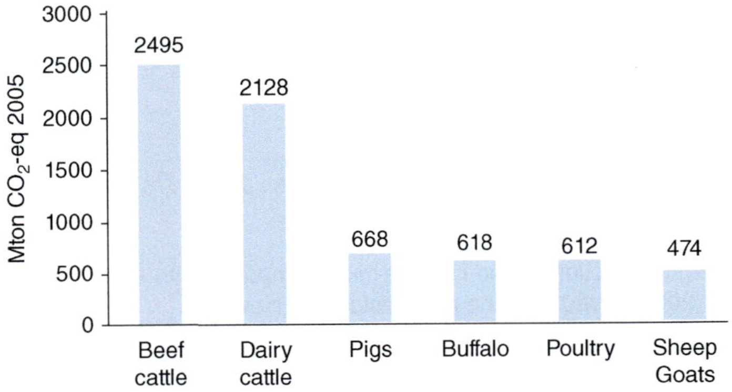

Fig. 10.7. Greenhouse gas emissions by species, according to the FAO. Adapted from Gerber *et al.* (2013).

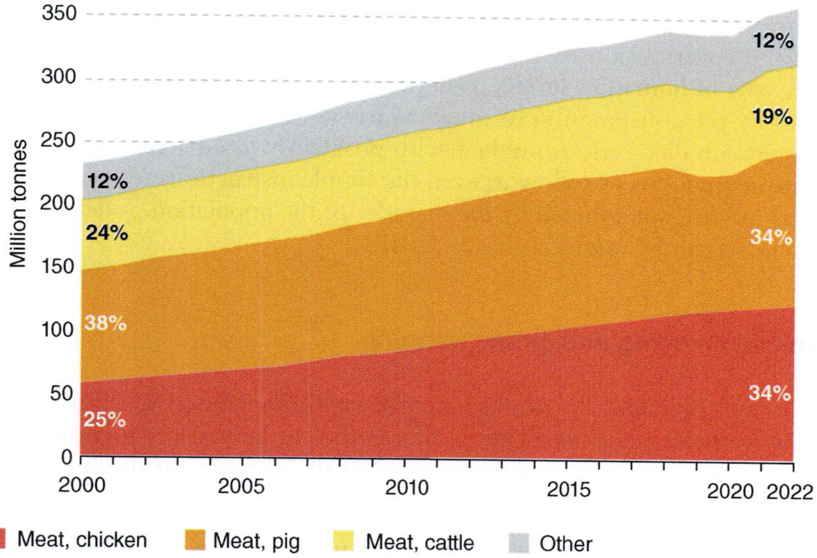

Fig. 10.8. World meat production in 2012. From Statistical Yearbook 2024, Figure 26, https://openknowledge.fao.org/server/api/core/bitstreams/d784864f-7f28-49d2-903e-6680d09a9d97/content/cd2971en.html#chapter-2_2 accessed 22 July 2025).

Table 10.2. Pollution caused by two alternative chicken production systems, expressed per ton of chickens produced. Adapted from Williams *et al.* (2006).

	Intensive production	Organic production
Primary energy used (GJ)	12,000	15,800
GWP_{100} kg 100-year CO_2 equivalent[a]	4,570	6,680
EP kg PO_4^{3-} equivalent[b]	49	86
AP kg SO_2 equivalent[c]	173	264

[a]GWP_{100}: global warming potential in 100 years.
[b]EP: pollution by nitrates and phosphates in the water and ammonia in the air, in units of phosphates (eutrophication).
[c]Acidification potential in SO_2 units.

The problem of the contribution of animals to global warming is a problem that falls outside the scope of this book. It can be argued, for example, that cattle use land that is not suitable for cultivation. Deciding whether it is reasonable to use these lands or abandon them to nature depends on many factors and very diverse circumstances, and it is difficult to generalize.

However, intensive production is frequently associated with a lack of animal welfare, which is not necessarily true. Welfare legislation has led to a marked improvement in the welfare of farm animals, something admitted even by the most conspicuous defenders of animals such as the philosopher Peter Singer, who acknowledged this fact in the second edition of the book that marked the beginning of the fight in defence of animals, *Animal Liberation* (Singer, 1990). Moreover, extensive production can occur in situations where the absence of well-being is dramatic, for example the harsh Australian winter for merino sheep, or the food shortage situation for beef cattle during the dry seasons in Brazil. Again, it is important to examine it on a case-by-case basis.

There are times when, with remarkable frivolity, vegetarianism is promoted as a solution to the contribution of livestock to global warming, without thinking that a considerable part of humanity suffers from protein malnutrition that can only be realistically covered by the consumption of meat. Moreover, if the population had to follow a balanced vegetarian diet, serious public health problems would arise due to ignorance or misuse of the instructions to follow it; even the simple instructions to consume less sugar and saturated fats are not adopted by the majority of the population. A detailed discussion of these problems can be found in Blasco (2013).

10.3.2 Animal breeding and global warming

What can animal breeding do to alleviate the problems caused by livestock in global warming? As we have mentioned before, more efficient animal production is one of the factors that can help mitigate this problem. Overall, increases in production efficiency help reduce global warming. Some examples are:

- *Improving food efficiency*. The conversion rate of feed to live weight, meat or egg production has been decreasing in pigs and poultry since the 1960s, as we saw in section 10.2.2. This implies that less and less feed is needed to produce the same amount of meat or eggs, with the consequent reduction in the carbon footprint; for example, Williams *et al.* (2006) calculated a 23% reduction in global warming potential (measured in t CO_2) after 20 years of selection in broilers. The first 'action' that the FAO (2019) proposed towards low-carbon livestock is improving efficiency.
- *Reducing the breeding stock*. By increasing litter size in monogastrics and small ruminants, fewer males and females are needed to produce the same amount of meat. This implies savings in emissions as there is less parental stock.
- *Reduction in the number of dairy cows*. By increasing milk production per cow, fewer cows are needed to produce the same amount of milk. In dairy cattle, the possibility of reducing the conversion rate of feed to milk is also being examined (see section 9.1.2.C).
- *Decrease in methane produced by ruminants*. Methane is the main greenhouse gas emitted by ruminants. Methane is produced in the rumen as a result of the fermentation of carbohydrates from plants ingested by the animal to produce easily assimilated volatile fatty acids. Currently, as discussed in section 9.1.2.C, rumen genomics and metagenomics studies are being carried out to detect genes or bacteria that can reduce methane production without affecting milk production or feed efficiency.

Main Concepts

- ***Somatic cloning:*** Obtaining individuals with the same genotype by substituting the nucleus of an embryo for the nucleus of a somatic cell suitably reprogrammed to act as an embryonic cell.
- ***Genetically modified animals:*** Animals whose DNA has been altered through molecular genetic procedures, including gene editing tools.
- ***Transgenic animals:*** Animals that have had a gene inserted into their DNA, often from another species.

- ***Knockout animals:*** Animals in which one or more genes have been inactivated by gene editing.
- ***Genetic editing:*** Techniques that allow the deletion of DNA bases and insert of new ones at selected sites.
- ***CRISPR (clustered regularly interspaced short palindromic repeats):*** A genetic editing technique whose efficiency, economy and versatility have made it prevail over other editing techniques.
- ***Microbiota:*** All the microorganisms of an animal, mainly bacteria, although archaea, fungi and protists may also be included. The most frequently examined microbiota is that of the digestive tract, mainly the rumen and ceca, or the microbiota of faeces. Studies on the microbiota of semen and the reproductive tract are also frequent.
- ***Metagenome:*** Genome of the microorganisms (mainly bacteria) that coexist in an animal.
- ***Metabolome:*** The set of metabolites of an animal at a specific time in a specific place (normally in the blood). It can also refer to the metabolites produced by the microbiota at a given time.
- ***Microbiome:*** The set of microorganisms of an animal, their genes and the metabolites they produce and that are found in their ecological niche.
- ***Long-term response:*** The response after many generations of selection. Currently, no stagnation of response has been observed in most productive traits.
- ***Global warming due to animal production:*** The increase in the temperature of the Earth caused by gases emitted by farm animals.

Self-assessment Questions (True or False)

1. Intensive animal production contributes to global warming by increasing it.
2. Animal genetic improvement cannot contribute to mitigating global warming.
3. Cloning produces individuals with highly variable production, as most of the variability in productive traits has environmental causes.
4. Cloning is a common technique used in animal genetic improvement.
5. The consumption of genetically modified animals is dangerous for human health.
6. Genetically modified animals do not play a major role in animal breeding programmes, essentially because most productive traits depend on many genes with little effect.
7. Long-term selection does not seem to have reached limits or stagnation in most of the productive traits.
8. It is not possible to modify several genes simultaneously by means of genetic editing techniques.
9. There are many examples of genetically modified animals used today in animal production.

References

Blasco, A. (2008) The role of genetic engineering in livestock production. *Livestock Science* 113, 191–201. DOI: 10.1016/j.livsci.2007.03.012

Blasco, A. (2013) *Animal Ethics: A Scientific Perspective*. Akal, Madrid, Spain (Kindle edition).

Boyle, E.A., Li, Y.I. and Pritchard, J.K. (2020) An expanded view of complex traits: from polygenic to omnigenic. *Cell* 169, 1177–1186. DOI: 10.1016/j.cell.2017.05.038

Burgess, D.J. (2020) CRISPR screens beyond Cas9. *Nature Review Genetics* 21: 273. DOI: 10.1038/s41576-020-0232-1

Caballero, A. (2020) *Quantitative Genetics*. Cambridge University Press, Cambridge, UK

FAO (2019) *Five Practical Actions Towards Low-carbon Livestock*. FAO, Rome.

FAO (2023) *Pathways Towards Lower Emissions. A Global Assessment of the Greenhouse Gas Emissions and Mitigation Options from Livestock Agrifood Systems*. FAO, Rome.

Fontanesi, L. (2016) Metabolomics and livestock genomics: insights into a phenotyping frontier and its applications in animal breeding. *Animal Frontiers* 6, 73–79. DOI: 10.2527/af.2016-0011

Gerber, P.J., Steinfeld, H., Henderson, B., Mottet, A., Opio, C. *et al.* (2013) *Tackling Climate Change Through Livestock. A Global Assessment of Emissions and Mitigation Opportunities*. FAO, Rome.

Gottardo, P., Gorjanc, G., Battagin, M., Gaynor, R.C., Jenko, J. *et al.* (2019) A strategy to exploit surrogate sire technology in livestock breeding programs. *G3: Genes, Genomics, Genetics* 9, 203–215. DOI: 10.1534/g3.118.200890

Hammer, R.E., Purse, V.G., Rexroad, C.E. Jr, Wall, R.J., Bolt, D.J. *et al.* (1985) Production of transgenic rabbits, sheep and pigs by microinjection. *Nature* 315, 680–683. DOI: 10.1038/315680a0

Havenstein, G.B., Ferket, P.R. and Qureshi, M.A. (2003) Growth, liveability, and feed conversion of 1957 versus 2001 broilers when fed representative 1957 and 2001 broiler diets. *Poultry Science* 82, 1500–1508. DOI: 10.1093/ps/82.10.1500

Hill, W.G. (1982) Predictions of response to artificial selection from new mutations. *Genetical Research* 40, 255–278. DOI: 10.1017/S0016672300019145

Hill, W.G. (2008) Estimation, effectiveness and opportunities of long term genetic improvement in animals and maize. *Lohmann Information* 43, 3–20.

Hill, W.G. (2016) Is continued genetic improvement of livestock sustainable? *Genetics* 202, 877–881. DOI: 10.1534/genetics.115.186650

Marks, H.L. (1996) Long-term selection for body weight in Japanese quail under different environments. *Poultry Science* 75, 1198–1203. DOI: 10.3382/ps.0751198

Niemann, H. and Kues, W.A. (2003) Application of transgenesis in livestock for agriculture and biomedicine. *Animal Reproduction Science* 79, 291–317. DOI: 10.1016/s0378-4320(03)00169-6

Palmiter, R.D., Brinster, R.L., Hammer, R.E., Trumbauer, M.E., Rosenfeld, M.G. *et al.* (1982) Dramatic growth of mice that develop from eggs microinjected with metallothionein–growth hormone fusion genes. *Nature* 300, 611–615. DOI: 10.1038/300611a0

Palmiter, R.D., Norstedt, G., Gelinas, R.E., Hammer, R.E. and Brinster, R.L. (1983) Metallothionein–human GH fusion genes stimulate growth of mice. *Science* 222, 809–814. DOI: 10.1126/science.6356363

Quinton, V.M., Wilton, J.M., Robinson, J.A. and Mathur, P.J. (2006) Economic weights for sow productivity traits in nucleus pig populations. *Livestock Science* 99, 69–77. DOI: 10.1016/j.livprodsci.2005.06.002

Renne, U., Langhammer, M., Wytrwat, E., Dietl G. and Bünger, L. (2003) Genetic-statistical analysis of growth in selected and unselected mouse lines. *Journal of Experimental Animal Science* 42, 218–232. DOI: 10.1016/S0939-8600(03)80004-4

Singer P. (1990) *Animal Liberation*. Jonathan Cape, London.

Stott, A.W., Coffey, M.P. and Brotherstone, S. (2006) Including lameness and mastitis in a profit index for dairy cattle. *Animal Science* 80, 41–52. DOI:10.1079/ASC40520041

Tait-Burkard, C., Doeschi-Wilson, A., McGrew, M.J., Archibald, A.L., Sang, H.M. *et al.* (2018) Livestock 2.0 – genome editing for fitter, healthier, and more productive farmed animals. *Genome Biology* 19: 204. DOI: 10.1186/s13059-018-1583-1

van der Steen, H.A.M., Prall, G.F.W. and Plastow, G.S. (2005). Application of genomics to the pork industry. *Journal of Animal Science*,83 (Suppl. E): E1–E8. DOI: 10.2527/2005.8313_supplE1x

van Vleck, L.D. (1999) Implications of cloning for breed improvement strategies: are traditional methods of animal improvement obsolete? *Journal of Animal Science* 77 (Suppl. 2): 111–121. DOI: 10.2527/1999.77suppl_2111x

Whiteside, S.A., Razvi, H., Dave, S., Reid, G. and Burton, J.P. (2015) The microbiome of the urinary tract – a role beyond infection. *Nature Reviews Urology*,12, 81–90. DOI: 10.1038/nrurol.2014.361

Williams, A.G., Audsley, E. and Sandars, D.L. (2006) *Determining the Environmental Burdens and Resource Use in the Production of Agricultural and Horticultural Commodities*. Main Report. Defra Research Project IS0205. Cranfield University and Defra, Bedford, UK.

Yu, L., Batara, J. and Lu, B. (2016) Application of genome editing technology to microRNA research in mammalians. In: Korman, M.S.D. (ed.) *Modern Tools for Genetic Engineering*. Intech Open, London.

Solutions

Solutions to self-assessment questions

1. F, 2. F, 3. T, 4. F, 5. F, 6. T, 7. T, 8. F, 9. F.

Index

Note: Page numbers in **bold** type refer to **figures**.
Page numbers in *italic* type refer to *tables*.